50p

Seals of the world

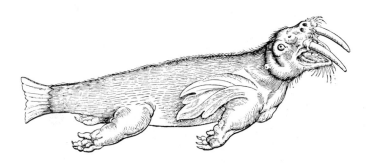

Seals of the world

Second edition

Judith E. King

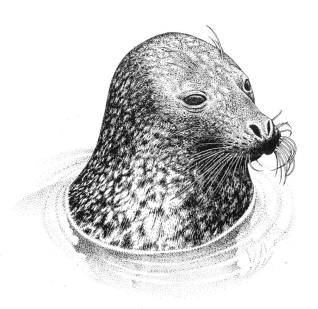

British Museum (Natural History)

Oxford University Press

© Judith E. King 1983
First published 1983 by
British Museum (Natural History) Cromwell Road,
London SW7 5BD
and Oxford University Press, Walton Street,
Oxford OX2 6DP

Oxford   London   Glasgow   Kuala Lumpur
Singapore   Jakarta   Hong Kong   Tokyo
Delhi   Bombay   Calcutta   Madras   Karachi
Ibadan   Nairobi   Dar es Salaam   Cape Town

Publication No 868

British Library Cataloguing in Publication Data

King, Judith E.
    Seals of the world. – 2nd ed.
    1. Seals (Animals)
    I. Title
    599.74'5   QL737.P64

    ISBN 0-565-00868-4
    ISBN 0-19-858513-6 (Oxford University Press)

Typeset by Rowland Phototypesetting Ltd.
Bury St Edmunds, Suffolk
Printed in England by Acolortone Ltd. and bound by
Cambridge University Press

Contents

Acknowledgements

It is still a pleasure to repeat my grateful thanks to the many people that were so helpful in the production of the original book which served as the foundation for the present edition. Particular thanks are due to the late Dr F. C. Fraser for the original idea and constant encouragement; to Professor R. J. Harrison for help in elucidating pinniped anatomy, to the late Mr A. C. Townsend for supervising the Latin and Greek names, and to Mr M. C. Sheldrick for the originals of some of the skeletal drawings.

I am deeply grateful to all the many people who did the original research and wrote the papers that I have consulted in the compilation of this book. Without them there would of course be no book. Some of these authors are mentioned in the list of references, but I hope the others will not take it amiss that they are not mentioned personally.

Many more people are deserving of thanks for helping me with this new book. To those people all round the world who responded to my letters asking for details, for photographs and for reprints – I am deeply grateful: to Charles Repenning and Clayton Ray; Murray Dailey; Gerald Kooyman – I acknowledge gratefully much needed help on the chapters on fossils, parasites and diving respectively: to the University of New South Wales who put up with seals in the midst of their marsupials, and of course to the British Museum (Natural History) for allowing me to bring the book at least nearly up to date. I also acknowledge the patience of my husband Basil Marlow during the protracted gestation of the book, and the kindness of Judith Simos who did the typing.

Preface

Increasing interest has been shown in the pinnipeds (seals, sea lions and walruses) over the past two decades. In 1960 the Zoological Record listed approximately 74 papers published that year on pinnipeds. In the ten years between and including 1970 and 1979 an average number of 125 pinniped papers were published each year (*J. Mamm.* – Recent literature of mammalogy). In 1972 an international symposium on the Biology of the Seal was held at the University of Guelph, Ontario. About 140 people interested in seals heard over 60 papers, which were later published in a large volume of 557 pages.

All aspects of seals are being studied at an astonishing rate. There have been spectacular advances, for instance, in the knowledge of seal physiology and their fossil history. One may quote the veteran biologist Victor B. Scheffer who in his opening address to the Guelph meeting said 'with continual fractionation of knowledge it is no longer feasible for one person to write a monograph on marine mammals'. He is right. I hope simply to summarize the subject to help those who have not the opportunity to dive into the sea of published papers to retrieve the fragments of information they need.

The writing of this book has a close similarity with the painting of the Forth Bridge – or Sydney Harbour Bridge – as fast as one gets to the end, the beginning is starting to look rusty. But this is the way with bridges and books. Since the publication of my *Seals of the World* in 1964 there has been such progress in the knowledge of seals that the book was beginning to look very rusty indeed. It could not be touched up, it had to be re-written. Consulting as many papers and books as I could – many times the number given in the references – I have tried to make an accurate summary of what is known about seals.

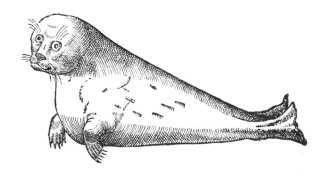

Introduction

Seals are known to most people, if only by no more than the entertaining and skilful balancing antics of the seal most commonly seen in circuses and zoos – the Californian sea lion. A few seals, usually of the local species, are to be seen in most zoos throughout the world.

The number of different 'sorts' of seal, though small when compared with the different sorts of beetle, or even with the different sorts of all mammals, is greater than many people appreciate. There are not just *a* seal, *a* seal lion and *a* walrus – but five genera of sea lions, two genera of fur seals, one of which includes eight species, ten genera of 'true' seals with nineteen species, and the walrus, though the numbers do vary a little according to different authors.

The names given to seals are curious. A large group may be called a herd, but if it is a breeding colony, it is a rookery. The adult males, or bulls, gather themselves harems of females, or cows. The young animal, until it is about four or five months old is a pup, and then, until it reaches its first birthday, is a yearling. A young walrus is a calf. Groups of pups are known as pods, and immature males are bachelors. Sealers terms are wig and clapmatch, which are used for breeding males and females, respectively. A bedlamer is a young Harp or Hooded Seal, under five years old. This name is a corruption of 'bête de la mer', the name given to the seals in the fifteenth and sixteenth centuries by the Breton settlers in Canada.

For anyone who may find a dead seal on a beach the following notes on measurement may be useful. The very simplicity of the spindle shape of a seal makes it difficult to determine exact points of measurement. Should the 'length' of a seal be taken to the end of its tail, or to the end of its outstretched hind flippers? So many authors do not say which they have used. Scheffer (1967) has given us standards to use – his 'standard length' is 'The straight line distance from snout to tip of tail flesh on the unskinned body, belly up, ideally with the head and vertebral column in a straight line'. If the seal body cannot be arranged neatly, then 'Curvilinear length is taken when the seal cannot be stretched belly up, as when frozen or stiff, or lying among rocks, and too heavy to be moved. It is the shortest surface distance from snout to tip of tail flesh along back, belly or side. It is usually measured with a flexible tape'.

The preliminary pages show some early illustrations of seals. The head of the walrus (p. 1), from C. Gesner's *Historiae Animalium*, 1558, bears a noticeable likeness to Durer's drawing (p. 72), and was probably taken from it. The posterior end of the animal is the artist's interpretation of Durer's note. The cheerful 'Sea Lion' (p. 8) with the 'seaweed tail' is from R. Brookes, *The Natural History of Quadrupeds*, 1763, where, as it mentions in the text that the male has a large snout or trunk, it is obviously the Elephant Seal that is depicted. G. Rondoletius, 1554, *De Piscibus*, published the seals (p. 9) and (p. 7), the latter probably representing the Common Seal, and the former the Mediterranean Monk Seal. The drawing of the 'Vitulus marinus' of P. Belon, 1553, *De Aquatilibus* (p. 6), is the earliest of those shown here, and is remarkable for accuracy, particularly in the details of the hind limbs and tail, which are so often shown in later drawings looking like a bunch of seaweed.

Common and scientific names

SEA LIONS
1. Northern, or Steller's Sea Lion
   Eumetopias jubatus (Schreber, 1776)
2. Californian Sea Lion
   Zalophus californianus (Lesson, 1828)
   Z. c. wollebaeki (Sivertsen, 1953)
   Z. c. japonicus (Peters, 1866)
3. Southern Sea Lion   *Otaria byronia* (Blainville, 1820)
4. Australian Sea Lion   *Neophoca cinerea* (Péron, 1816)
5. Hooker's Sea Lion   *Phocarctos hookeri* (Gray, 1844)

FUR SEALS
6. Guadalupe Fur Seal
   Arctocephalus townsendi (Merriam, 1897)
7. Galapagos Fur Seal
   Arctocephalus galapagoensis (Heller, 1904)
8. Juan Fernandez Fur Seal
   Arctocephalus philippii (Peters, 1866)
9. South American Fur Seal
   Arctocephalus australis (Zimmerman, 1783)
   A. a. gracilis (Nehring, 1887)
10. Subantarctic Fur Seal
    Arctocephalus tropicalis (Gray, 1872)
11. Antarctic Fur Seal
    Arctocephalus gazella (Peters, 1875)
12. South African Fur Seal
    Arctocephalus pusillus pusillus (Schreber, 1776)
13. Australian Fur Seal
    A. p. doriferus (Wood Jones, 1925)

14. New Zealand Fur Seal
    Arctocephalus forsteri (Lesson, 1828)
15. Northern, Pribilof, or Alaska Fur Seal
    Callorhinus ursinus (Linnaeus, 1758)
WALRUS
16. Walrus
    Odobenus rosmarus rosmarus (Linnaeus, 1758)
    O. r. divergens (Illiger, 1815)
NORTHERN PHOCIDS
17. Grey, or Atlantic Seal
    Halichoerus grypus (Fabricius, 1791)
18. Common, Harbour, or Spotted Seal
    Phoca vitulina Linnaeus, 1758
    P. v. concolor DeKay, 1842
    P. v. stejnegeri Allen, 1902
    P. v. richardsi (Gray, 1864)
19. Larga Seal   *Phoca largha* Pallas, 1811
20. Ringed Seal, or Floe rat
    Phoca hispida Schreber, 1775
    P. h. krascheninikovi Naumov & Smirnov, 1936
    P. h. ochotensis Pallas, 1811
    P. h. botnica Gmelin, 1785
    P. h. saimensis Nordquist, 1899
    P. h. ladogensis Nordquist, 1899
21. Caspian Seal   *Phoca caspica* Gmelin, 1788
22. Baikal Seal   *Phoca sibirica* Gmelin, 1788
23. Greenland or Harp Seal, Saddleback
    Phoca groenlandica Erxleben, 1777

Living pinnipeds

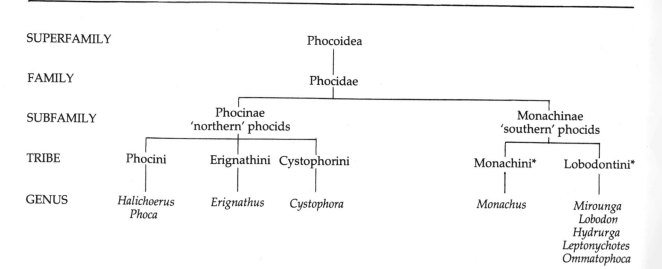

* The otariid subfamily names and the monachine tribe names are now not generally used – see text.

Common and scientific names

24. Banded or Ribbon Seal
      Phoca fasciata Zimmerman, 1783
25. Hooded or Crested Seal, Bladder nose
      Cystophora cristata (Erxleben, 1777)
26. Bearded Seal   *Erignathus barbatus* (Erxleben, 1777)
      E. b. nauticus (Pallas, 1811)

SOUTHERN PHOCIDS
Monk seals
27. Mediterranean Monk Seal
      Monachus monachus (Hermann, 1779)
28. West Indian Monk Seal
      Monachus tropicalis (Gray, 1850)
29. Hawaiian or Laysan Monk Seal
      Monachus schauinslandi Matschie, 1905

Antarctic seals
30. Weddell Seal
      Leptonychotes weddelli (Lesson, 1826)
31. Ross Seal   *Ommatophoca rossi* Gray, 1844
32. Crabeater Seal
      Lobodon carcinophagus (Hombron & Jacquinot, 1842)
33. Leopard Seal   *Hydrurga leptonyx* (Blainville, 1820)

Elephant Seals
34. Southern Elephant Seal
      Mirounga leonina (Linnaeus, 1758)
35. Northern Elephant Seal
      Mirounga angustirostris (Gill, 1866)

Total population numbers

The figures below have been rounded off, but the text should be consulted for details.

	millions		thousands
Lobodon	50	*Otaria*	240
P. hispida	6–7	*A. tropicalis*	214
P. groenlandica	2·5	*Ommatophoca*	150
Callorhinus	1·7	*Zalophus*	99
		Halichoerus	82
	thousands	*P. sibirica*	50
A. p. pusillus	850	*M. angustirostris*	48
Hydrurga	800	*A. forsteri*	40
M. leonina	700	*A. p. doriferus*	25
P. caspica	600	*Phocarctos*	6
Leptonychotes	500	*Neophoca*	5
Erignathus	500	*A. galapagoensis*	1–5
P. largha	400	*A. townsendi*	1
A. gazella	400	*M. monachus*	1
Cystophora	365		
P. vitulina	360		below one thousand
A. australis	320	*A. philippii*	800
Eumetopias	300	*M. schauinslandi*	700
Odobenus	250	*M. tropicalis*	0
P. fasciata	240		

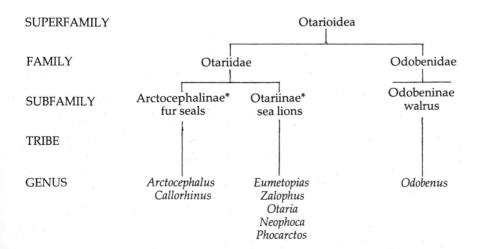

SUPERFAMILY		Otarioidea		
FAMILY		Otariidae		Odobenidae
SUBFAMILY	Arctocephalinae* fur seals	Otariinae* sea lions		Odobeninae walrus
TRIBE				
GENUS	*Arctocephalus* *Callorhinus*	*Eumetopias* *Zalophus* *Otaria* *Neophoca* *Phocarctos*		*Odobenus*

* The otariid subfamily names and the monachine tribe names are now not generally used – see text.

SECTION 1

The diversity of pinnipeds

Chapter 1 *Classification*

Although they are no longer regarded as an order in their own right (see chapter on fossils), all pinnipeds are sufficiently like one another, and sufficiently different from all other groups of mammals, as to be instantly recognizable. There is a high degree of aquatic specialization which shows all over the body, from its general streamlined shape with all the attendant characters such as the reduction of pinnae and external genitalia, to the characteristic flippers (see chapter on shape).

In the skull the enlarged orbits, the indistinct or absent lacrimal bones, the reduced incisor number (never more than $\frac{3}{2}$) and the near-homodont condition of the cheek teeth are among the pinniped characters (Burns & Fay, 1970, Anderson & Knox Jones, 1967). Among the fine points of the anatomy of the ear Repenning (1972) notes that 'As distinctive of pinniped carnivores as their flippers is the great size of their round window membrane relative to the size of the oval window', and he also notes the presence of the round window fossula (= fossula fenestra cochleae, see chapter on hearing) which is not present in other carnivores but is present in all pinnipeds. The distensible cavernous tissue is also noted as being among the pinniped characters.

SUPERFAMILY CHARACTERS

Phocoidea	Otarioidea
Marine, freshwater, estuarine	Marine only
Hind flippers cannot be turned forward	Hind flippers can be turned forward
All flippers furred on both surfaces	All flippers with naked palms and soles
Hind flipper nails of same size on all 5 digits	Hind flipper nails larger on middle 3 digits, rudimentary on digits 1 and 5
Mammae with 2 or 4 teats	Mammae with 4 teats
Epidermis dark in colour	Epidermis light in colour

Skull characters listed in chapter on skull and skeleton

SUPERFAMILIES

The pinnipeds are conveniently divided into two superfamilies – Phocoidea and Otarioidea, whose more obvious differences may be listed (see Superfamilies table). The Phocoidea includes the 'true' or 'earless' seals, while the Otarioidea includes the sea lions, fur seals and Walrus. Some of the skull characters that separate these two groups are listed in the chapter on the skull, and differences in the various parts of the skeleton are also mentioned in that chapter.

FAMILIES

Three families of living pinnipeds are recognized – Phocidae (earless seals), Otariidae (fur seals and sea lions) and Odobenidae (walrus) – the last two forming the Otarioidea. The usual characters that are used for distinguishing the families are listed (see Family table), though of course this is only a selection. Characters 10–17 are taken from Repenning (1972), where much more detail is given. Detailed characters of the skull and also the skeleton of the Otarioidea at superfamily, family and subfamily levels are given in Repenning & Tedford (1977), and the fossil members are included.

Family Odobenidae

The family Odobenidae is divided into two subfamilies, but one of these (Dusignathinae) includes only fossils (see chapter on fossils). Living walruses belong to the subfamily Odobeninae and are in the single genus *Odobenus*.

Family Otariidae

The living otariid seals are often divided into two subfamilies: Otariinae – sea lions, and Arctocephalinae – fur seals. Scheffer (1958) followed this trend, saying that he felt that 'The living members at least, fall naturally into two fairly distinct groups'. The usual, more obvious differences between the two groups are the blunter nose and sparse underfur of the sea lions, compared with the sharper nose and abundant underfur of the fur seals.

Most sea lions have a larger body size than most fur seals. Most sea lions have the third upper incisor larger, with a circular cross-section, and usually five upper post canines, compared with the shorter third upper incisors with an oval cross-section and the usual complement of six upper post canines of fur seals (Repenning *et al.*, 1971). Fur seals have the tip of the baculum very narrow, whereas it is broader in sea lions (Kim *et al.*, 1975). The behavioural differences between sea lions and fur seals are not very distinct. The karyotypes of two sea lions at least (*Zalophus*, *Eumetopias*) are distinct from those of *Arctocephalus* and *Callorhinus*, though all have the same number.

The differences between the sea lions and fur seals however, are not regarded as being of very great significance, and Repenning (1976) suggests that there is no need for subfamilies to be recognized within the Otariidae. In a later, more comprehensive publication, Repenning & Tedford (1977) enlarge on this opinion and use the subfamilial names within inverted commas as a 'convenience in expression' only. Mitchell (1968) notes the presence of hybrids between *Arctocephalus* and *Zalophus* and asks 'Is this proper between members of putative subfamilies of mammals?'

The two genera of fur seals, *Arctocephalus* and *Callorhinus* are, however, regarded as worthy of generic separation. Many characters do not hold good all the time, but *Callorhinus* is separable from *Arctocephalus* by three characters. It lacks fur on the dorsum of the fore flipper (the fur of the limb stops abruptly at a line across the wrist), the facial angle is abrupt (less than 125° in *Callorhinus* and more than 125° in *Arctocephalus*), and the premaxillary bones, in direct lateral aspect, always thin to less than half their antero-posterior dimension midway between the incisors and the first contact with the nasal bones (Repenning *et al.*, 1971). The shape of the adult baculum in *Callorhinus* with its 'figure of eight'-shaped apex is different from all other otariids and this, together with other unusual facts of its life suggest that it is 'a specialized offshoot of the Arctocephaline stem' (Kim *et al.*, 1975). Both genera of fur seals share a rather generalized sucking louse (*Proechinophthirus*) but have different species of the louse, indicating that they have evolved separately for some time (see chapter on parasites).

The bacula of *Arctocephalus* and the sea lion *Zalophus* are not dissimilar and are regarded as the most primitive of those found in otariids. *A. p. pusillus*, as noted in the chapter on fur seals, is the most sea-lion-like of the fur seals, and in this respect the hybrid pups produced between these two animals should be remembered (see chapter on reproduction).

Little is known of the relationships between the different genera of sea lions. They are a relatively

recent group, having evolved from the main *Arctocephalus* line probably less than 3 million years ago (see chapter on fossils), and all of them share the same species of sucking louse, which is found only on sea lions (see chapter on parasites). The bacula show characters that may have some significance in showing relationships (Morejohn, 1975, Kim *et al.*, 1975). The changing shape of the apex of the developing baculum of *Eumetopias* shows that it passed through stages of resembling first that of *Arctocephalus* or *Zalophus*, then *Otaria* and *Neophoca*, then *Phocarctos*, before assuming the final shape characteristic of *Eumetopias*.

Family Phocidae

The Phocidae used to be divided into three subfamilies – the Phocinae, Monachinae and Cystophorinae based mainly on the incisor count ($\frac{3}{2}$, $\frac{2}{2}$, $\frac{2}{1}$ respectively) and the fact that the Elephant Seals (*Mirounga*) and the Hooded Seal (*Cystophora*) had an inflatable proboscis. More detailed comparison of the skull and skeleton (King, 1966) suggested that *Cystophora* would fit into the Phocinae, and *Mirounga* into the Monachinae. A slight readjustment of some of the cranial characters was made by Burns & Fay (1970), but the subfamilies of the Phocidae remain as the Phocinae – or Northern phocids, and the Monachinae – or Southern phocids, although the geographical common terms should not, of course, be taken too precisely. (See Phocid subfamily table.)

SUBFAMILY PHOCINAE The subfamily Phocinae has previously been divided into two tribes – Phocini and Erignathini (Scheffer, 1958, King, 1964). The addition of *Cystophora* to this subfamily, and a more detailed comparison of large numbers of skulls (Burns & Fay, 1970) has altered this situation. The latter authors recognize the tribes Phocini, Erignathini and Cystophorini, the first including Grey, Harbour, Ringed, Harp and Ribbon Seals, while the Bearded and Hooded Seals are in their own tribes (see also chromosome section p. 136). The most useful characters for distinguishing the tribes are listed (see table), but reference to the detailed work of Burns & Fay (1970) is essential to understand the variation involved. The Erignathini are regarded as the most primitive of the arctic phocids because of the resemblances to the Monachinae (Burns & Fay, 1970, King, 1966) (and see sections on *Erignathus* and on chromosomes). *Cystophora* is sufficiently different to merit a separate tribe, but resembles the other arctic phocids more than does *Erignathus*.

No subtribes are recognized within the subfamily as there is too much variation and intergradation.

FAMILY CHARACTERS

Phocidae	Otariidae	Odobenidae
1. No external ear pinna	Small pinna present	No external ear pinna
2. Tail distinct and free	Tail distinct and free	Tail enclosed in web of skin
3. Guard hairs without medulla	Guard hairs with medulla	Guard hairs without medulla
4. Tip of tongue notched	Tip of tongue notched	Tip of tongue rounded
5. Testes internal	Testes scrotal	Testes internal
6. Upper canines normal in size	Upper canines normal in size	Upper canines enlarged to form tusks
7. No grooves on upper incisors	First and second upper incisors with transverse groove	No grooves on upper incisor
8. Lower incisors 1 or 2 on each side	Lower incisors 2 on each side	Lower incisors absent
9. Lower jaw symphysis not fused	Lower jaw symphysis not fused	Lower jaw symphysis fused in adults
10. Bulla inflated	Bulla little inflated	Bulla moderately inflated (though appears flat externally)
11. Round window opens externally	Round window opens into middle ear cavity	Round window opens into middle ear cavity
12. Ossicles large	Ossicles not enlarged	Ossicles large, but like otariid in shape
13. Tympanic membrane large	Tympanic membrane small	Tympanic membrane large
14. No internal acoustic meatus	Have round acoustic meatus	Have wide, shallow acoustic meatus
15. Basal whorl of cochlea runs transverse to orientation of skull	Basal whorl of cochlea runs postero-lateral to orientation of skull	As in otariids (and other carnivores)
16. Tentorium reduced, not touching petrosal	Tentorium strong, not touching petrosal	Tentorium strong, pressed against petrosal
17. Mastoid not fused to jugular process	Mastoid large, fuses with jugular process	Mastoid large, usually not fused to jugular process
18. Supra orbital processes absent	Supra orbital processes present	Supra orbital processes absent

Within the tribe Phocini, *Halichoerus* is the most distinct in the shape of its snout. A detailed comparison of skull characters (Burns & Fay, 1970) shows that in the other four genera usually included in the Phocini – *Phoca*, *Pusa*, *Pagophilus* and *Histriophoca* – while their skulls are quite distinct and recognizable, the variation is such that in every character they inter-grade. The above authors are of the opinion that the skull differences 'do not quite attain the generic level' and they use *Phoca* for all of them, and use the 'usual' generic names as subgenera. Electrophoretic findings from the red cells and blood sera support this use of *Phoca* (McDermid & Bonner, 1975).

PHOCID SUBFAMILY CHARACTERS

Phocinae	*Monachinae*
1. Lateral swelling of mastoid process forms an oblique ridge at an angle of about 60° from the long axis of the mastoid bone as a whole	Ridge of mastoid process absent.
2. Petrosal widely exposed in foramen lacerum posterius	Petrosal not usually widely exposed in foramen lacerum posterius
3. Foramen lacerum posterius extending anteriorly along medial edge of bulla (except *Erignathus*)	Foramen lacerum posterius small, not extending along medial edge of bulla
4. Spine of scapula well formed	Spine of scapula reduced, acromion process knob-like
5. Projecting ledge on cuneiform of carpus	No projecting ledge on cuneiform
6. Metacarpals of digits 1 and 2 approximately similar in size	First metacarpal noticeably larger than others
7. Metacarpal head with well developed palmar ridge	Metacarpal heads without palmar ridge
8. Ilium everted, excavated laterally (except *Erignathus*)	Ilium less everted, without lateral hollow (includes *Erignathus*)
9. Large claws on fore and hind flippers	Reduced claws on hind flippers

PHOCINAE – TRIBE CHARACTERS (skull only)

1. Least interorbital width in adult skulls in the anterior half of the interorbital septum
2. Petrobasilar vacuity continuous with the posterior lacerate foramen and extending anteriorly usually beyond the posterior carotid foramen
3. Incisors $\frac{3}{2}$ (or $\frac{2}{1}$)
4. Least depth of jugal greater than the least width between the orbits
5. Least depth of jugal less than $\frac{1}{3}$ of its greatest diagonal length
6. Roots of the upper incisors greatly compressed laterally
7. The process from the posteromedial edge of the bulla in the vicinity of the carotid foramen absent

Of these seven characters the tribe Phocini differs from all other pinnipeds in respect of character 1.

The tribes Phocini and Cystophorini agree in characters 2, 4, 5, 6, 7, while the tribe Erignathini is usually different.

The tribes Phocini and Erignathini agree in the number of incisors $\frac{3}{2}$, while the Cystophorini have $\frac{2}{1}$.

(Taken from Burns & Fay, 1970 – who should be referred to for the detailed comments.)

According to Burns & Fay (1970) these four genera (or subgenera) show no obvious lines of relationship other than common ancestry from a more generalized form, 'While *Histriophoca*'s nearest structural relative is *Pagophilus*, the latter shows equal or closer relationship with *Phoca*'. *Histriophoca* is regarded as the most divergent from a common ancestor of the '*Phoca*' group, *Pagophilus* has great resemblances to *Pusa*, *Phoca* and *Histriophoca*, but is probably closer than any to the ancestor of this group.

SUBFAMILY MONACHINAE

Scheffer (1958) discusses briefly the early history of this subfamily. He notes that the monk seals were frequently separated from the Antarctic phocids at subfamily level but does not consider that this is warranted – 'In fact the only important difference between *Monachus* and the southern phocids, keeping in mind the great variation within the latter group, is one of geography' (Scheffer, 1958). He does, however, create two new tribes – Monachini for the three species of *Monachus*, and Lobodontini for the four genera of Antarctic phocids (see table for Scheffer's tribe characters). With *Mirounga* now included in the Monachinae, Scheffer's tribe characters do not hold good, though *Mirounga* is obviously better fitted for the Lobodontini rather than the Monachini. It is the opinion of Hendey & Repenning (1972) that 'Recognition of any tribal subdivision of the Monachinae now seems pointless'.

TRIBE CHARACTERS*

Tribe Monachini
'Nasal processes of premaxillary broadly in contact with the nasals; post canines wide and heavy, crushing type. Mammary teats 4. Embryonal (and new born) pelage jet black; adult pelage never spotted; vibrissae smooth. Breeding only in subtropical waters'.

Tribe Lobodontini
'Nasal processes of premaxillary barely touching (*Leptonychotes*) or not touching the nasals; post canines not crushing type. Mammary teats 2. Embryonal (and new born) pelage white (grayish in *Leptonychotes*); adult pelage spotted in some species; vibrissae very faintly beaded. Breeding only in antarctic waters'.

* Taken from Scheffer (1958) – but does not include *Mirounga*.

The structure of the bony parts of the ear indicates that *Monachus* is the least specialized genus of the living monachine seals (Hendey & Repenning, 1972, Repenning *et al.*, 1979, Repenning & Ray, 1977). The origin of *Monachus schauinslandi* and *Monachus tropicalis* from a common ancestor is noted in the chapter on fossils. The very primitive nature of *M. schauinslandi* and the very early date at which it must have separated from its ancestral population is also mentioned. *M. schauinslandi* has certain skull characters in common with *M. tropicalis* and different from *M. monachus* (King, 1956) which are in line with its ancestry. It has also some characters which indicate its primitive nature. *M. schauinslandi* has the fibula unfused to the proximal end of the tibia – a primitive feature, unique among phocids (Ray, 1976). It also has a foramen in the pubis for the obturator nerve, between the acetabulum and the obturator foramen which is not present in any other phocid, living or fossil (King & Harrison, 1961, Ray, 1976, Repenning & Ray, 1977). Repenning & Ray

(1977) also note that the bony structure of the ear is more primitive in *M. schauinslandi* than in *M. tropicalis*, and also is more primitive 'than that of any known living or extinct species of monachine seal and probably of any phocoid seal'. In the structure of the posterior vena cava with its complicated duplication and persistence of a complex network of anastomotic channels *M. schauinslandi* is again primitive (King & Harrison, 1961). The Hawaiian Monk Seal has been described as 'the modern representative of the most ancient of living phocoid lineages, and as such might be characterized not altogether improperly as a "living fossil".' (Repenning & Ray, 1977.)

The origin of *Mirounga* is noted in the chapter on fossils, and the skull differences of the two species in the chapter on elephant seals.

The relationships of the Antarctic phocids are not yet known. In considering the relationships of the Late Pliocene South African monachine *Homiphoca* (see chapter on fossils), Muizon & Hendey (1980) also compare the cranial and dental characters of the four Antarctic seals. They make an informal separation of the four genera, based in part on the following characters:

	Lobodon and *Hydrurga*	*Leptonychotes* and *Ommatophoca*
1.	High specialized cheek teeth with well developed accessory cusps	Reduced cheek teeth, small or absent accessory cusps
2.	Posterolingual cusps well developed on upper cheek teeth	Cusps not developed
3.	Molars well developed	Molars reduced
4.	Long snout	Short snout
5.	High occiput	Low occiput

Chapter 2 *The living species*

SEA LIONS

Northern sea lions
Steller's Sea Lion Eumetopias jubatus

DISTRIBUTION

The Northern, or Steller's Sea Lion, is an abundant, widely distributed sea lion of the cooler regions of the north Pacific. It is present on both sides of the Pacific, ranging from Hokkaido in the west to the southern waters of California, and is usually found on rocky islets or on the open sea coast. Animals are found on Hokkaido, where they are said to be abundant, especially on the Shiretoko Peninsula (Nishiwaki, 1972), on the Kuril Islands, and round the shores of the Sea of Okhotsk, Sakhalin and Kamchatka, and on the Commander Islands.

There are breeding colonies off Sakhalin, on Ion and Yamskiye Islands in the northern sea of Okhotsk, at Cape Shipunski and in the Gulf of Kronotski on the eastern coast of Kamchatka. The Pribilof Islands form the northernmost breeding ground. After the breeding season, males may move north and reach St Lawrence Island near Bering Strait.

The centre of abundance is probably on the Aleutian Islands (Kenyon & Rice, 1961) where a survey in 1959 and 1960 revealed some 98 rookeries and an estimated population of about 100 000 animals (see map), and a survey published in 1978 estimated as many as 200 000 animals (in Braham *et al.*, 1980). There has, however, been a significant population decline in the Eastern Aleutian Islands (Fox Islands) since 1957. At that time the population on these islands exceeded 50 000, but a survey in 1977 estimated a population of less than 25 000. There is at the moment no explanation for this decrease though some correlation with the pathogen *Leptospira pomona* has been suggested. This organism has been identified in *Eumetopias*, and infection with it has resulted in reproductive failure and reduction in numbers of *Zalophus* (Braham *et al.*, 1980). Breeding colonies of large size are also to be found on Kodiak Island in the Gulf of Alaska, and on Chicagof, Baranof and Prince of Wales Islands off the southern Alaskan coast. Further south, the British Columbia population has its main rookeries on Scott Islands where about three quarters of the Canadian population are to be found. In Californian waters there is an adult breeding population of about 1000 individuals on Ano Nuevo Island which is about 65km south of San Francisco

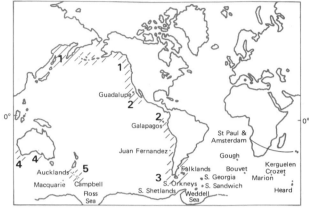

World distribution of sea lions,
1 Eumetopias, 2 Zalophus, 3 Otaria, 4 Neophoca,
5 Phocarctos.

Distribution of Steller's Sea Lion.

(Bartholomew, 1967), and breeding extends as far south as San Miguel Island in the Californian Channel Islands. The total Steller's Sea Lion population, estimated in 1960 by Kenyon and Rice (1961) was thought to be between 240 000 and 300 000.

There is some indication of seasonal movement, particularly by the males. Both on San Miguel and Ano Nuevo females and young are present throughout the year, but adult bulls are only on the islands in the breeding season (June–July), moving north at other times of the year (Orr & Poulter, 1967).

DESCRIPTION

Steller's is the largest of the sea lions, adult males having a nose to tail length of about 3m and weighing approximately 1 tonne. Adult females are smaller, about 2·2m in length and 270kg in weight. Both sexes are a slightly variable yellowish buff colour, and adult males develop a heavy muscular neck bearing a mane of longer, coarser hair. Newborn pups are about 1m in length, weigh 18–22 kg, and have a thick grey-brown or chocolate brown coat which changes to dark brown after six months, and to the lighter colour of the adult by the end of the second year. Albino pups with pink flippers and eyelids have been seen, and they appeared to have some difficulty in seeing clearly.

BREEDING

The pups are born between the middle of May and the middle of July, most births taking place between 5–16 June (Pitcher & Calkins, 1981). They stay close to their mothers for a week or so, but when several weeks old they gather in groups and play and sleep together. They play in rock pools, engaging in mock battles, but will not voluntarily go into the open ocean. By about September they have progressed to swimming in surge channels between the rocks, and then gradually progress to the ocean as their swimming skills increase (Orr & Poulter, 1967).

The pup may start to suckle within half an hour or so of being born, and during its first few days spends a large part of its waking hours in this activity. It has been noted that the mother remains calm while the pup is suckling, but becomes agitated, moving round the pup, lifting it and vocalizing when it stops suckling (Sandegren, 1970). Although probably supplementing their diet with small invertebrates as they grow older, the pups continue suckling at least until the next pup is born, and may occasionally continue for 18 months. Animals up to 39 months old have been seen either suckling or with milk in their stomachs. Occasionally females may suckle more than one individual at a time – two subadults (1–3 years old), or one subadult and a pup, or two pups have been seen to be suckled simultaneously. Pitcher and Calkins (1981) note – 'One large female nursed a smaller female that, in turn, was nursing a subadult. Milk was seen to flow from nipples of both females.' Sometimes the newborn pup may die and the yearling continue to suckle, and occasionally, in disturbed situations the newborn pup is rejected or ignored in favour of the yearling (Sandegren, 1970).

The dominant breeding bulls arrive at the rookery in early May, their numbers reach a maximum by early July, and by the end of August they have gone back to sea. Aggressive behaviour occurs between these big bulls and they frequently have scars on their heads and necks. The bulls maintain their territories and groups of cows gather round them. Conditions on the rookery are quieter when there are a greater proportion of adult males and females. It seems to be the subadult males that are responsible for most of the disruptive behaviour in the colony (Harestad & Fisher, 1975).

The females seem to have favoured spots for giving birth – gently sloping areas just above high tide level, well protected from waves and sun, and cows seem to be attracted to areas where other cows have given birth. The pup may be born head or hind flippers first, and occasionally the female will assist the birth by pulling at the pup with her teeth. The umbilical cord usually breaks spontaneously, and the placenta may appear with the pup, or after an interval. The female seems to take an interest in the placenta, staring at it, sniffing and nipping it, and slapping it with a fore flipper – this behaviour being interpreted as part of the general caretaking of the young (Sandegren, 1970).

For about the first two days after the pup is born the female is very aggressive towards other cows and is very active towards her pup. Frequently she will seize the pup by its neck and lift it up, only to drop it on the ground, doing this many times soon after birth. The pup shows signs of irritation and it may be that this performance is directed towards stimulating it into activity. The female makes a high-pitched slightly rattling sound to her pup, and the pup gives a low-pitched hoarse bleat. The bulls also make a rattling sound like a distant two-stroke motor bike. The bulls appear to ignore the pups.

Throughout the breeding season the females have a very definite type of display directed to the bulls. The female will crawl to the bull, vocalizing, salivating, and with vibrissae erected. She will wind her body round that of the bull, rubbing her head and neck against him and biting him. The bull reacts to the display by pressing his head on her back and sniffing at her genitalia. This display does not always mean that the cow is receptive to the bull, and copulation does not always follow (Sandegren, 1970).

Oestrous occurs between 10 and 14 days after parturition, and copulation takes place at this time, the peak being between 7 June and 4 July. The blastocyst implants in late September or October after a delay of about 3·5 months. Copulation usually takes place on land, though sometimes in shallow water. Males become sexually mature between 3 and 8 years old. In females the first pregnancies occur between 2 and 8 years old, the average age at the first pregnancy being 4·9 years (Pitcher & Calkins, 1981).

These sea lions are gregarious animals, and females and young animals in groups of several hundred individuals can often be seen forming compact rafts while floating on the water. These rafts drift about and sometimes all the animals forming them will dive simultaneously.

MORTALITY, PREDATORS AND PARASITES

Wounds on the fore part of the body are usually caused by other sea lions, but wounds on the hind end may be caused by sharks and killer whales. The most usual cause of death of young pups is drowning, sometimes caused by high waves that wash them into the water. Young pups who fail to avoid them may also be crushed by bulls.

Parasites such as trematodes in the intestine and bile duct, cestodes in the intestine, nematodes in the stomach, and larval nematodes in the lungs have been recorded. A massive infection of nematodes will cause ulcers which may lead to massive stomach bleeding and death. Verminous pneumonia from the larval nematodes may also cause death. Pups get infected with lice, and adults usually have mites in the nasal cavities.

FEEDING

Diet consists of squid, and also a wide variety of fish, almost anything available being taken such as herring, halibut, sand lances, flounders, rockfish, greenling, sculpin, pollack, cod and lampreys. Some of the fish are of commercial importance, and a certain amount of salmon is taken from the nets with consequent damage to fishing gear. The amount of salmon eaten by both sea lions and Harbour seals on the coast of British Columbia is estimated to be about 2·5 per cent of the annual commercial catch, and at this level is regarded as of negligible importance (Spalding, 1964). Sea lions watched feeding for 14 hours at the mouth of the Roque River, Oregon were seen to take 87 per cent of lampreys to 2 per cent of salmon – the remainder of their food being unidentifiable. These lampreys were up to 60cm long, were shaken violently by the sea lion and swallowed whole. Lampreys are parasites of salmon, so by taking such large numbers the sea lions are benefiting the salmon (Jameson & Kenyon, 1977).

Although the diet is mostly fish and squid, this sea lion occasionally takes Ringed Seals, Northern Fur

Steller's Sea Lion, Eumetopias jubatus *on Ano Nuevo Island, California. The difference in size between the large adult male and the smaller adult female is clearly apparent. Photo. J. E. King.*

Seals and sea otters (Gentry & Johnson, 1981). Young male sea lions have been seen killing and eating Northern Fur Seal pups up to about five months old. This has been known to happen since 1967 in the inshore waters of St George Island, Pribilofs, but not, strangely enough at St Paul Island only 65km away. The sea lions swim up under the fur seal pups, catch them by the abdomen and eviscerate them with a quick shake of the head. They then eat the carcass at sea, attended by scavenging birds. This predation occurs between August, when the fur seal pups first enter the water, and November when most of them have left the island. In 1975 the sea lions were estimated to have killed between 2700 and 5400 fur seal pups, this being 3·4–8 per cent of the pups alive at the time. The fur seal pups are regarded as a supplementary part of the young sea lions' diet, rather than an important part, and the predation does not add significantly to the level of mortality of the fur seals.

EXPLOITATION

Products of the sea lion were formerly used by Aleutian natives; skins for boat coverings, harness, waterproof clothing and boots, the meat eaten and the fat used for fuel. The carcasses are less in demand now, although the very thick hide may be used to a certain extent for leather, and the meat may go to fox and mink farms.

Californian Sea Lion Zalophus californianus

DISTRIBUTION

The Californian Sea Lion is one of the best known sea lions, seen in almost every zoo, and known to thousands as a performer in circuses. As its name suggests, it occurs mainly in Californian waters, but is also to be found on the Galapagos Islands and possibly in Japanese waters. They are shore living and coastal animals, rarely being found further than about 16km out to sea.

The three populations have been given separate subspecific names which, at least, serve to distinguish them geographically. The Californian animal is *Zalophus californianus californianus* (Lesson, 1828), the Galapagos animal *Z.c. wollebaeki* (Sivertsen, 1953), and the Japanese animal *Z.c. japonicus* (Peters, 1866). Although geographically distinct, it is difficult to tell the animals apart by their skulls, particularly as there are very few available from the Galapagos and Japan. Compared with others of similar age, skulls from the Galapagos are smaller and narrower, but so far, no differences have been demonstrated between the skulls of the Japanese and Californian animals (Sivertsen, 1953, King, 1961).

There is virtually no information on the Japanese animals and it is uncertain whether any still remain in this area. They apparently once lived on at least some of the small Japanese islands such as the Izu Islands, Oki Island and Take-shima (= Liancourt Rocks, 37° 15'N, 131° 52'E) (Nishiwaki & Nagasaki, 1960). After the Second World War the latter island was claimed by Korea and there has since been no information on sea lions. Only rumours now exist that there might still be sea lions sheltering under the Korean cliffs (Nishiwaki, 1973).

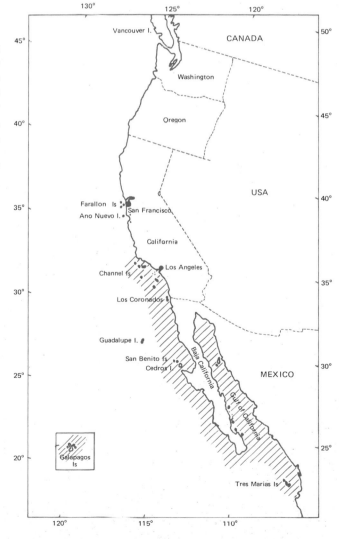

On the Galapagos Islands sea lions are plentiful, and occur on all the islands of the group. An estimate of 20–50 000 animals was made in 1963 (Brosset, 1963).

The sea lions are abundant in the waters of southern California and Mexico (see map). They live and breed on the islands in the Gulf of California and along the Pacific Mexican coast probably as far south as Islas Tres Marias. Breeding colonies may also occur on the Pacific coast of Baha California and on the associated islands such as Guadalupe, San Benito and Cedros Island. In the waters of southern California the sea lions breed on the Californian Channel Islands group in the vicinity of Los Angeles, the main breeding rookeries in this group being on Santa Barbara, San Clemente, San Nicolas and San Miguel. The latter island is the most northerly breeding place although the occasional pup is born on the Farallone Islands.

A census of the Channel Islands sea lion population in June 1964 – in the breeding season when the greatest numbers are present – gave a total of over 34 000 animals. A later count of animals on San Nicolas in June 1969 gave a population of about 11 000 on this island and this indicated that the total Channel Islands population was much the same in 1969 as it was in 1964 (Odell, 1971). Counts of animals in the Guadalupe–San Benito area in June 1968 gave a total of over 15 000 sea lions with most of them being on Islas Benito, Cedros and Natividad (Brownell *et al.*, 1974).

After the breeding season most of the adult and subadult males leave the breeding rookeries and move northward. They are regular winter visitors on the coast of southwest Vancouver Island where the numbers increase to a peak of nearly 500 in February, and then decrease from about May onwards as they return south again (Bigg, 1973, Bartholomew, 1967). On Vancouver Island they share some of the haul out spots with Steller's Sea Lions, but they then use the inner parts of the reefs while *Eumetopias* remains on the outer rocks.

Ano Nuevo Island, just south of San Francisco, is another place where large numbers of adult male sea lions congregate in autumn and winter. Some 13 000

A large group of male Californian Sea Lions congregating on the Ano Nuevo Island, California. The seals also occupy the now deserted light house keeper's buildings in the background. Photo. J. E. King.

sea lions may be present on this small island of about 6·5 hectares which they share with *Eumetopias*, *Mirounga* and *Phoca*, and it has been calculated that on some of the more crowded beaches only about 0·6m² is available for each animal (Orr & Poulter, 1965).

On Ano Nuevo there seem to be two brief population peaks – a lower one about the middle of May, and a higher one at the beginning of September. After the May peak the population declines until the middle of June when there are practically no sea lions on the island, and then starts to build up again after the end of July. After the September peak the numbers gradually level off to the winter population of about 1000 animals. The virtual absence of animals in June and July is, of course, due to the breeding season further south, and animals resting on the island on their way to and from places further north would account for the peaks in population numbers (Orr & Poulter, 1965). Females and young animals either remain on the breeding grounds throughout the year or move southward.

DESCRIPTION

Adult males are about 2·4m nose to tail length and weigh about 300 kg, adult females are smaller and slimmer, about 1·8m in length and 100 kg in weight. Both sexes are a dark chocolate brown colour, although there may be variations in shade. The adult male has a very noticeably raised forehead due to the extremely high sagittal crest on the skull. As noted in the chapter on the skull, this crest may be about 4cm in height, it starts growing in the animals fifth year and is fully developed by the tenth year (Orr *et al.*, 1970). There may be some lighter coloration of the hair over the top of this crest in adults.

BREEDING

Most observations have been made on the animals of the Californian Channel Islands, particularly on the colony on San Nicolas. Here, some bulls have been seen defending territories during May, June, July and August, but such behaviour is at its peak in late June and early July. Most of the territories are along the

Adult male Californian Sea Lions, Zalophus californianus *on Ano Nuevo Island, California. The typical high forehead of the adult male can be seen. Photo. J. E. King.*

water's edge, so the bulls always have access to the sea to keep cool. They will fight to defend their territories, though once established, fights are less frequent. The bull patrols his territory, in water and on land, barks almost incessantly to advertise his presence, and makes formalized movements to establish his boundaries. Each bull will hold a territory for about two weeks and will presumably then go to sea to feed and fight for a new territory on returning. Thus individual territories are held by a succession of different bulls.

The females are gregarious, but indifferent to the territorial boundaries of the males, and the males do not herd them or restrain them. If conditions, such as storms or tides dictate that a group of females moves, then the males will modify their territories. Activity on the rookery is the same, or possibly more pronounced at night when conditions are cooler. *Zalophus* is one of the noisiest of sea lions and barks continually.

On San Nicolas most pups are born during June. They are about 75cm nose to tail length, weigh about 6 kg and are a rich chestnut brown colour. The female is protective towards her pup for about two or three days, maintaining contact with it and tugging it or lifting it should it move away. If she needs to go to the sea or to a pool to cool off she takes the pup with her, maintaining the contact by pulling the frequently reluctant pup into the water. It is situations like this that have given rise to the story that sea lions teach their pups to swim. After this brief period of parental attention, the female then starts to leave her pup for increasing periods.

The pups gather in groups and spend much of their time sleeping, playing, or making increasingly bold exploratory trips round the rookery. By late July they are playing in the tide pools.

Very young pups suckle frequently throughout the day. However, by the time the pup is about three weeks old it will suckle for about half an hour once a day, and less frequently as it grows older. As with other sea lions suckling continues at least until the next pup is born.

The female comes into oestrous about two weeks after the pup is born. She will usually solicit the attentions of the bull by rubbing her body against his and making submissive movements. Copulation seems to be actively terminated by the female pulling away and biting at the neck of the male (Peterson & Bartholomew, 1967).

The natural history of the sea lions on the Galapagos Islands is not well known but it appears that the breeding season there lasts for several months possibly because of the equable climate. In 1963–64 the first newborn pups were seen on 25 September and pups continued to be born until the end of February. Harem activity and patrolling by the bulls extended from August until March. This occurred on South Plaza Island, off the east coast of Indefatigable (Santa Cruz), but a newborn pup was seen at the end of June on the west side of Albemarle and other young pups seen there in September and October (Snow *in litt.* Orr, 1967). Fur seals (*Arctocephalus galapagoensis*) are also on the Galapagos Islands. They usually occupy the rocky parts of the coast, whereas *Zalophus* is usually on the smoother beaches, but the two animals exist together apparently without animosity.

MORTALITY, PREDATORS AND PARASITES
Killer whales and sharks take sea lions, and fishermen and so-called 'sportsmen' may shoot them. The sea lions suffer from various maladies, both in captivity and in the wild. Lungworms frequently cause death from verminous pneumonia, and seal pox virus causes a skin disease in free-living and captive animals. Lesions of the aorta, severe vertebral column and hind leg deformity (Morejohn, 1969) and an epizootic of leptospirosis are among pathological conditions that have been reported, the latter causing some reproductive failure and reduction of numbers. These sea lions have a high level of helminth parasitism with cestodes and trematodes in the intestines, hookworms in the intestines of unweaned pups, and nematodes in the lungs and gut. The presence of the opal-eye fish as an intermediate host for the nematode *Parafilaroides decorus* is mentioned in the chapter on parasites. Mites are present in the trachea, and in the nasopharynx, and lice are present particularly round the snout and anus of pups.

FEEDING
Californian Sea Lions feed mainly on squid and octopus, but may also take any available fish such as herring, hake and anchovies. Although present in salmon fishing areas they seem to show a greater preference for squid rather than salmon, and are known to take lampreys which are parasites of salmon.

EXPLOITATION
Sea lions possess the normal pinniped playfulness and have been seen, while under water, to chase and catch their own exhaled bubbles of air. Circus training is accomplished by constantly rewarding the sea lions with fish. Individual methods of trainers probably vary, but ball balancing may be achieved by constantly throwing a ball at a sea lion until it is accidentally balanced, or by holding a ball on the animal's nose, until it realizes what is meant.

Balancing is of course the main forte of the performing sea lion. An accomplished animal can balance a wineglass full of water on its nose and then go down some steps and into a moat of water. It will then swim over a hurdle, and finally get out of the water again – all without spilling the water in the balanced wineglass. It may well then applaud its own performance by enthusiastic clapping with its fore flippers.

Sea lions are intelligent animals and have a good

memory, and although it may take a year before a trick is ready to be shown in public, the sea lion will be able to demonstrate it perfectly later, even after a complete rest of three months. The performing life of a sea lion may last eight to twelve years.

Southern sea lions

Southern Sea Lion Otaria byronia

DISTRIBUTION

The Southern Sea Lion is found on both Atlantic and Pacific coasts of South America, approximately from Uruguay to Peru (see map). Large colonies perhaps totalling 50 000 animals are on Isla de Lobos and associated Uruguayan Islands (Vaz Ferreira, 1950) though the animals on some islands are being drastically reduced in numbers because of persecution by local fishermen (Pilleri & Gihr, 1977). Sea lions probably occur at any suitable place along the coast, with concentrations in various areas. One of these is the Peninsula Valdes, Argentina where a population of over 14 000 was estimated in 1975 (Ximenez, 1976). Some 70 other colonies along the Argentinian coast had a population of over 150 000 animals in 1954 (Carrara, 1954). There was a large population on the Falkland Islands, estimated at nearly 400 000 in 1937 (Hamilton, 1939b), but aerial surveys in 1965 and 1966 showed a drastic reduction in numbers and the Falkland Island population was then estimated to be only 30 000 and the situation has not changed since then (Strange, 1972, 1979, Laws, 1973). Sea lions are to be found right down to the tip of South America on Tierra del Fuego and Staten Island.

Sea lions have been protected by the Chilean government since 1965 though there is no recent information on the status of these animals along the Chilean coast (Laws, 1973). There is no good evidence for these sea lions ever having lived on the islands of Juan Fernandez. Sea lions breed along the Peruvian coast about as far north as Isla Lobos de Tierra (6° 30′S) but the greatest numbers occur between latitudes of approximately 13–15°S. The breeding groups occur now on precipitous rocky islets and similar relatively inaccessible spots such as the Paracas Peninsula and the nearby islands of the Chinchas and Ballestas groups. Formerly beach sites were used by this sea lion but the animals were too vulnerable to indiscriminate hunters, and have either been hunted out from such places or have deserted them for more isolated beaches. As many 36 650 skins of sea lions and fur seals

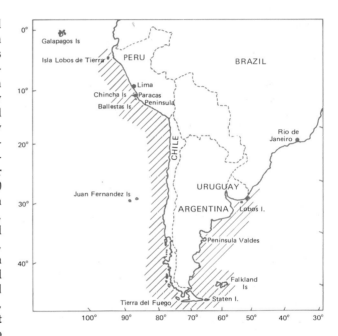

obtained on a short stretch of the Peruvian coast were bought by a dealer in 1941–2. Sea lion hunting was prohibited in 1959 and although animals are undoubtedly still killed illegally, the population is building up again and in 1966 the Peruvian population was estimated as being about 20 000 (Grimwood, 1968). About 400 adults and pups were seen on a shingle beach at the foot of a cliff on the Ballestas Is. in March 1974. A partially decomposed body of an adult male *Otaria* found on the Galapagos Islands in 1973 seems to be the only record of this sea lion on these islands, but was probably just a straggler (Wellington & de Vries, 1976).

DESCRIPTION

Adult male *Otaria* are about 2·3m nose to tail length and weigh about 300 kg when in peak condition; adult females are 1·8m and weigh 144 kg. There is a certain amount of variation in colour, though a brown colour prevails in both males and females. The adult bulls are generally very dark brown with the mane of slightly longer coarser hairs on the heavy muscular neck a slightly lighter colour. The belly of both males and females is a dark yellow. Females are dark brown with the back of the head and neck dull yellow, or sometimes with the whole head and neck yellow (Hamilton, 1934). In both sexes occasional lighter animals of a more golden colour are seen and this variation in colour is very characteristic of *Otaria* and has been known for several centuries. Hamilton (1934) notes that 'In Pigafetta's account of Magellan's voyage (1520) mention is made of "sea-wolves of many colours" being seen on the coast of South America . . . on the north side of the entrance to the Rio de la Plata'.

After the moult from the natal coat, the first adult coat is dark grey, developing a reddish tinge during the first year. After this, young animals of both sexes are reddish brown, deepening in colour with the in-

creasing age, particularly in the male. *Otaria* adult males have the characteristic thickened neck and chest region as in other sea lions, but the apparently enormous heavy head with its broad, rather upturned snout gives a profile which is typical only of this sea lion.

BREEDING
At birth the pups are 85cm in nose to tail length and are black, fading to a dark chocolate in colour. Pups may be born from about 25 December to 15 January, although most are born at the beginning of January. The period of lactation seems to be long, as in other sea lions; some pups still suckling until the next pup is born, but starting to eat solid food sometime before this.

In the non-breeding season there is no segregation of any part of the herd, but as the summer approaches the breeding animals congregate, surrounded by a fringe of immature sea lions and idle bulls. All suitable stretches of the coast are occupied, and it is only

A group of Southern Sea Lions, Otaria byronia, *a male with several females and pups. The difference in size between the sexes is clearly seen as are the mane, heavy head and upturned snout of the male. Photo. G. Pilleri.*

because portions of it are unoccupiable that the herd is divided up into rookeries, each consisting of several harems. The average number of cows per harem is nine, but the harems are frequently so close together that their limits are known only to the harem bulls, who are very quick to recognize and dissuade trespassers. Although ready to fight idle bulls, round the edge of his territory, a harem bull will not leave his station, even if a very high tide should almost cover him. Disruptions of the harem system may sometimes occur when groups of subadult males gang up and attack the harem bulls, abducting females and pups (Vaz Ferreira, 1965). The females mate again a few days after they have given birth, and are then allowed by the males to return to the sea, from which they land at intervals to suckle the pups. After all the available cows have been mated there is a gradual disintegration of the harems. The males have by then spent at least two months defending their territory, fighting and mating, all without feeding, and almost without sleeping, and by February they are extremely thin and spend much time sleeping. During the next six months they feed and recover their strength, and even by August are beginning to take an interest in the females again.

As the females return to the sea the pups wander about and tend to gather in groups or 'pods'. They spend much of their time playing and sleeping, but although they play at the edge of the water they do not venture any distance out until coaxed by their mothers. Even then, they try to get out of the deeper water by climbing on their mothers' back. The male becomes sexually mature in the sixth year, and it is then that the mane starts to appear, and the female during the fourth year, probably producing the first pup in the fifth year. Not all the animals moult at the same time, and various stages can be seen between April and August (Hamilton, 1934).

MORTALITY, PREDATORS AND PARASITES
There are relatively few natural enemies of this sea lion though leopard seals may occasionally take a pup.

A group of Southern Sea Lions, Otaria byronia *on a beach on the Ballestas Islands, Peru. Most of the animals are females and their pups. Photo. J. E. King.*

Killer whales are known to take large numbers of pups from the colonies on the Peninsula Valdes, and possibly from other places too. When the pups are about a month old and playing at the edge of the sea, killer whales have been seen to charge in close to the beach, creating panic amongst the sea lions, and taking perhaps twenty sea lion pups in an hour (Wilson, 1975a). In the same area killer whales have been seen patrolling the coast, and one tossed a full grown sea lion 6–8m in the air with its powerful tail flukes, and this performance was repeated several times with the same sea lion, recalling the actions of a cat with a mouse (Bartlett, 1976).

The death rate of pups is, as usual, fairly high, as they may die of starvation, by being crushed by the large bulls, or from serious roundworm infection. Small cuts on the hands of sealers dealing with these, and all other commercially used seal carcasses, often become infected with erysipelas, and the frequency of this, and associated finger infections has lead to them being regarded as an occupational disease known as 'seal finger'. Similar 'whale fingers' occur among whaling men. A seal pox virus has been found in wild *Otaria* resulting in a nodular skin condition of the animal with areas of alopecia (Wilson & Poglayon-Neuwall, 1971).

FEEDING

The main food of the sea lions seems to be squid and the crustacean *Munida* but undoubtedly other organisms such as small fish are eaten. On the Falkland Islands these sea lions have been seen to eat both young pups and adult females of the South American fur seal *A. australis* (Gentry & Johnson, 1981). The penguins of the area, such as Rockhopper, Gentoo and Magellanic penguins are not infrequently taken by sea lions (Hamilton, 1946, Boswall, 1972). Stones are swallowed, sometimes up to 9kg of pebbles and sharp-angled stones being found in a single animal.

EXPLOITATION

From at least the sixteenth century onwards these sea lions have been known and made use of by man for their meat, hide, and the oil obtained from the blubber. Hamilton (1934) notes that the sea lions were an important source of 'refreshment' for travellers down the South American coast. Simon de Alcazaba for instance killed a few hundred sea lions in 1535, and Drake's people 'kylled some seyles for owr provysyon'. At the present time the government of Uruguay kills a certain number of sea lions commercially. No killing is allowed in Argentina.

Australian Sea Lion Neophoca cinerea

DISTRIBUTION

The Australian Sea Lion is found only in Australian waters. It occurs on many of the offshore islands from Houtman's Abrolhos in Western Australia to Kangaroo Island in South Australia. On Houtman's Abrolhos sea lions are not now common, but there are populations of 30–50 on some of the islands between Geraldton and Perth. Carnac Island off Fremantle supports about 30 animals, and although they are present at such places as Eclipse Island and Cheyne Beach, greater numbers are present on several of the islands of the Recherche Archipelago, where Wedge and Daw Islands for instance support about a hundred seals. The total population in Western Australia has been estimated as 700 sea lions (Abbott, 1979, Marlow & King, 1974).

The islands of South Australia from those of Nuyts Archipelago, the Investigator Group, those in and around the southern end of Spencer Gulf, to Kangaroo Island have been surveyed, and a total of about 2300 sea lions counted (Ling & Walker, 1977). Including animals away at sea they suggest that the total popula-

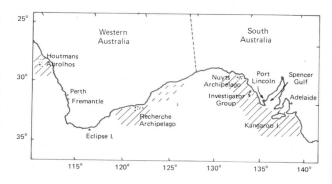

tion is probably between 3000 and 5000 (Inns *et al.*, 1979). Of the South Australian localities, Seal Bay on Kangaroo Island and Dangerous Reef, a small rocky island off Port Lincoln probably have the greatest numbers, Seal Bay with 300–400, and Dangerous Reef with 600–700, though counts vary considerably with the time of year.

Neophoca does not normally occur east of Kangaroo

Island now, though a skull was collected from Cape Barren Island in the Furneaux Group in Bass Strait in 1973 (Guiler, 1978). Matthew Flinders too noted sea lions on some of the islands in this group in 1798 when he was going to rescue the crew of the wrecked *Sydney Cove* (Flinders, 1814, Marlow & King, 1974). Flinders also saw sea lions on Waterhouse Island, off the north Tasmanian coast, and Wood Jones mentions identifying their bones from aboriginal kitchen middens in Tasmania (in Marlow & King, 1974). A pre-1900 specimen has been recorded from King Island in the Bass Strait (Guiler, 1978) and there is a fossil skull from Queenscliff, Victoria that has been dated to the Last Interglacial (in Marlow & King, 1974).

Neophoca figured quite considerably in the discovery of Australia. The early Dutchmen, wrecked on the Abrolhos, named one of the islands there 'Robbeneiland', and also used the seals for food. Matthew Flinders also killed seals for food during his voyage of exploration in the *Investigator* (Flinders, 1814). While on Kangaroo Island in April 1802 one of his sailors was bitten by a sea lion:

'Richard Stanley (entered at Madeira), the man who returned lame, having been simple enough to attack a large seal with a small stick, in an incautious manner, was seized upon by the seal and much bitten in the leg. As he attacked the seal without any useful object in view, he was not undeserving of some punishment for his malignity; he was paid severely, for it is probable that he will allways (sic) be more or less lame from it . . .' (Austin, 1964).

After this Francois Péron in his exploration of the south coast of Australia saw sea lions on Kangaroo Island and later published their scientific name which still stands today. His Excellency Captain George Grey, Governor of South Australia met sea lions 'as large as donkeys' at Rivoli Bay, S.A. in 1844. He shot one, and its skull is today still in the collections of the British Museum (Natural History), (Angas, 1847, Ray & Ling, 1981).

DESCRIPTION

Adult males are bulky animals, which makes them appear larger than they actually are. In nose to tail length they are about 2–2·5m and possibly weigh 300 kg, while the more slender females are slightly shorter at about 1·7–1·8m and weigh about 80 kg. Adult bulls are a rich chocolate brown in colour with the powerful neck region clothed in slightly longer rougher hair to give a mane. On the vertex of the head and continuing down the nape of the neck the hairs are white. Adult females are a silvery grey to fawn dorsally, with a creamy ventral surface. Female seals are much the same colour throughout their lives, but males have age-related changes. After the moult of the natal coat

and for about the first two years of life young males resemble females in colour, and the sexes are difficult to tell apart unless the genital region can be seen. With increasing age the males tend to get larger and more bulky than the females, the chest becomes slightly spotted, and the muzzle becomes darker. In later age groups the body darkens and there is a slight lightening in colour on the top of the head. Adult males that have not yet reached full breeding status have a dark coat, white nape, and a distinctive white ring round the eye, and do not yet have the heavy shoulders of the harem bull (Marlow, 1975).

BREEDING

Much still remains to be discovered about the breeding time of *Neophoca*. In contrast to other pinnipeds the pups are apparently not born regularly, at the same time each year. Limited observations on Dangerous Reef and the South Neptune Is. showed that most of the pups were born in October but, referring to the pupping time Marlow (1975) says 'In some years it may commence before the beginning of October and extend through late December into early January.' Observations on the Kangaroo Island colony have shown that in 1975 most pups were born in June, though births were recorded throughout the year and well into 1976. In 1976 the main season was in October, but again the pups were produced over many months. Accordingly, an 18-month breeding cycle has been suggested for *Neophoca* (Ling & Walker, 1978). A marked female gave birth on 29 September 1976, was seen afterwards on numerous occasions, but did not produce another pup until the beginning of February 1978. A second marked female also produced a pup in late September 1976, and the subsequent pup in mid March 1978 (Ling & Walker, 1978).

On Dangerous Reef in particular, several cows have been watched while giving birth, and the presence of fresh placentae made the birth date of other pups certain. Fifteen cows whose pupping date was known were subsequently seen copulating between four and nine days later. There is thus presumably a post partum oestrous which occurs about seven days after the birth of the pup (Marlow, 1975).

The relationship between a post partum oestrous and a postulated 18-month breeding cycle has yet to be determined. If the delay in implantation is the normal 2·5–3 months, then the active gestation period is some 14 months; conversely if the active gestation period is the normal eight or nine months, the delay before the blastocyst is implanted must be considerably prolonged. When compared with the arrangements in other pinnipeds neither of these explanations sounds likely.

Adult females come ashore about three days before

the pup is born and may join other females on the beach. The adult bulls actively herd their cows, and attempt to head them off should they try to move out of his circle of influence. They are quite ruthless in this, and can be very rough with the cows. At this time territorial fights take place between the bulls, and frequently a cow may be coerced into a rival's harem when the attention of the rightful owner is distracted.

Cows about to pup are restless, straining and turning to investigate their genital area. The amniotic sac appears and ruptures, and the female becomes even more restless. She frequently flails her hind end about, but little damage appears to be suffered by the pup which may be born head or hindflippers first. The pup shakes its head and begins vocalizing to its mother, an important factor in later recognition. If the pup has been produced in an exposed area, the cow will attempt to move it to a better position – picking it up very clumsily by any part of its body.

The placenta is produced between 30 minutes and six hours after birth. It is usually ignored by the cow and eventually dries up or is eaten by gulls (Marlow, 1975).

Newborn pups are about 70cm in nose to tail length and weigh about 6–8 kg. They are chocolate brown in colour, and their fur is softer and thicker than that of the adult. When they are about two months old their natal coat, which by now has faded to a gingery colour, starts to moult, and is replaced by one that is silvery grey dorsally and cream ventrally. Pups may start suckling about an hour after birth and may continue for at least a year, although they probably start some independent feeding before this. Cows about to give birth have been seen to be still suckling the large pup of the previous breeding season, and on one occasion at least the newly born pup was ignored by the cow for nearly a day while the juvenile animal continued to suckle.

For 14 days the pup has close contact with its mother, suckling, lying next to her or clambering over her. Then the cow goes to sea, leaving the pup in a sheltered position, and returning at intervals to feed it.

Australian Sea Lions, Neophoca cinerea *on Dangerous Reef, South Australia. The adult male has the pale head. The adult female has a young pup lying close to her. Photo. B. J. Marlow.*

A pup Neophoca cinerea *that has reached the stage of adventurous exploration of its home island – Dangerous Reef, South Australia. Photo. J. E. King.*

As the pup becomes older it becomes more adventurous, playing in pools and perfecting its swimming, play-fighting with other pups, until it is about 3 months old when it is well able to enter the sea.

Neophoca is an aggressive animal, and pups quietly sleeping or walking about are liable to savage attacks by bulls, subadult males and adult females. Pups may be bitten, or picked up, shaken and tossed, when they may sustain lethal injury from a fractured skull or perforated lungs, the latter caused by the big canines of the adult.

FEEDING

Without killing a seal at sea it is frequently almost impossible to get an idea of the food eaten. Obviously *Neophoca* will eat such local fish as are available, and fish bones and squid beaks have been found in the faeces. Seals have been seen taking the livers from school sharks caught in nets (Inns *et al.*, 1979). Stones are frequently found in the stomach, and piles of such stones may often be found in the appropriate internal position in a dried up carcase.

PREDATORS AND PARASITES

Lice and mites occur externally on *Neophoca*, and nematodes, acanthocephalans and cestodes occur internally. Cestodes are perhaps the most obvious in a sea lion colony as some animals may occasionally have a mass of such parasites protruding from the anus. Broken off fragments of cestodes are soon removed by silvergulls (Marlow, 1975). In the water round Dangerous Reef S.A. in particular, record specimens of white pointer sharks (*Carcharodon carcharias*) have been taken. Many sea lions bear wounds, sometimes of a characteristic curved shark-jaw shape with impressions of the triangular teeth, and missing hind flippers show evidence of escape from the shark.

EXPLOITATION

These sea lions are protected in Australia, though fishermen kill some that they consider to be interfering with the fishing. The easily visible colony on Kangaroo Island is a tourist attraction and it has been estimated that they are an economic asset to South Australia, and could, if well managed, continue as such (Stirling, 1972b).

Hooker's Sea Lion, New Zealand Sea Lion
Phocarctos hookeri

DISTRIBUTION

Hooker's Sea Lion is an animal of very restricted distribution. Its main centre is on the Auckland Islands some 322km south of New Zealand, but there is also a breeding colony on Campbell Island, and a few pups may be born on the Snares Islands. Stragglers, probably mostly young males reach Macquarie Island to the south, and Stewart Island and the New Zealand mainland to the north.

Although these sea lions may be found at any suitable place round the coast of the main Auckland Island, and on any of the small associated islands in the Auckland group, such as Rose Island, their main breeding places in the Auckland Islands are on the

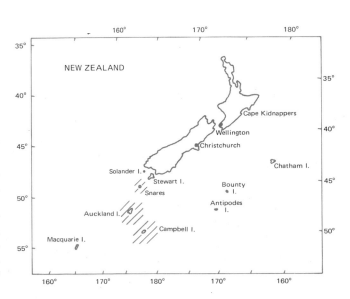

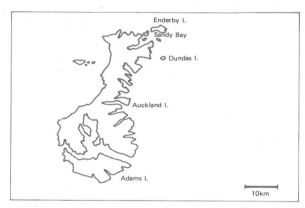

Details of Auckland Islands.

wide sweep of Sandy Bay on Enderby Island, and on the nearby Dundas Island.

During much of the year the animals are scattered out at sea, on the beaches and indeed over most of Enderby Island at least, where they may be found resting deep in the forest or on the grass on the high cliffs. During the breeding season they are more concentrated on the beaches and are easier to count. At the height of the breeding season there are about 1000 sea lions on Sandy Bay – 200 males, 400 cows and 400 pups (Marlow, 1975). There is a larger colony on Dundas Island where about 1700 pups were counted in January 1978, an estimate was made of 3550 sea lions on the island (Falla *et al.*, 1979). The population on Enderby Island seems to have been fairly stable for nearly 40 years, but over this period the Dundas colony has increased enormously. About 350 adult females have been estimated to be on the rest of the Auckland Islands, and about 50 on Campbell Island. A total of the animals associated with breeding groups has been suggested as 6000, but as this figure takes no account of immature animals, it is obviously far from the final count (Falla *et al.*, 1979).

On the Snares 47 sea lions (36♂) were seen in one day in January 1971, but only a very few pups are born there (Crawley & Cameron, 1972). Trampled vegetation and tracks into the bush indicate that beaches in the Port Pegasus area of southern Stewart Island are used as regular hauling out grounds. Relatively small numbers (±20) of sea lions have been seen, and these are all males, mostly subadults with a few adult bulls (Wilson, 1979).

On the New Zealand mainland stragglers are sometimes seen at Kaikoura north of Christchurch, and in the region of the Otago Peninsula on South Island. Sea lions used to breed on the west coast of South Island in 1863, though they have probably not done so since the turn of the century (Berry & King, 1970). A jaw from Cape Kidnappers at the southern tip of Hawke Bay, North Island, previously thought to be of considerable

palaeontological antiquity, is now regarded as being from a midden of negligible age (Weston *et al.*, 1973). This jaw is from a juvenile male animal, and the skull of a young female *Phocarctos* is known from a Maori midden from a few kilometres from Cape Kidnappers (Berry & King, 1970), perhaps indicating breeding sites on North Island at earlier times. Further examples of *Phocarctos* from North Island are 3 teeth from a fourteenth century site on the Coromandel Peninsula and fragmentary remains of two animals from a site of similar antiquity near Wellington (Smith, 1978, 1979).

The Auckland islands were discovered in 1806 and at that time fur seals and sea lions were abundant there. Enderby Island became a centre for sealing operations in the area, sealing reached its height in about 1822, and following the usual pattern the seals were virtually cleared out by 1830. The sealers made occasional visits thereafter and took what seals they could find.

Captain Thomas Musgrave of the *Grafton* was wrecked on the Aucklands after prospecting for tin on Campbell Island. He and his crew remained there for 20 months from January 1864 to September 1865. During this time the sea lions formed a valuable part of their food, and indeed Musgrave wrote his journals in seals blood (Musgrave, 1866, Marlow & King, 1974).

In 1881 fur seals and sea lions were protected by law, though poaching still continued. From about 1888 the sea lion colony on Sandy Bay gradually built up to its present numbers.

There were two settlements on the Auckland Islands – Maoris from the Chatham Islands stayed there from 1842 to 1856 and then the Southern Whale Fishery Company leased the islands from the Crown. Charles Enderby, the chief commissioner to the Company, established a settlement there with English colonists from 1849 to 1852, and pastoral leases were granted for a while, with sheep being run there. The climate, however, is against such activity, and human settlements did not last long. Between 1880 and 1927 New Zealand Government vessels called in occasionally at the Auckland Islands and established huts provisioned with food and clothing. Signposts were erected indicating the direction to the nearest food depot. They also liberated cattle, sheep and goats to join the already introduced rabbits and pigs. All these precautions were to facilitate the survival of sailors from the many spectacular shipwrecks that occurred on the rocky coasts as a result of the wild weather in the region (Taylor, 1971, King & Marlow, 1974). After 1927, the regular visits to the islands were discontinued, but coast watching parties were established during the Second World War, and there have been about four scientific expeditions since (Yaldwyn, 1975).

DESCRIPTION

Adult bull Hooker's Sea Lions are about 2–2·5m in nose to tail length, though their bulk makes them appear larger, and females are about 1·6–2m in length. The bulls are black with a mane of rougher, longer hair on their well-developed neck and chest. The profile of the face is blunter and more rounded than in *Neophoca*, and the muzzle appears shorter. Adult females are virtually identical in colour to *Neophoca*, being silvery grey dorsally and cream ventrally. As with *Neophoca*, the young males are coloured like females, darkening with age until the adult black colour is attained (Marlow, 1975).

BREEDING

The breeding season starts at the beginning of December when the adult males space themselves out along the sandy beach. The territory defended by each male is virtually a 2m circle on which it rests and from which it makes short aggressive charges. Storms and other disturbances cause major reorganizations and it is obvious that the bulls are not defending a piece of territory as such, but a personal space around themselves, which of course moves with them. They will defend their space for the whole of the breeding season without going to sea to feed or cool themselves, and sometimes without females at all. This presumably indicates that it is the defence of the territory that motivates the animal, and in this respect it is different from *Neophoca*, which is also motivated by territory but will desert the chosen area if no females appear (Marlow, 1975). Territorial defence by *Phocarctos* is much more ritualized, and there are relatively few serious fights.

The cows start to concentrate on the beach at the beginning of December. On emerging from the sea they first run the gauntlet of the solitary bulls that line the beach just above the water line, though these bulls are very rarely successful in capturing a cow. In 1972, the cows gathered in groups and the number on the beach rapidly increased from four on 1 December, to 276 on 21 December. Each emerging cow galloped across the sand to join an existing group, all lying almost piled up on each other. The bulls do not herd

Hooker's Sea Lion, Phocarctos hookeri *on Enderby Island, Auckland Islands. This adult male shows its blunt rounded face and heavy neck, and also a wound at the corner of its mouth. Photo. J. E. King.*

cows at all, and are usually very gentle with them, allowing them to move about the beach without any restraint.

Compared with *Neophoca*, the pups of *Phocarctos* are born over a very restricted time. In 1972–3 pups were born between 6 December and 7 January (Marlow, 1975). They are produced head or hind flippers first, and the placenta follows usually about half an hour later. The large numbers of skuas (*Catharacta lonnbergi*) on the Auckland Islands very soon clear up the placentae, and in fact a group of squabbling skuas is a good indication that a birth has taken place. The cows may sniff at the placenta but usually ignore it. A single cow was seen to pull a portion of the placenta off and swallow it, but it is probable that this unusual action was the result of stress caused by the attentions of the skuas (Marlow, 1974).

The new born pups are 75–80cm in nose to tail length, and are chocolate brown in colour with a lighter area on the nape of the neck and top of the head which frequently extends as a stripe along the dorsal surface of the muzzle to the nose. Most of them begin to suckle within about half an hour of birth. As *Phocarctos* cows have been seen suckling large pups that are probably nearly a year old, it is probable that the lactation period extends for about a year as in other sea lions, with the pup starting to feed on solid matter well before the end of this period. Sleeping cows have not infrequently been seen with two pups suckling at the same time but when the cow wakes she sniffs at the pups and threatens the intruder. Pups have been seen making the rounds of sleeping females, and yearlings too are not above trying to get an easy meal this way.

For two or three days after the birth the pups lie quietly by their mothers, but after this they are much more active than *Neophoca* pups. Except when suckling, the pups tend to gather in large groups or pods, and spend much time sleeping, playing and play-fighting. With increasing age they move more freely about the beach and the grassy area at the back of it, and by about six weeks of age, are playing at the edge of the sea or in rock pools. Shortly after this they will enter the sea with their mothers. When they are about two months old most of them are swimming strongly and are able to visit neighbouring islands, so the main breeding area on Sandy Bay is again deserted.

Oestrous and copulation take place about seven days after the birth of the pup. The pups moult their natal coat when they are about two months old, and the adults start to moult at the end of February, when the breeding season is over.

Phocarctos is far less aggressive than *Neophoca* and although females with pups will snap at other females and strange pups, only very rarely will they toss a pup. The males, even the adult bulls are very tolerant towards pups, seeming to ignore them, though occasionally a pup might get trampled upon during a fight.

MORTALITY, PREDATORS AND PARASITES

There is virtually no information about diseased animals, and wounds obviously due to sharks are rare. Maybe there are fewer sharks in the latitude of the Auckland Islands, though an occasional missing flipper might be attributable to these fish.

Most deaths of *Phocarctos* pups are due to starvation. Rabbits were introduced to the islands, probably by settlers in about 1850, and they have made large areas of burrows at the back of Sandy Bay. Pups wriggle into these burrows to shelter from the wind and rain, and if they get in too deeply, they are unable to turn around or to move backwards, and eventually starve (Marlow, 1975, King & Marlow, 1974). Lost pups too will starve since normally no other cow will suckle them. Lice are present externally, nematodes are found in the gut, and encysted cysticerci have been found in the blubber.

FEEDING

Little is known about the food eaten though penguins have been reported as being chased and eaten. Fish and cephalopods are taken, and also various crustaceans, such as the swimming crab *Nectocarcinus antarcticus*, crayfish *Jasus lalandei*, and prawns such as *Munida* and *Nauticaris* (Yaldwyn, 1958). As most food is digested while at sea, evidence of food eaten is usually obtainable only from indigestible remains that are vomited up, and even then one has to be very quick to reach these specimens before they are removed by gulls and skuas. Stones are also ingested.

EXPLOITATION

Phocarctos is totally protected, and the normally uninhabited Auckland Islands are a flora and fauna reserve.

FUR SEALS

Southern fur seals *Arctocephalus*

Although not all members of this genus live in the southern hemisphere, the group as a whole is frequently known by this title, distinguishing them from the Northern fur seal *Callorhinus*.

Studies on the different species of *Arctocephalus* have only rarely dealt with more than one species at a time. Animals of this genus are so widely spread over the world that it has always been difficult to get reasonable numbers of specimens of each species, and to assemble enough skulls in one place so that comparative work could be done. Repenning *et al.* (1971) have managed this task, and their arrangement of the species is accepted here. Eight species are recognized, and the distribution and biology of each one will be detailed here.

The inter-relationships of the eight species are not yet known, and studies on the skulls give only possible pointers to this complicated subject. *A. pusillus* is the largest fur seal, and seems to be most distinct. The shortness of the maxillary process of the zygomatic arch, and the tendency towards single rootedness of the first upper molar are characters in which this fur seal is rather more like sea lions than other *Arctocephalus*. Repenning *et al.* (1971) suggest that possibly all the other species of *Arctocephalus* have 'a relationship to, and possibly origin from, the South American fur seal, *Arctocephalus australis*, and seemingly have one common trait, the evolution of simplified cheek tooth patterns'. Later work, however, on the Miocene fossils *Thalassoleon* and *Pithanotaria*, which have simple teeth, now suggests that those modern otariids with well-developed cusps, as in *A. australis* and *A. pusillus* are more advanced (Repenning & Tedford, 1977). These authors note – 'Those living species associated with continental shores, South America, Africa, and Australia, now appear to be the most advanced of the

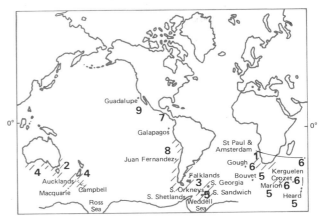

World distribution of fur seals of the genus Arctocephalus, 1 A.pusillus pusillus, 2 A.pusillus doriferus, 3 A.australis, 4 A.forsteri, 5 A.gazella, 6 A.tropicalis, 7 A.galapagoensis, 8 A.philippii, 9 A.townsendi.

living fur seals, and insular species, with simple cheek tooth crowns, appear to retain a primitive condition.' The Juan Fernandez and Guadalupe Fur Seals, *A. philippii* and *A. townsendi* have not infrequently been separated as a distinct genus *Arctophoca*, and in fact the two fur seals have on occasion been combined specifically. A comparison based on large numbers of specimens has not yet been possible, but Repenning *et al.* (1971) working on all the available material are of the opinion that 'the *philippii–townsendi* complex is in some ways distinctive, but classing these 2 species (only provisionally held to be distinct) in a separate genus, *Arctophoca*, seems unwarranted' and 'whereas our sample of neither *A. townsendi* nor *A. philippii* is large enough to defend the reliability of their differences statistically, it is suggestive enough to warrant their specific separation until such time as larger samples may prove the observed differences irrelevant.'

Guadalupe Fur Seal *Arctocephalus townsendi*

DISTRIBUTION

At the present time this fur seal is known only from a small part of the eastern coast of Guadalupe Island, Mexico (latitude 29°N) (see map), where it occurs in rocky areas at the base of high cliffs and in caves. Although breeding only on Guadalupe, a few individual fur seals sometimes reach San Miguel and San Nicolas to the north in the Californian Channel Is-

lands, and Cedros Island, Baja California to the south. The total population is believed to be between 500 and 1000 animals. This fur seal is totally protected both by Mexican and American law (Kenyon, 1973b).

The history of the sealing along the Pacific American coast mirrors that elsewhere. Vast quantities of skins were taken towards the end of the eighteenth century

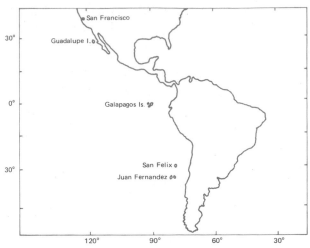

Distribution of Guadalupe, Galapagos and Juan Fernandez fur seals.

without regard for the continuance of the species involved, or even for the identity of the animals – they were all 'fur seals' and thus commercially valuable. The skins taken off the Farallon Islands, some 130 000 in 1808 and 1809 were probably *C. ursinus*, while those taken from Juan Fernandez were *A. philippii*. However, the skins taken from the San Benito Islands, Baja California, and from Guadalupe were almost certainly *A. townsendi*, although of course the sealers left no specimens to prove it.

It is believed that the bulk of the population of fur seals on Guadalupe was killed approximately during the period 1800–20. The Russians knew Guadalupe, and it is possible they brought over Aleuts from Alaska to do the sealing, but other people may well have been involved also. On Guadalupe today many stretches of smooth polished lava rock show where thousands of fur seals once moved and rested, and ruins of the sealers' stone huts still remain. Between 1876 and 1894, the last commercial catches of the Guadalupe Fur Seal were made.

Guadalupe Fur Seal, Arctocephalus townsendi *showing the long pointed nose. An adult female and pup. Photo. Ardea.*

No scientific specimens were collected until 1892 when Dr Charles Townsend brought four skulls back to America. These were described as a new species, *A. townsendi* in 1897, but it was not until about 1928 that any scientist knew what the animal looked like. In that year two bulls were received by the San Diego Zoo, the animals having been discovered and captured by fishermen, who thought there were about 60 animals on the island (Hubbs, 1956). A few animals are reported to have been seen in the 1930s but the observations were not recorded.

In 1949 a lone male fur seal was seen on San Nicolas Island, and this led to Dr Carl Hubbs and his colleagues planning an expedition to Guadalupe in 1950, but on this occasion they saw no fur seals. Another visit in 1954 revealed a small breeding colony in a cave, and subsequent visits have shown a growing population.

DESCRIPTION

As might be expected, there is at the moment a lack of detailed information on this seal. Adult males are about 1·8m in nose to tail length and weigh about 136 kg. In colour they are a dusky black, the head and shoulders appearing greyish because of the lighter tips to the guard hairs, and the animal appearing a grizzled grey when dry. They are thus a very typical

'*Arctocephalus* colour', and merge well with the black blocks of lava on the beaches. Both the Guadalupe and Juan Fernandez Fur Seals are distinguished from other members of the genus *Arctocephalus* by their extremely long pointed noses. 'This elongation is not only in the non-osseous nose and its supporting cartilages, but is accentuated by a bulbous terminal rhinarium and by conspicuously downward-facing nostrils that produce a shark-like muzzle' (Repenning *et al.*, 1971).

BREEDING

Pups are born in June. During a visit to Guadalupe in the last two weeks of June all pups were in the black natal coat, and two fresh placentae were noted (Brownell *et al.*, 1974). Bulls seen in May appeared to be holding territories as they were reluctant to abandon their positions. They were in or near caves or recesses in the rock, and these seem to be favoured situations as animals may go at least 25m into the caves. Loosely organized harems with about ten cows per bull seem to be formed (Peterson *et al.*, 1968a).

VOCALIZATIONS

The sounds made by *A. townsendi* have been described as a bark, a roar and a cough in different situations. These sounds are different from those produced by *Callorhinus* and *Zalophus*, and similar to those of *A. gazella* (Peterson *et al.*, 1968a).

Galapagos Fur Seal
Arctocephalus galapagoensis

DISTRIBUTION

The Galapagos Fur Seal, as its name would suggest, is found only on the Galapagos Islands which lie approximately between the equator and 1°S, some 10° off the coast of Equador, to which country the island group belongs. Fur seals are present on at least ten islands of the group, and the population is probably considerably more than a thousand individuals (Orr, 1973) and some estimates even suggest as many as 5000 (Clark, 1979). The fur seals share the Galapagos Islands with the sea lion (*Zalophus californianus wollebaeki*), but whereas the sea lion occupies the beaches, the fur seal seems to prefer the rocky area with sea caves.

Fray Tomas de Berlanga, Bishop of Castilla del Oro, who was on his way to Peru, stopped to take on water at the Galapagos Islands in March 1535, and thus discovered the islands. Throughout the nineteenth century both fur seals and sea lions were taken, as for

example in 1816 when Captain Fanning took 8000 fur seals and 2000 sea lions. Captain Reed took 6000 fur seals between 1872 and 1880 (King, 1954), but the bulk of the commercial sealing was over by about 1900. Naturalists from Stanford University, on the schooner *Julia P. Whalen* under the command of Captain Noyes spent ten months (1898–9) round the Galapagos, during which time the crew did some sealing. Numbers of fur seals were obviously declining as they collected only about 200 skins on this trip, and had only achieved the same total on several previous annual visits. On the Stanford University trip the naturalist Edmund Heller collected the type specimen which he later named and described (Heller, 1904).

Although very reduced in numbers the Galapagos Fur Seal population was never exterminated, and a few specimens were caught and presented to San Diego Zoo in 1932–3 (Townsend, 1934). It was not,

Galapagos Fur Seal, Arctocephalus galapagoensis, *from James Bay, James Island, a poorly known animal. Photo. Robert T. Orr.*

however, until 1957 that a colony of reasonable size was rediscovered, and since that time the population has continued to increase (Orr, 1973).

The Galapagos Fur Seal is regarded as a full species, but its close relationship to the South American Fur Seal *A. australis* cannot be doubted. These two animals are geographically quite close, and the northerly directed Peru Current that washes both the Pacific shores of South America and the Galapagos Islands probably took the fur seal to the Galapagos Islands at some time.

DESCRIPTION

This is the smallest of the fur seals. Few animals have been measured, but the nose to tail length of the adult male is about 1·5m (Heller, 1904, Townsend, 1934). Judging by skull measurements the female is only slightly smaller.

In colour, 'The front and sides of the muzzle as well as the underparts of the body are a light tan, contrasting with the grizzled grey-brown fur of the back and sides. The ears are light tan except along the margins. The posterior vibrissae are dark while those situated more anteriorly on the muzzle are light proximally, becoming dark distally' (Orr, 1973). The nose is short and fleshy with a small button like rhinarium.

BREEDING

There is virtually no information on the life history of this fur seal. Like other fur seals they prefer the rocky and more inaccessible parts of the coast and are to be found in sea caves. Heller (1904) notes that he found pups of various ages in December, whereas Brosset (1963) recorded new born pups in August. Pups of *Zalophus*, the only other pinniped on the Galapagos, may be produced in almost any month of the year, a possible response to the equable climate, so perhaps the fur seal has an equally long pupping period.

PREDATORS AND PARASITES

An epizootic of unknown origin affected the Galapagos *Zalophus* in 1970–1 and it seems that some of the fur seals also died (Orr, 1973).

FEEDING

Local fish are presumed to be the food of this fur seal.

Juan Fernandez Fur Seal
Arctocephalus philippii

DISTRIBUTION

This fur seal is restricted to islands off the coast of central Chile (see map). It is present on the islands of the Juan Fernandez group, latitude 33°S (Isla Más a Tierra,* Isla Santa Clara, Isla Más Afuera*), and on the Isla San Felix group (Isla San Ambrosio and Isla San Felix) which are about as far off shore as Juan Fernandez, but about 7° latitude further north. The most recent report on the population of these islands (Aguayo, 1973) suggests that there are 700–800 fur seals on the Juan Fernandez group. Only two fur seals were seen on San Ambrosio at the latest count in 1970, but Aguayo suggests that these probably (and hopefully) represent a slightly larger population. The fur seal is protected by the Chilean government.

In the interval between the discovery of the islands by Juan Fernandez in 1563, and their later re-discovery and exploitation by sealers some 230 years later, fur seals were extremely abundant. The explorer William Dampier noted that at the time of his visit in 1683 'Seals swarm as thick about this island, as if they had no other place in the World to live in'. He also mentioned that they had 'fine thick short Furr', and that 'Large Ships might here load themselves with Seals Skins, and Trane-oyl' (Dampier, in Hubbs & Norris, 1971). Unfortunately this latter suggestion came true and the fur seals were taken at such a rate that by 1824 the numbers had been drastically reduced and the animals were virtually extinct commercially. Some dates and figures will illustrate the process:

1683	Visit by Dampier. Seals very numerous.
1687–90	Men from the *Bachelors Delight* left on Más a Tierra to cure seal skins.
1704	Alexander Selkirk marooned on Más a Tierra – hundreds of 'sea lions' roaring.
1792	Crew of the ship *Eliza* killed 38 000 seals on Más Afuera and started the China trade by selling the skins in Canton for 16 thousand dollars.
1797	Reported to be 2–3 million seals on Más Afuera. Three million taken to Canton in 7 years. Crews from 14 ships all killing seals on the island at the same time.
1798	Captain Fanning of the *Betsey* took 100 000 skins to Canton from Más Afuera, and estimated that he had left 5–700 000 animals still there.

* (The official Chilean names for these islands are: Isla Robinson Crusoe for Isla Más a Tierra, and Isla Alejandro Selkirk for Isla Más Afuera)

Juan Fernandez Fur Seal, Arctocephalus philippii, *a drawing showing the pointed nose. Drawing by Helmut Diller. Bruce Coleman Ltd.*

1801	A report that a ship carried a single cargo of a million skins to the London market from Más a Tierra.
1807	Captain Morrell noted that the sealing business on Más Afuera was now 'scarcely worth following'.
1824	Juan Ferandez group virtually abandoned by fur seals.

The same sequence of events occurred on the San Felix group, but after the sealing was over, it was not until 1970 that two animals were seen. The above list of dates is abbreviated from King (1954) and Hubbs & Norris (1971) who quote the original references.

The fur seals were, however, not completely exterminated. It was in fact not until 1864 that Dr R. A. Philippi, then Director of the Natural History Museum in Santiago, Chile, collected from the Juan Fernandez group the skull and skin that Peters named after him as a new species (Peters, 1866). In 1870 Captain Gaffney saw 300–400 animals on Más Afuera, and took 19. He also saw a few fur seals on Más a Tierra. Skulls were received by the British Museum (Natural History) and by the Australian Museum in 1883 and 1884 respectively, probably as exchange material from the Chilean Government (Repenning *et al.*, 1971). Skottsberg explored the Juan Fernandez islands in 1916–17 and took a cestode, obviously from a recently dead animal.

The expedition by Dr Carl Hubbs to Guadalupe in 1954 which resulted in definitive observations of *A. townsendi* that was thought to be almost extinct, stimulated this zoologist's interest in island populations of *Arctocephalus*. His discussions in 1958 with people familiar with Juan Fernandez, and who had reported seeing seals, led to increased hope. In December 1965 two independent groups of people saw up to 200 seals on Más Afuera, and photos were taken of young pups. Further observations of fur seals were made in 1966, and in 1968 Dr Kenneth Norris and colleagues went to Más a Tierra where they saw and photographed the fur seals and took a specimen for definite identification (Hubbs & Norris, 1971). Chilean naturalists have since been keeping check on the population (Aguayo, 1973).

DESCRIPTION
The Juan Fernandez Fur Seal is one of the least known of the genus *Arctocephalus*. In many respects it resembles the Guadalupe Fur Seal, but as mentioned earlier the two fur seals are regarded as being distinct. In photographs, however, the two fur seals appear remarkably similar and both have the long fleshy nose that distinguishes them from other *Arctocephalus* (see *A. townsendi*). Both fur seals are docile animals and appear to have similar habits – hauling out on lava rock at the base of cliffs, and inverting themselves in the water, head down, so that the hind flippers are left swaying gently above the surface of the sea (Hubbs & Norris, 1971).

The nose to tail length of an adult male was 2m, and this animal was estimated to weigh 159 kg. 'The bulls have a rather slim, pointed snout, a heavy mane of silver-tipped guard hairs, and blackish-brown, shiny fur on the posterior body parts and belly. All the males are scarred around the neck and fore flippers by bites, probably from rival males' (Hubbs & Norris, 1971).

BREEDING
The pups are born in early summer, about the beginning of December.

South American Fur Seal
Arctocephalus australis

DISTRIBUTION
The South American Fur Seal occurs, as its name suggests, on the coasts and offshore islands of South America between, approximately, Lima and Uruguay. In Peru it is found along the more southerly parts of the coast, the northern limit probably being the breeding colony at Paracas Peninsula south of Lima (Grimwood, 1968). In 1968 the Peruvian Department of Fisheries estimated the Peruvian population of this fur seal to be about 12 000 animals (Vaz-Ferreira, 1979). Little is known about the animal in Chilean waters, but 40 000 were counted in 1976 (Vaz-Ferreira, 1979).

In the Falkland Islands the fur seal numbers are increasing gradually, but the colonies have not yet increased to a level where sealing could again be contemplated. A count in 1973 gave a figure of 14 000–16 000 animals (Strange, 1979). The Argentinian population is not well documented, but in 1954 two colonies on Staten Island totalled 2300 animals, and on Escondida Island there were 400 animals (Carrara, 1954).

In Uruguayan waters the main breeding colonies are on Lobos Island, and on the Torres and Castillo Grande Islands just north of the River Plate Estuary, but visiting non-breeding animals may be found at other places. The Uruguayan population is increasing

and in 1972 there were estimated to be 252000 animals (Ximenez, 1973), and 101470 pups were born in the 1972–3 season (Vaz-Ferreira, 1979). The world population of this fur seal is thus about 323000 animals (see Exploitation, below). There is no migration, but the seals disperse widely.

DESCRIPTION

There is some evidence that mainland animals tend to be rather smaller than those on the Falkland Islands. The condylobasal length of the skull of an adult male from the mainland is 230mm, while a Falkland Island skull of similar maturity may be 255mm (King, 1954). Similarly, Uruguayan adult males weigh 136 kg, while Falkland males may weigh 159 kg, but really adequate comparative work has not been done in this field. If the two groups are entitled to separate subspecific names, then *A. australis australis* (Zimmerman, 1783) is available for the Falkland Island population, and *A. australis gracilis* (Nehring, 1887) for the mainland population. Fur seals on the Galapagos Islands are recognized as a separate species (*A. galapagoensis*).

With the acceptance of a possible difference in size between mainland and island populations, adult males in general are about 1·9m in nose to tail length and weigh 159 kg, while adult females are smaller at 1·4m and 48 kg. The males are blackish grey with longer hairs on the neck and shoulders, the females and most of the immature animals are rather variable in colour, but the dorsal surface is usually greyish black, shading to a lighter colour ventrally (Vaz-Ferreira, 1950).

BREEDING

The adult bulls take up their positions in November and defend their territories until the end of the breeding season. The males are polygamous, but do not gather the females into harems. Adult but non-breeding males who have lost their fights or who are not yet strong enough to fight the territorial bulls usually gather to the seaward side of the breeding groups. Immature non-breeding males reach their

South American Fur Seal, Arctocephalus australis. *An adult male in tussock grass terrain. Photo. Bill Vaughan.*

maximum numbers after the breeding season has ended. They play and fight together, but do not maintain territories.

The pups, clad in soft black fur are born in November and December, weighing 3–5 kg at birth. The adults mate in late November and December, probably about 6–8 days after parturition, judging by other species of *Arctocephalus,* and the breeding groups have started to break up by the beginning of January. The female suckles her pup for about a year, and implantation of the blastocyst takes place in March–April (Vaz-Ferreira, 1979).

Although occurring in very much the same areas as the Southern sea lion, the two animals tend to keep separate, the sea lions preferring the sandy beaches while the fur seals live in more rocky places. Except at the breeding season they tolerate each other's presence, and although on occasion the later breeding sea lion may occupy the same site as the fur seal, they are not encouraged to come too near when breeding is in progress, and in any conflict the fur seal usually wins. However, the sea lions have been seen to eat fur seal pups and adult females (Gentry & Johnson, 1981).

PREDATORS AND PARASITES
Apart from man the chief enemies of the fur seal are probably killer whales and sharks. A fur seal has been seen with the whole of one hind flipper and half of the other bitten off, probably by a shark, and a shark has been caught with five young fur seals in its stomach (Vaz-Ferreira, 1950). The usual populations of internal parasites are found – mites in the respiratory system, nematodes and cestodes in the gut.

FEEDING
Little detail is available on the food eaten by this fur seal but remains of cephalopods, crustaceans, lamellibranchs and sea snails have been found in stomachs of animals caught on land. Stomachs of seals caught in fishing nets at sea contained such fish as *Engraulis anchoita, Trachurus lathami, Cynoscion striatus* and *Pneumatophorus japonicus* (Vaz-Ferreira, 1979).

EXPLOITATION
Uruguay is the only South American country exploiting this fur seal. The harvesting is managed by the government, and forms an important resource for the country. Sealing takes place in the winter months (July–August) and young males form the basis of the catch, being used for their pelts and oil. The average annual harvest is about 12 000 animals (Vaz-Ferreira, 1979).

Subantarctic Fur Seal Arctocephalus tropicalis

DISTRIBUTION
The Subantarctic or Amsterdam Fur Seal lives on the isolated subantarctic islands that lie north of the Antarctic Convergence.* It is found on the islands of the Tristan da Cunha group, on Marion, Prince Edward, Crozet, Amsterdam and St Paul.

The fur seals of the Tristan da Cunha group (Tristan, Inaccessible, Nightingale, Gough) have their headquarters on Gough Island. About 600 fur seals were present on Inaccessible in 1956 (Swales, 1956) and about 90 on Nightingale Island (in Bester, 1980). On Tristan there are about 500 seals, but few (20) pups are born (in Bester, 1980).

Sealers worked the islands at the end of the eighteenth century taking fur seals and elephant seals, but by the 1830s most of the seals had been taken and a sealing party visiting Gough Island in 1892 found very few animals (Holdgate, 1965, Wace & Holdgate, 1976). A nucleus of fur seals must have remained, however, and given rise to the present colony. On Gough Is-

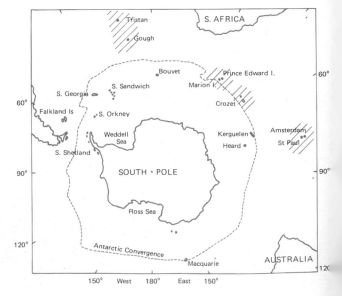

land, 350km from Tristan, fur seals are found on the storm-beaten rocky west coast beaches at the base of cliffs. Members of the Gough Island Scientific Survey visited the island in 1956, and counted 13 000 fur seals, most of them on the west coast (Swales, 1956). A visit on behalf of the University of Pretoria in 1973, counting seals on the east coast estimated a two-fold increase but as the counts were made at different times of year they are not really comparable (Shaughnessy, 1975). A later count up to 1977–78 gives an estimated total population of 2 000 000 animals, so that 93 per cent of the world population of this seal is on Gough Island (Bester, 1980).

The Prince Edward Islands, consisting of Marion Island and the smaller Prince Edward Island, lie 2300km southeast of Cape Town. Although normally lying about 290km north of the Antarctic Convergence,* the Convergence itself does not remain in a fixed position and may be displaced by about 80km north or south. Thus at times the Convergence would lie closer to the islands and would influence the local temperatures (Condy, 1978).

Sealing on the Prince Edward Islands started about the end of the eighteenth century and both elephant seals and fur seals were taken. Although sealing ves-

sels called at the islands it appears that fur seals were not over abundant there. The last fur seals taken from these islands were a cargo of nearly 800 skins taken as recently as 1921. In 1952 there were about 500 fur seals on Marion Island of which about 160–170 were breeding males (Rand, 1956b). By 1975 the total population on Marion Island was estimated as close to 7000 seals, with another 2000 on Prince Edward Island (Condy, 1978). As is usual with fur seals they tend to live on the rugged rocky exposed west coasts of the islands.

The Crozet Islands still had fur seals at the end of the first quarter of the nineteenth century, though possibly they were never as abundant as on Amsterdam Island. Sealing reduced the numbers still further and by the 1860s there were virtually none left. Fur seals were not recorded on these islands again until 1962 when a single animal was seen. Another was seen in 1969 and four more in the next two years, but so far there has been no major recolonization of these islands by fur seals as has happened at so many other places. Although the recent animals are *A. tropicalis* there is

* Antarctic Convergence – the meeting place where the cold Antarctic water sinks below the warmer subantarctic water.

Subantarctic Fur Seal, Arctocephalus tropicalis. *An adult male on Gough Island, showing the pale chest and face.* Photo. P. D. Shaughnessy.

some evidence that the original inhabitants were not. The original Crozet seals apparently had better quality fur than those on Amsterdam Island and could have been *A. gazella*. This would agree with the colder climate on these islands at the time of the sealing gangs (Despin *et al.*, 1972).

At the end of the eighteenth century fur seals were so numerous on the islands of St Paul and Amsterdam that a visiting Dutchman in 1754 had to beat a path through them with a club before he could land (Paulian, 1964). The animals were later killed commercially and traded with China, and by the end of the 1870s, there were few left. Recent (1970) estimates of the population give 4868 on Amsterdam Island and 353 on St Paul Island (in Bester, 1980).

The total population of *A. tropicalis* is therefore probably something in excess of 214 000 animals with increasing numbers of animals in the breeding colonies on all islands except possibly at the moment, the Crozets.

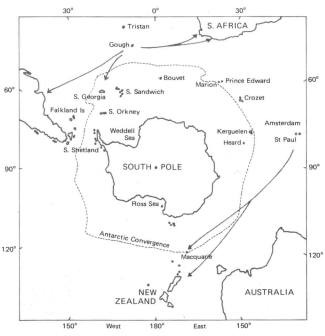

Wandering individuals of A. tropicalis.

Individuals of the Subantarctic Fur Seal have been found far away from the breeding colonies, and some of their wanderings indicate that the Antarctic Convergence is no real barrier to travel. In August, 1976 an adult male was found alive at Tramandai, Rio Grande do Sul in southern Brazil (Castello & Pinedo, 1977). Five males have been recorded at South Georgia between 1972–76 in December–March. One of these was

defending a territory and was treated by the surrounding *A. gazella* in a normal manner; another was accosting cows of *A. gazella*, but was not seen to mate (Payne, 1979b). There are 22 records of *A. tropicalis* reaching the southern coast of South Africa between 1966 and 1979. Most of the animals were males and were found in winter and early spring (May–September) (Shaughnessy & Ross, 1980). It is most probable that the South African animals came from the rapidly expanding Gough Island colony, and their trip to South Africa would be assisted by the West Wind Drift. The South Georgia and Brazil visitors also presumably came from Gough, although the direction of the current would be against them. A young male *A. tropicalis* was seen at Macquarie in March, 1959, and stragglers of this species have also been seen on the west coast of South Island, New Zealand (Csordas, 1962). (See also the section on New Zealand Fur Seal.) Presumably these animals came from Amsterdam Island, some 5000km away.

These wandering individuals of *A. tropicalis* would be found in areas that are normally inhabited by other species of fur seals – by *A. australis* in Brasil, *A. gazella* on South Georgia, *A. pusillus* in South Africa and *A. forsteri* in Macquarie and New Zealand. They can, however, be recognized by their distinctive colour, crest, voice, teeth and characters of the skull (King, 1959a, Condy, 1978, Shaughnessy & Ross, 1980).

DESCRIPTION

Adult males have a nose to tail length of 1·5 to 1·8m and may weigh up to 140 kg. Payne (1979b) notes the dimensions of two of the South Georgia visitors – an 11-year-old male (aged from canine sections) had a nose–tail length of 166cm and weighed 115 kg; an eight-year-old male was 157cm and weighed 95 kg. Adult females are smaller – 1·2–1·3m in nose–tail length and up to 50kg in weight. Condy (1978) notes that *A. tropicalis* bulls and cows are short and compact in appearance with short thick necks, short black hairless ears and short broad foreflippers when compared with *A. gazella*. The foreflippers are also proportionately shorter in *A. tropicalis* when compared with those of *A.p. pusillus* (Shaughnessy & Ross, 1980). The latter authors also note the large prominent eyes, the higher pitched voice, and the strong musky odour by which *A. tropicalis* may also be distinguished from *A.p. pusillus*. Payne (1979b) also notes that the call of *A. tropicalis* is quite distinct from that of *A. gazella*.

The colour of the adult animals is very characteristic. In the adult males the body is a dark greyish brown with the belly a more ginger colour. The chest, nose and face to above the eyes are white to orange in colour. The mystacial vibrissae are very long and white. The guard hairs on the crown of the head are black with white tips, about 75mm in length and form an upright brush-like crest which is more obvious when the animal is agitated. The crest seems to develop between the ages of six and ten years. The colouring of the adult females is very much as in the adult males – the yellowish chest and face being noticeable, and the rest of the body grey-brown with a more ginger belly. Females do not have the crest, but the whiskers are long and white.

BREEDING

The general biology of *A. tropicalis* seems to be much the same throughout its range. Apart from the cows and young pups who remain, most of the fur seals spend most of the winter and early spring (c. June–September) away from the breeding grounds, probably out at sea. In September the first of the adult males start to haul out on the breeding grounds and stake out their territories. In October and November the rest of the seals haul out, the younger animals keeping away from the breeding beaches. Between the end of November and February the pups are born, about five to six days after the cows arrive, and copulation takes place eight to twelve days after the birth of the pup. By mid-December the newly mature cows have mated, and they, and many of the immature animals return to the sea. The cows stay with their pups until after copulation has taken place, then they spend increasing periods at sea. After the breeding season is over, the adults moult and by about the end of April there are very few adults on the breeding beaches (Swales, 1956, Condy, 1978, Rand, 1956b, Paulian, 1964).

The pups at birth are about 60cm nose to tail length and weigh about 4·5 kg with the males tending to be a little (5cm) longer and a little (90 g) heavier than the females. They are black with chestnut coloured underfur and black mystacial whiskers. Between the ages of eight and twelve weeks they moult this natal coat for one of adult appearance. They suckle for ten to eleven months, probably until the next pup is born. The cow, at least for a short period is solicitous of the welfare of the pup, moving it into the shade of rocks and defending it if necessary.

FEEDING

A. tropicalis is a fairly general feeder, eating squid, Nototheniid fish and krill, their importance probably being in that order. Paulian (1964), however, notes that the fur seals on Amsterdam Island, at least in summer, eat primarily squid, but also take Rock-

hopper penguins, and he found no fish remains in the stomachs.

PREDATORS AND PARASITES

The usual range of parasites is present – lice on the head and neck, mites in the nasal passages, cestode cysts in the blubber, and adult cestodes in the intestine, nematodes in the oesophagus and stomach, and acanthocephalans in the small intestine (Shaughnessy & Ross, 1980). Although there is no actual evidence it seems very likely that the sharks of the area would be able to take pups, and killer whales would also be a danger to them (Paulian, 1964).

POSSIBLE INTERBREEDING

The stragglers of *A. tropicalis* at such places as South Georgia and South Africa have already been mentioned. It was also noted that on South Georgia one of these *A. tropicalis* males was holding a territory in the midst of the normal population of *A. gazella*. On Marion and Prince Edward Islands, in the normal range of *A. tropicalis* quite considerable numbers of *A. gazella* have been found, and both species are breeding on these islands. In addition to the 7000 *A. tropicalis* on Marion Island, there are 200–300 *A. gazella*, and an unknown number on Prince Edward Island (Condy, 1978).

Not only are both species present and breeding on Marion Island, but there is said to be some evidence of possible interbreeding. Adult bulls have been seen with the crests and light faces of *A. tropicalis*, but they were large animals and had the large flippers and less colourful coat of *A. gazella*. Photographs were taken, but no specimens of these possible hybrids, so the question must remain open for the moment. *A. gazella* bulls have been seen with *A. tropicalis* cows in their harems, and conversely *A. gazella* cows have been seen in the harems of *A. tropicalis* bulls (Condy, 1978).

The development of *A. gazella* pups on Marion Island seems to be about two to three weeks ahead of the *A. tropicalis* pups, and the former have a much shorter lactation time. There is also possibly a slight ecological separation as there seems to be a tendency for *A. gazella* to gather on the smoother areas of the rocky beaches and on the grassy slopes. The difference in the nature of the calls would help in species recognition.

The populations of *A. tropicalis* are expanding, but not at the rate of those of *A. gazella*. The rate of increase of *A. tropicalis* is taken to be more 'normal' than that of *A. gazella* which is apparently reacting to the increased amount of krill available as food as a result of the decline in baleen whales (Condy, 1978).

Antarctic Fur Seal Arctocephalus gazella

DISTRIBUTION

The Antarctic or Kerguelen Fur Seal was first described from a female animal collected on Kerguelen in
1874 (Peters, 1875). It has often previously been
known as the Kerguelen Fur Seal, but as Kerguelen is
the only part of its range where it is not known to
breed at the moment, and the centre of the population
is on South Georgia, Antarctic Fur Seal is a more
suitable name.

This seal occurs on islands south of the Antarctic
Convergence – on South Georgia, South Sandwich,
South Orkneys, South Shetlands, Bouvet, Kerguelen,
Heard and McDonald. There is also a breeding colony
north of the Convergence, on Marion Island. The
colony of *A. gazella* on South Georgia is probably the
fastest growing pinniped colony known, and it is
possible that this has some correlation with the greater
availability of its major food supply – krill – and the
decline of the baleen whales (Condy, 1978).

The number of seals in the South Georgia colony has
changed from 'pretty numerous' as reported by Captain James Cook in 1775 to virtual extinction by the
1820s, to a grand total now of something approaching
400 000 animals.

Bonner (1968) and Payne (1977, 1978) summarize
the history of this seal in South Georgia. After discovery of the islands in 1775 the sealing started in 1790 and
reached a climax in the season of 1800–1801 when
112 000 fur seals were taken. By 1822 the fur seal was
commercially extinct and it was estimated that nearly
1·25 million fur seals had been killed. After this virtually nothing was heard of the fur seals on South Georgia
until a single one was taken in 1915, and in 1919 five
were seen on Bird Island off the northwest tip of the
South Georgia mainland.

Bird Island is the centre of the South Georgia
population. The fur seals there were estimated to
number about 60 in 1933, and by 1956 had increased to
about 12 000. By 1961 the annual production of pups
on Bird Island was 9900 and small colonies were
starting on the mainland. The annual pup production
climbed to 11 500 in 1963, 25 000 in 1971 and 90 000 in
the 1975–77 season. The last figure is for all colonies on
the South Georgia mainland and associated islands
and would be equivalent to a total population of
369 000 fur seals in March, 1976 (Payne, 1979a).

Although Bird Island has been the centre of the
present increase in fur seals at South Georgia, the
remnant of the population left over from sealing days
is thought to have been on the Willis Islands, and to
have spread to Bird Island from there. By 1960 a few
pups were being born at Elsehul on the mainland, and
by 1972 large stretches of the mainland coast had been

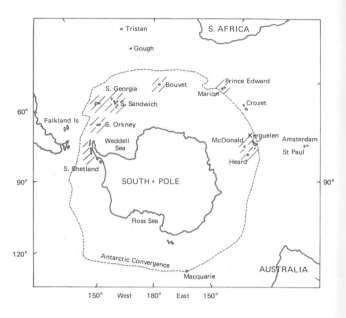

occupied and the pup production was greater than on
Bird Island where nearly all the suitable areas were
occupied (Payne, 1977).

Colonization of the other islands of the Scotia Arc
has presumably taken place from the flourishing Bird
Island colony, the South Sandwich and South Orkney
groups, being nearer to South Georgia, having greater
numbers of fur seals than the South Shetlands. The
colony on Bouvet was probably never completely exterminated and is recovering slowly. Although fur
seals are breeding on Heard and McDonald, the bulk
of the animals there, and on Kerguelen, are summer
visitors. Presumably they come from Bouvet, though
nothing is known of their origin.

The South Sandwich Islands are not thought ever to
have had a very large population of fur seals. Six
thousand were taken in 1875–77, but less than a hundred skins the following year in 1878. In 1881 no fur
seals were found, and only 400 skins taken in 1891.
There are are no further records until January, 1960
when about 400 animals were seen on Visokoi Island,
including harem bulls and black pups. In March, 1962,
800 fur seals were seen on many of the islands of the
group and the breeding population seems to be increasing (O'Gorman, 1961, Holdgate, 1963).

In the South Orkneys the fur seals were exterminated by the sealers and there are no records of seals
returning until fairly recently. In 1947 one animal was
seen on Signy Island, and in 1973 counts of 100 fur

seals on Signy were not uncommon. These were mostly adult males, seen in autumn (February–April). Pale coloured animals like some of those on South Georgia have been seen on Signy and a South Georgia tagged animal has been sighted. On Michelsen and Powell Islands in the S. Orkneys the fur seals have increased from 111 in 1956 to 923 in 1971, with about 60 pups. Over the S. Orkneys in February, 1971 a total of 2035 fur seals was counted, of which the far greater proportion (93 per cent) were males (Laws, 1973).

On the South Shetland Islands 42 fur seals were seen on Livingstone Island in 1958 and two pups were born there the following year. By 1966 the count had risen to 200 animals on Livingstone, and 300 on some of the other islands with breeding noted on Elephant and Livingstone Islands (Laws, 1973, O'Gorman, 1961, Aguayo, 1970). Between 1966 and a census made in the 1972–73 season there was a great increase in numbers. On Livingstone for example there were 3024 animals in 1973 and it was estimated that there were 6–7000 fur seals on all the islands of the South Shetland group (Aguayo, 1978).

The Norwegian island of Bouvet was discovered in 1739 but seldom visited. It was visited sporadically by sealers, but has never been a major source of skins. The last skins were taken in 1927 when the *Norvegia* collected 800. The visits of the *Norvegia* to Bouvet during 1927–29 resulted in the first scientific collection of fur seal specimens from the islands, together with an account of the animals there. It was estimated that there was a breeding colony of about 1200 animals in 1928 (Sivertsen, 1954). Biologists who visited Bouvet in 1964 noted about 500 fur seals and estimated an annual pup production of about 150–180. The beaches on Bouvet are too small to support a large fur seal population, but the volcanic platform that appeared between 1955 and 1958 increases the area suitable for hauling out (Holdgate *et al.*, 1968).

On Heard Island between 1948–51 small numbers of fur seals were seen, but there was no breeding. In 1963 about 500 fur seals were seen, and two very young pups. In March, 1969 there were estimated to be 3000 fur seals on the island. Some suckling pups were seen but it was thought that most of the fur seals were

Antarctic Fur Seal, Arctocephalus gazella. *A big adult male sits beside a pair of reclining females. Photo. T. S. McCann.*

summer visitors from breeding grounds somewhere else (Budd, 1970). Further increases were noted in January, 1971 where in a single area the population had almost doubled, but although there were a few pups the population was mostly non-breeding bulls (Budd, 1972).

On McDonald Island, 38km west of Heard Island, the first landing was made in 1971 and on two of the main beaches 85 males, 68 females and 46 black pups were seen. Other parts of the island could also harbour fur seals but the population was considered too small to be the reservoir for the increase in numbers at Heard Island (Budd, 1972).

Fur seals used to be more abundant on Kerguelen but are thought to have been exterminated by the sealers. There is little recent information. Paulian (1953) saw and caught a single adult male on the eastern side of the island in 1951 and suggested there might be more on the more favourable west coast. Angot (1954) saw eight small animals in 1952. In 1967, 143 animals were reported (Budd & Downes, 1969), but so far no births have been recorded. Only a relatively small area of the coast has been surveyed for seals, so whether the increase in numbers is due to breeding on Kerguelen or to visitors coming in from somewhere else is not known.

Although normally resident south of the Antarctic Convergence, there are instances of *A. gazella* being found north of it. A pup, tagged in January, 1973 was found in September of that year, as bones, on Tierra del Fuego (Payne, 1979b). Allowing four months for lactation, the pup must have then have swum the 2000km against the current of the West Wind Drift in a maximum time of five months. A collection of skulls of *A. tropicalis* made by R. W. Rand on Marion Island in 1952 contained one skull that was different from the rest (King, 1959a). This skull came from a lactating female which makes it very probable that it was rearing a pup in the near vicinity (Payne, 1979b). Further evidence of *A. gazella* breeding on Marion Island was obtained from Condy's observations there between 1973 and 1976. As noted in the section on *A. tropicalis*, 200–300 *A. gazella* are on Marion Island, and an unknown number on Prince Edward Island (Condy, 1978). Possible interbreeding between the two species is also noted under *A. tropicalis*.

DESCRIPTION
The nose to tail length of adult male *A. gazella* is about 1·8–1·9m, and they weigh about 140 kg. Adult females are about 1·3m in length and rarely exceed 50kg in weight (Payne, 1979a). The males are grey-brown over most of the body with a grizzled mane. The chest and neck may be a silvery grey or much the same colour as the back, and the face is a slightly darker grey. They do not have the very definite yellow face and chest of *A. tropicalis* and do not have the crest of hair on the top of the head. Compared with *A. tropicalis* the body appears longer and less bulky and the neck more slender. Adult females are grey-brown dorsally with the belly, chest and neck white to grey. They are overall much lighter in colour than *A. tropicalis* (Condy, 1978, Bonner, 1968). When compared with *A. tropicalis*, the foreflippers of *A. gazella* are longer in relation to the body size, the eyes are smaller, and narrower, and the animals do not have the conspicuous white mystacial whiskers (Condy, 1978, Payne, 1979b). A colour variation that occurs in both sexes and all ages is the virtually complete absence of pigment in the guard hairs. The animals appear white, though they are not albinos, as pigment is developed in the underfur and in the naked skin areas (Bonner, 1968).

As both *A. tropicalis* and *A. gazella* may occur on the same beaches some mention of the distinguishing characters of their skulls should be made. *A. tropicalis* has a narrower skull with a narrow palate, a short narrow rostrum and long nasal bones. The upper cheek teeth are of normal size and do not get excessively worn. *A. gazella* has a wider skull and palate, a short broad rostrum and short flaring nasal bones. The upper cheek teeth in particular are very characteristic – they are small, with no accessory cusps, darkly stained, worn very smooth on their lingual surfaces, and the last two are extremely tiny (King, 1959 a and b, Repenning *et al.*, 1971).

BREEDING
Most work has been done on the biology of the South Georgia animals (Bonner, 1968, McCann, 1980). The mature bulls probably start to haul out in the middle of November, and nearly all of them have established their territories by the beginning of December. All the available territory is soon divided up so that the late comers have to fight for possession, slashing at each other with their teeth. Particularly fierce aggression is shown against new animals arriving from the sea and attempting to make their way through the established territories of other bulls. Territorial fighting results in a hierarchy of dominance, and the most successful bulls will have their territories at the water's edge. Territories extend from the beach inland into the tussock, and also from the favoured spots just above high water mark seawards so that some bulls maintain a territory of a minute piece of rock that is submerged except at low water. It is the younger, less experienced bulls that have these less favoured spots either far in the tussock or partially submerged. Territorial bulls give off a pungent musky, sweetish odour during the breeding season.

The cows come in towards the end of November, though most of them arrive in early December. They are gregarious, and prefer sheltered dry shingle beaches. The bulls do not gather cows, but do their best to stop them leaving the territory, an occupation that gets more difficult as the number of cows increases.

The pups are born within a day or two of the arrival of the cows, and most are born at the beginning of December. They are about 65cm nose to tail length, 6kg in weight, and are dark brown or black in colour with a grey-brown belly. Immediately after birth the pup starts searching for food and establishing a vocal bond with its mother. The cow remains with her pup for about eight days, and she then comes into oestrous and mates. She then goes to sea for feeding and after this alternates trips to sea with a few days on shore to allow the pup to suckle.

Lactation lasts for 110–115 days (four months) (Payne, 1979a), so that towards the end of March pups are beginning to be weaned. Their birth weight of about 6 kg increases during lactation and at weaning male pups weigh about 17 kg and are about 3·5 kg heavier than female pups. This indicates a rate of increase of 98 g a day for males – a very fast rate of growth for a fur seal. This is correlated with the short lactation time compared with most other fur seals (Payne, 1979a). The pups moult the dark natal coat for the more silvery yearling coat from January onwards, the process taking about a month, and about half the pups have completed the moult by the end of February. By the middle of April relatively few pups still remain at the rookery, and in fact large groups of pups can be seen swimming out at sea from the middle of March onwards.

After the middle of January the rookery structure disintegrates, territorial bulls go to sea to feed and the younger bulls attempt to take over the territories. By the middle of May very few animals are on shore. The bulls are sexually mature at three to four years, but are not able to hold territories until they are at least eight years old. Females produce their first pup at three or four years old.

FEEDING

The main food of the fur seals at South Georgia, and probably in other places of its range is krill (*Euphausia superba*), which is extremely abundant south of the Antarctic Convergence. Other food items such as local fish and squid are also eaten, particularly by animals feeding close inshore, but krill is the main diet. The curious wear on the lingual surface of the cheek teeth, particularly the upper ones, where the inner surface may be worn flat or even concave, has already been mentioned. There is no obvious reason for this wear, but it may possibly be caused by the hard exoskeleton of the krill as it is moved around the mouth (Bonner, 1968).

MORTALITY, PREDATORS AND PARASITES

The parasite infestation of these seals is said to be relatively low and little is known about causes of death. Fibrous cysts have been found in the abdomen, and an affliction of the coat resembling mange has been noted (Bonner, 1968), but neither of these appeared to be serious. In the vicinity of South Georgia at least, killer whales are not common. Barnacles (*Lepas australis*) have been found firmly attached to the guard hairs of the back of breeding cows. The barnacles of course die when the seals haul out during the breeding season.

South African or Cape Fur Seal
Arctocephalus pusillus pusillus

DISTRIBUTION

The Cape Fur Seal is abundant in the waters of South Africa and South West Africa (Namibia) where there are some 20–23 colonies, some on the mainland, but most on offshore rocks and islands. The colonies stretch between Black Rocks on the eastern side of Algoa Bay, about 48km east of Port Elizabeth, to a small, relatively new colony at False Cape Fria (c. 19°S) on the coast of South West Africa.

In South African waters some of the colonies are at Black Rocks, Algoa Bay, Seal Island Mossel Bay, Geyser Rock Dyer Island, Seal Island False Bay and Elephant Rock near the mouth of Olifants River. In the waters of South West Africa some of the colonies are at Sinclair Island, Wolf Bay and Atlas Bay on the mainland about 56km south of Luderitz, and the Long Islands just off the coast in this area, the Luderitz Bay islands, Hollamsbird Island, Cape Cross and False Cape Fria. Most of these colonies carry 500–1000 bulls, some over 3000 (Long Island) and Wolf and Atlas Bays have some 5000 bulls between them. In South West African waters some 150 000 pups are born annually. This represents 68 per cent of the total numbers of

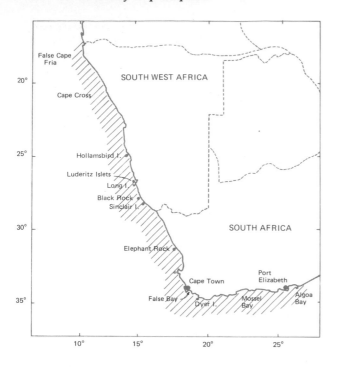

Cape Fur Seals born which makes the total pup production about 220 000 animals. The total population of *A. pusillus*, including seals of all ages is estimated to be 850 000, (Rand, 1972, Shaughnessy, 1979a and 1979b).

Although they occur mainly in the inshore waters and do not make regular migrations, these fur seals do nevertheless range widely in small groups while feeding. Adult bulls have been found as far as 160km out at sea in the trawling grounds, and many individuals that have been tagged have been recovered far from their breeding grounds. The longest recorded distance is between Seal Island in False Bay and Cape Cross (*c.* 1500km), and some 16 seals have made this voyage, the youngest being only ten months old (Shaughnessy, 1979).

DESCRIPTION

As already mentioned, *A. pusillus* is the largest of the fur seals and has some characters that are rather more like sea lions than fur seals.

Adult males are about 2·3m nose to tail length, and weigh between 200–350 kg according to the time of year. Adult females are smaller, 1·5–1·8m in length and about 120kg in weight. The fully adult bull has a coarse outer coat of dark blackish grey colour, lighter below. The females are an indeterminate brownish grey colour, lighter brown ventrally. At birth the pups have a velvety black coat formed of short curled hairs. This coat is moulted during March and April, when they are about four to five months old, the actual

moult taking about four to five weeks. The new coat is an olive grey colour, the black hairs being tipped with varying amounts of white, and the thick brown under-fur develops. This coat is moulted a year later for one that is a silvery grey colour, although the bright colour of the new coat fades during the summer so that the dorsal surface appears darker, and the ventral lighter. The next, and subsequent moults take place in January and February (Rand, 1956a).

BREEDING

Although these seals do not migrate, just before the breeding season starts there is a definite decrease in the numbers of animals on the rookery. From about mid-October onwards the adults start coming ashore, the bulls establishing territories, fighting and maintaining them for about six weeks without leaving to feed. Cows also frequently establish claims to a personal bit of territory where they will produce their pups (Rand, 1967).

Pups are born during the end of November and beginning of December, most of them are born during a 34-day period with a median date of 1 December. Some, however, may be born between the end of October and the beginning of January. The pups are about 60–70cm in nose to tail length, and weigh 4·5–7 kg, with the males tending to be heavier than the females. Suckling starts very soon, within an hour, or sooner, after birth.

For about a week the mother stays with her pup, but then goes to sea for a day or two, gradually increasing the time spent away until by the second month the pup may be left alone for two weeks. By April when the pups are 4·5–5 months old, the males may weigh 22 kg, and the females 17 kg (Rand, 1956a).

The pup, although born with most of its milk teeth, sheds these by May–June, and doubtless helps this to happen by its habit of biting at miscellaneous floating objects. After this, the pup supplements its milk diet with small crustaceans and fish, although lactation continues until October, when the females have a more prolonged period at sea. If a newborn pup dies, that of the previous year may continue suckling for some time into its second year. With increasing age the young animal becomes more adventurous and by the time it is seven months old it is capable of spending two or three days away at sea. Later, during their first and second years the young seals spend much time at sea and come on land mainly during the annual moult in December and January. The young males in particular are active, noisy animals, threatening each other, lunging and snorting as they settle their order of hierarchy. They also play, particularly in the evenings, sliding over the rocks and scampering over the sand (Rand, 1967).

Mating takes place about six days after the birth of the pup and there is about a four-month delay before the blastocyst becomes implanted in April or May. The females are sexually mature and mate for the first time at the end of their third year. Males are also believed to be sexually mature in their third or fourth year, but are unable to hold a harem until several years later.

PREDATORS AND PARASITES

It is possible that killer whales may take fur seals at sea, but there is little real evidence of this and the seals take little notice of underwater playback of killer whale sounds that have been recorded. There are, however, many reports of 'great white sharks' taking seals. Black backed jackals (*Canis mesomelas*) may take pups and placentae, and strand wolves (= Brown hyaena, *Hyaena brunnea*) scavenge carcasses from the rookery. The usual population of internal parasites is present – acanthocephalans and cestodes in the intestine, nematodes in the stomach, and mites in the nasopharynx and trachea. The pig tape worm (*Taenia solium*) has been found in this fur seal and this, together with an old record of its occurrence in the Mediterranean Monk Seal, are the only records of this parasite being found in a marine mammal. How this fur seal became infected is not known, but it could have ingested the mature cestode proglottids in a fish that had acquired them from human faeces. The adult male fur seal was found ill and convulsing and was later found to have cysticerci in the brain, heart, lung and liver, and also in the skeletal muscle (De Graaf *et al.*, 1980).

FEEDING

The feeding movements and habits of this seal have been studied in some detail. During the five to six weeks of the breeding season the harem bulls are ashore continuously, feeding little, if at all; but during the rest of the year they spend two or three weeks of every month on feeding expeditions. Cows spend slightly more time on land owing to the demands of the pups and the time spent at sea varies between one and three weeks a month. When the pups are about five months old they begin to take in other substances besides milk. Although some of these are inedible,

South African Fur Seal, Arctocephalus pusillus pusillus *at Wolf Bay, South Africa. An adult male occupies the foreground. Photo. P. D. Shaughnessy.*

such as pieces of wood, small crustaceans and fish are also caught, and by the time it is weaned the young seal is reasonably proficient at catching its own food and gradually attains the adult diet. This latter includes small shoaling fish such as maasbankers (*Trachurus trachurus*) and pilchards (*Sardinops ocellata*), but almost any available fish may be eaten.

In the last 20 years the populations of maasbanker and pilchard have decreased while the number of anchovies has increased, so probably the feeding habits of the seals have been modified.

Cephalopods also form a large proportion of the food eaten; rock lobsters are taken, although not in great numbers; and various small crustaceans, some probably from fish stomachs and some parasitic upon the fish, are also found in stomach contents. Stones are also found in stomachs, but by far the greater number of stones is found in the yearlings, and it is suggested that the stones allay hunger pangs while the mother is away at sea. Older yearlings that can augment the milk diet are less often found with stones. Smaller fish are usually swallowed whole, and under water, while larger ones are brought to the surface and mouthfuls of flesh torn from them. The seals normally hunt and feed singly except where a shoal of fish attracts a number of seals. It is thought that most feeding is done during the daytime, those animals that can do so returning to the rookery at night, the others sleeping at sea. The seals are surface feeders, rarely going deeper than 45 metres. It is not believed that the seals interfere to any serious extent with the commercial fisheries of the coast.

EXPLOITATION

Sealing began with the discovery of the Cape but the herds were soon over exploited and the market disappeared. In the eighteenth century the same cycle was repeated, and although the South African Government took over control in the present century, it was from the commercial point of view, and research into the biology of the animals was not started until 1946. With an increasing understanding of the life history the management of the herds has improved and the wasteful methods of previous centuries are unlikely to be repeated.

At the present time sealing is under the control of the Sea Birds and Seal Protection Act of 1973, and permits are issued for the taking of seals. A quota system in operation allows a maximum of 40 per cent of the pups born to be harvested, but the situation is under continuous review.

Sealing takes place in summer (October–November) and again in winter (July–August), different classes of animals being hunted. The bulk of the summer catch is composed of the surplus bulls. These are shot, and are taken more for their blubber than for their pelts, which are apt to be scarred. The season lasts only four to five weeks, as the condition of the bulls rapidly deteriorates. About 2000 bulls are taken. Winter sealing is directed to the pups. These are commercially most valuable from the age of seven to ten months when their underfur reaches its greatest length and best texture; the skin is free from scars, and the blubber is thick. About 60–80 000 pups are taken annually and the blubber is used as well as the pelt (Shaughnessy, 1976 and 1979a).

ECOLOGY

The Cape Fur Seal is an abundant animal and is increasing in numbers. A controlled harvest does the main stock no harm. There is competition between the fur seals and sea birds for breeding space on the offshore islands, and the seals have already displaced the birds from one colony. They are probably also responsible for the white pelican vacating its nests on more than one site. Other animals such as the Cape gannet, Jackass penguins, cormorants and small cetaceans also share the food supply. Uncontrolled numbers of fur seals would be detrimental to these other animals and would also cause even more nuisance to the human fishermen, who would then retaliate so that unknown numbers of seals would be shot at sea (Shaughnessy, 1976).

Australian Fur Seal
Arctocephalus pusillus doriferus

Repenning *et al.* (1971) note the skull characters in which *A. pusillus* differs from other species of the genus, and comment that in spite of the great distance between the populations of South Africa and Australia they were unable to find any really good distinguishing characters between skulls from the two areas. They do, however, maintain the separation of the two groups at subspecific level 'on the basis of an observed difference in the length of the crest that unites the mastoid process with the jugular process of the exoccipital: it is proportionately greater in our specimens from Australia than in specimens from Africa' (Repenning *et al.*, 1971, Marlow & King, 1974).

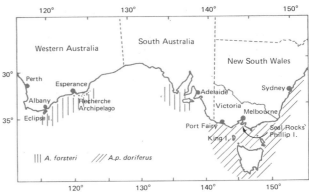

Distribution of fur seals in Australian waters.

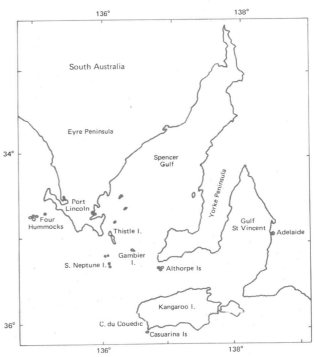

Enlargement of the area in South Australia where A.forsteri *is found.*

DISTRIBUTION

The Australian Fur Seal is an abundant animal in Bass Strait, and at suitably rocky and exposed places on the coasts of Tasmania, Victoria and southern New South Wales. The total population numbers about 20–25000, and about half of these are on the two largest colonies at Lady Julia Percy Island and Seal Rocks off Phillip Island. Other Victorian breeding colonies are on Kanowna Island off Wilsons Promontory, and on the Skerries. In New South Wales there are small and relatively unknown colonies on Montague Island and on Seal Rocks near Port Stephens (32° 28'S, 152° 33'E) (Warneke, 1979). In Bass Strait there are colonies of 1500 on Judgement Rocks and Reid Rocks, and slightly smaller colonies of 500–750 on West Moncoeur Island, Moriarty Rocks and Tenth Island. In the south of Tasmania there are about 900 seals on Needle Rocks in the Maatsuyker Group and 100 on Pedra Blanca. Smaller groups of seal occur on any of the many rocks and islets round the coast, and a total of about 6500 has been estimated for the Tasmanian population (Pearse, 1979). These fur seals do not migrate and are completely protected. Their voice is mentioned in the section on the New Zealand Fur Seal.

DESCRIPTION

Adult males are 2–2·2m in nose to tail length, and weigh 218–360 kg. They are dark grey-brown all over with a well developed mane of coarse hair on their necks and shoulders. The slightly lighter tips to the hairs of the mane gives the animal a paler area over the back of the neck. When clean and wet the bulls appear more steely-grey. Adult females are 1·2–1·7m in length and 36–110kg in weight. They are silvery grey dorsally with a creamy yellow throat and chest, and a chocolate brown abdomen. Immature animals of both sexes are grey-brown dorsally and yellow ventrally. The apparent colour varies enormously depending a lot on the cleanliness of the animals. The general impression of a breeding rookery, where large numbers of the seals spend much time ashore is of an over-all ginger-brown colour.

The pups at birth are about 62–80cm in nose to tail length and weigh 4·5–12·5 kg with male pups being about 1 kg heavier than females. The newborn pups are shiny dark brown to black dorsally and grey or light brown ventrally. After about three months they moult and assume the colour of the immature animals.

BREEDING

The timing of the breeding season is very much as in the South African Fur Seal. The breeding colonies are never totally deserted, but the occupants are mostly females and their growing young. There is an influx of big territorial bulls towards the end of October. They take up their territories and defend them against rivals, the competition and subsequent commotion becoming fiercer as the prime sites are taken and more animals arrive.

As in most other otariid fights, the two bulls front up to each other, the defender trying to push the aggressor out of his territory. Slashing attacks with the canines are made and the opponents grab hold of mouthfuls of skin and shake vigorously. A successful attack on a foreflipper may result in the overbalancing of the opponent, and a fight may go on for several hours. If there is eventually no outright winner and the combatants are exhausted, the disputed territory may be divided between them. Although the bulls jealously maintain their territories, cows and young animals are not herded into harems and can move freely where they wish.

The pups are born between the beginning of November and the end of December, but most of them between 20 November and 7 December. Each pup stays close to its mother until she mates about five or six days after parturition. The adult females then go to sea to feed, returning about once a week to feed the pup. This continues for about eight months, and then the pup starts to take a certain amount of solid food as well, and will go to sea with its mother. The young pups congregate in pods, sleeping, playing, and swimming in tide pools on hot days. Although an unusual occurrence, and possibly the result of some stress, a female has been seen to tear at the placenta and eat portions of it.

The big territorial males do not usually leave their chosen sites to feed until the end of the breeding season. Once they have left the breeding colonies very little is known of their whereabouts, and it is assumed that they spend most of the time out at sea, but there are of course many rocks and islets where they could rest. Immature seals are rather sedentary and do not move further than about 150km from their birthplace (Warneke, 1975).

Australian Fur Seal, Arctocephalus pusillus doriferus. *An adult male cooling his hindquarters, beside three females with young pups. Taken on Seal Rocks, Victoria. Photo. B. J. Marlow.*

Lactation lasts for eleven to twelve months, normally until the next pup is born, though sometimes a pup may be suckled through a second or even a third year if some accident befalls the current pup (Warneke, 1979). There is a delay in the implantation of the blastocyst of approximately three months. The females are sexually mature when they are between three and six years old, and the males between four and five years. At this age, however, the males are not physically capable of strenuous fights with the large territorial bulls in order to try and establish their own territories; and long-term tagging has shown that it is not until they are eight to twelve years old, or even older, that they are capable of achieving the status of territorial bull. Even then their breeding careers seldom last more than three seasons, though a maximum of six seasons is known.

MORTALITY, PREDATORS AND PARASITES

Killer whales and white sharks are probably the main enemies of this fur seal. Both sharks and fur seals may sustain injuries from the spines of sting rays. In one recorded occurrence a nine-month-old fur seal died of a pericardial infection and septicaemia due to a sting ray spine that punctured both the oesophagus and the pericardium and penetrated the wall of the right ventricle (Obendorf & Presidente, 1978). The stomach and intestine of these fur seals house their quota of cestodes, nematodes and acanthocephalans. There may be large quantities of mites in the nasopharynx, trachea and bronchi with consequent overproduction of mucus which can lead to respiratory problems and even death from asphyxia.

FEEDING

Squid (*Notodarus*, *Sepiotheuthis*) and octopus form a large proportion of the food eaten, but many different fish are also taken depending on season and availability.

Barracouta, whiting, flathead, mullet and parrot fish, are amongst those taken, and rock lobsters are also eaten. Shoaling surface and midwater fish and also sedentary bottom-dwelling fish are taken. Probably the whole of the continental shelf area is available to this seal as a feeding area, and recoveries from nets indicate that it can dive to at least 120m (Warneke, 1979).

EXPLOITATION

These fur seals were probably first recorded as having been seen by George Bass, when in the early part of 1798 he explored the coast of Victoria and discovered Westernport. He managed to extend his six weeks' supply of stores to nearly double that time "with the assistance of occasional supplies of petrels, fish, seal's flesh and a few geese and black swans, and by abstinence" (Flinders, 1814), and thought there was the possibility of commercial exploitation of the seals of Wilsons Promontory. One of the survivors of the wrecked *Sydney Cove* has reported fur seals in Bass Strait (Abbott & Nairn, 1969) and Flinders, on his trip to the Furneaux Islands to rescue the cargo of this ship, also commented on the number of fur seals he had seen. This resulted in the *Nautilus* leaving Sydney in October, 1798 for the first sealing expedition in this area, achieving a cargo of some nine thousand skins, and the beginning of the sealing industry here' (Marlow & King, 1974). By 1820 the original herds had been seriously depleted, though a certain amount of sealing continued in Victoria until the seals were protected there in 1891.

New Zealand Fur Seal *Arctocephalus forsteri*

DISTRIBUTION

The New Zealand Fur Seal is usually an animal of exposed rocky coasts where offlying reefs give some protection from heavy seas. Most of the breeding colonies in New Zealand are on Open Bay Islands and along the rocky Fiordland coast south of this, round to Solander Island, Ruapuke Island and Stewart Island, though there are also colonies at Cape Foulwind and to the north of Heaphy River. Non-breeding groups are plentiful round much of the coast of South Island, and in winter large colonies form in the south of North Island. There are also smaller groups, mostly on the west coast and as far north as Three Kings Islands. Seasonal movements, made mainly by the males are between the breeding grounds and hauling out

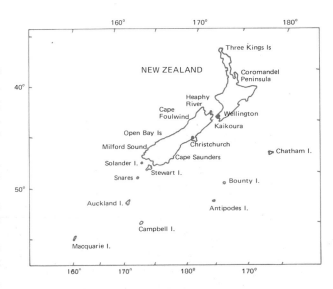

grounds in other places. There seems to be a tendency for the seals to make a general movement north so that there are, for instance, many males at Cape Saunders in February, at Kaikoura in June, and at Wellington in July, after which there is a movement south again for the breeding season (Crawley & Wilson, 1976).

Breeding colonies do not now occur on the coasts of North Island, and even non-breeding groups are infrequent on the northeast coast. However, archaeological material from layers probably no later than the fourteenth century on the Coromandel Peninsula, just east of Auckland, suggest that there used to be breeding colonies in this area (Smith, 1978).

Breeding colonies and non-breeding groups are also found on most of the islands in the New Zealand area – on the Snares, Auckland Islands, Campbell Island, Macquarie Island, Bounty Island, Antipodes Island and Chatham Island. On Macquarie Island the breeding population is small, but there is a large visiting population of perhaps 650–1000 animals, mainly young non-breeding males that haul out to moult during summer (Johnston, 1972). On the rocky western coast of the main Auckland Island small colonies are present and the terrain seems to be suitable for further expansion. About 1000 fur seals are estimated to be on these islands (Wilson, 1974). On Antipodes Island there seems to be no evidence of breeding, but there is a non-breeding group of about 1000 seals. The following list of the numbers of fur seals in the breeding areas of New Zealand is taken from Crawley & Wilson, 1976:

Westland	6 600
Fiordland	8 600
Solander Island	5 000
Stewart Island	3 150
Snares Islands	1 150
Auckland Islands	1 000
Campbell Island	2 000
Bounty Islands	5 500
Macquarie Island	650
Chatham Islands	2 150
Total:	35 750

The above authors suggest that the figure for the total New Zealand population could be as high as 40 000.

New Zealand Fur Seal, Arctocephalus forsteri. *An adult male, wet from the sea. Photo. E. H. Miller.*

New Zealand Fur Seals are also present in Australian waters (King, 1969a). There are breeding colonies on Eclipse Island and on some of the islands of the Recherche Archipelago, Western Australia; on the islands of Four Hummocks and South Neptune, and on Cape du Couedic on Kangaroo Island, South Australia. The South Neptune colony is the biggest in Australian waters with about 1300 animals (Ling & Walker, 1976). Up to 300 seals have been counted at Cape du Couedic, but it is believed that many of these have come from the South Neptune colony. There are perhaps 400 fur seals in Western Australia, and one might estimate a total of about 2000 New Zealand Fur Seals in Australian waters (see map p. 55).

There is no geographical barrier between the South Australian colonies of *A. forsteri* and the large colonies of *A. pusillus doriferus* off the Victorian coast. *A. forsteri* occasionally reach Seal Rocks off Phillip Island, the main stronghold of *A.p. doriferus*, even attempting to hold a territory there, and Seal Rocks animals are known to turn up in South Australian waters. Analysis of the vocalizations of these two fur seals have been compared with those of other species of *Arctocephalus* from the literature (Stirling & Warneke, 1971). The unusual situation of two vocal types within the genus has been reported, *A. forsteri* and *A. gazella* from South Georgia having similar calls, and *A.p. doriferus* being quite different apparently with a reduced vocal repertoire. This would presumably assist the females in recognition of the correct males should they be in the wrong locality.

Straggling New Zealand Fur Seals outside their normal range are occasionally found in the Sydney region. In August 1972 and September 1973 three young animals, estimated to be about eight months old, were stranded in New Caledonia, very much further north than any of the usual haunts of this fur seal. From the currents and weather conditions at the time, Rancurel (1975) suggested an Australian origin for these young animals, but their exact place of origin is not known. They are, however, definitely *A. forsteri* and not *A.p. doriferus* as distinguishing characters of their skulls have shown (King, 1976).

In 1925, Le Souef found *c.* 100 *A. forsteri* on Saltpetre Rocks, west of King Island in Bass Strait, and a skull confirms his identification (King, 1969a). *A. forsteri* does not now normally occur east of Kangaroo Island, but there is some evidence of archaeological material (dated to about 1500 years ago) of this fur seal on Hunter Island, between the Tasmanian mainland and King Island (O'Connor, pers.comm.). At the present time the Bass Strait islands are occupied by the Australian Fur Seal (*A.p. doriferus*) and Reid Rocks just south of King Island is a major breeding ground for this fur seal.

After the voyages of Captain Cook made the presence of the New Zealand fur seal known, it suffered from the same depletion to the commercial world as did the other fur seals. George Forster, the botanist on Cook's second voyage, sketched a 'sea bear' that was killed in Dusky Sound, South Island in 1773. In spite of the slaughter of fur seals that took place during the first two or three decades of the nineteenth century, when tens of thousands of animals were killed, European scientists knew no more about the New Zealand fur seal than was evident in Forster's drawing. Not until 1871 were skulls collected from Milford Sound and their description published (Cumpston, 1969, Abbott & Nairn, 1969, Marlow & King, 1974).

The history of the fur seals on Macquarie Island (Csordas & Ingham, 1965, Cumpston, 1969) indicates that after the discovery of the island in 1810, fur seals had been virtually totally removed by 1820. Although there was an annual visitation of the yearling seals, probably from the Auckland Islands, the first definite evidence of breeding was not obtained until 1955, and the breeding population is still small. It is not definitely known that the Macquarie fur seal of pre-sealing days *was A. forsteri*. Sealers apparently distinguished the fur seals of Macquarie and also of Antipodes Island as the 'upland seal', and regarded them as different from the New Zealand and Australian animals. Possibly the 'upland seal' was *A. tropicalis* – the Subantarctic Fur Seal. A young male seal of the appropriate colour for this latter animal was seen on Macquarie in 1959, and it is possible that other widely travelling young male Subantarctic Fur Seals may occasionally reach the west coast of South Island (Csordas, 1962). If *A. tropicalis* was the original inhabitant of Macquarie Island, it has obviously now been replaced by *A. forsteri*. Examination of the blood serum of modern fur seals from Macquarie, New Zealand and Australia has shown that the transferrin types of the Macquarie and New Zealand *A. forsteri* are similar, but are different from those of the Australian *A. forsteri*, though the significance of this genetic difference is not at the moment apparent (Shaughnessy, 1970).

DESCRIPTION

The nose to tail length of adult males reaches about 2 m, and they may weigh 200 kg. Adult females reach 1·5 m in nose to tail length and 90 kg in weight. 'The coat colour of adults merges from a dark grey-brown dorsally to a lighter grey-brown ventrally. The thick under fur is a rich chestnut colour; the guard hairs are coarse and dark grey, often with white tips which impart a silvery sheen to the dry fur' (Crawley & Wilson, 1976).

BREEDING

Outside the breeding season, between March and

September, the breeding rookeries are occupied mostly by subadults, yearlings and cows, but from mid-October to late December the number of adult bulls increases, while the number of yearlings decreases. The main influx of bulls occurs about two weeks before most of the cows arrive, and they spend this time establishing their territories. Adult cows arrive in large numbers at the end of November and through December, having just had a period of intensive feeding at sea.

The pups are born a few days after the females come ashore, between late November and mid-January, with a peak in mid-December. Both head and tail presentations are recorded, the former births taking about two minutes, while the latter take about six minutes. The placenta appears about an hour later, though the timing is very variable. After the birth the mothers are more placid than they were before, but the pups are relatively active – coughing, looking around, shaking themselves and calling. This and other mother–pup behaviour has been studied by McNab & Crawley (1975).

For about ten days the female stays with her pup, and about eight days after giving birth will mate with a territorial bull. Most copulations are in December. She then leaves her pup and goes to sea for a feeding trip of three to five days, returning to spend a further period of two to four days with the pup. As the pup gets older the cow's feeding trips become longer, and the temporarily abandoned pups gather in pods. During January there is a general breakdown of the harem system as the males, who have remained on land for perhaps ten weeks, without food, depart to sea. Many of the adult females are also at sea, and the land population is now increased by an influx of subadults (Crawley & Wilson, 1976).

At birth the pups are about 55 cm long and weigh 3·5 kg. Their fur is black in colour, but after the first moult at about two months, it changes to a silvery grey and they are then called yearlings. By the time they are about six weeks old the males tend to weigh about 0·6 kg more than the females, but there is no appreciable difference in length. The rate of weight gain varies considerably with the age of the pup. Between birth and 60 days male pups are gaining 74 g a day, but the greatest amount can be as much as 159 g per day between 50 and 60 days, and the lowest as little as 7 g per day between 60 and 240 days. An average daily gain for both sexes between birth and 240 days is 24 g (Crawley, 1975). As is usual in most otariids the pups are suckled for about a year. Aspects of the behaviour of the New Zealand Fur Seal have been studied by Miller, 1974, Stirling, 1970 and Gentry, 1973.

The above details are based on fur seals in New Zealand. On the Australian South Neptune Islands the timing of the breeding season is similar, with births being recorded from 29 November to 22 January with the peak at about 20–30 December. Recorded weights of newborn pups vary between 4·1 and 4·4 kg. Most copulations occur at the beginning of January (Stirling, 1971a & b).

FEEDING

The New Zealand Fur Seal has, predictably, been accused of taking fish that the human fishermen wanted, and thus interfering with their commercial fishing. Investigation of stomach contents, however, gives little justification for this view. The food of the fur seals consists very largely of squid and octopus, with some barracouta and other fish being taken. Fish, however, forms a minor part of the diet, and it is largely non-commercial fish that are eaten. Lampreys, rock lobster and crab are eaten, and penguins are also taken (Street, 1964, Marlow & King, 1974, Crawley & Wilson, 1976).

PREDATORS AND PARASITES

Little is known about the parasites and diseases of this fur seal, but a healed break in the tibia and fibula was noted in an animal that stranded near Sydney.

EXPLOITATION

Permits have been required for killing the New Zealand Fur Seal since 1875. A limited amount of sealing under licence has been allowed, but now normally only those required for research purposes are taken.

The Northern Fur Seal
Northern Fur Seal Callorhinus ursinus

DISTRIBUTION

The Northern, Pribilof, or Alaska Fur Seal might almost be called *the* fur seal as it is the animal that is used most extensively for seal fur coats; the herds are 'farmed' carefully and most aspects of the seal have been well studied. When speaking of its range one must distinguish between the more restricted breeding areas and the considerably larger migration areas. The main places where these seals breed are on St George and St Paul Islands of the Pribilof Islands in the eastern Bering Sea; Copper and Bering Islands of the Commander Islands in the western Bering Sea; and on Robben Island off Sakhalin (see map).

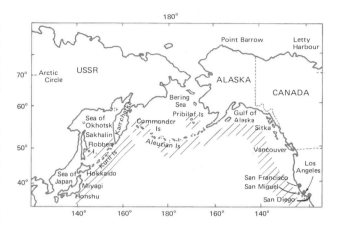

The stock of fur seals on the Kuril Islands was thought to be extinct, but over recent years a colony has been re-establishing itself and increasing in numbers. Two groups of rocks, Kamennye Lovushki and Srednev Rocks in the Kuril Islands hold the majority of this population (Johnson, 1975).

Another recently discovered breeding colony of *Callorhinus* is on San Miguel Island, California. In July 1968 a colony of about 100 individuals, including new-born pups, was discovered there, and it was thought that it had already been established for three to four years (Peterson *et al.*, 1968b, Howorth, 1976). Before they took to San Miguel as 'home' many of the animals had been tagged as pups on the Pribilof Islands and Commander Islands (Johnson, 1975). Now, like all the other colonies of this fur seal the numbers have increased.

The total present population of *Callorhinus* may be tabulated as follows (Lander, 1979):

Pribilof stock	1 300 000
Commander stock	265 000
Robben stock	165 000
Kuril stock	33 000
San Miguel stock	2 000
	1 765 000

The populations of *Callorhinus* are protected and managed by the International North Pacific Fur Seal Commission which was set up in 1957 with representatives from USA, Canada, USSR and Japan. The USSR has control of the Commander Island, Robben Island and Kuril herds, while the USA manages the Pribilof and San Miguel herds. Canada and Japan each receive a percentage of skins from both the USA and the USSR (NOAA, 1977).

During winter and spring large numbers of the seals migrate south, returning again to their breeding grounds in summer. The limits of the southerly migra-

tion are at about latitude 35°N on the Japanese side, and at San Diego, California, 33° 10′N on the American side.

Very occasionally a young seal may wander north, but less than half a dozen have been recorded from the Arctic coast, the furthest being from Letty Harbour, North West Territories 69° 50′N, 124° 24′W, and from Bathurst Inlet (long. 108°W). When moving south the seals are widely dispersed in the north Pacific between the Aleutians and Vancouver Island and then keep normally up to 48–100 km off shore, although the deep Alaskan inlets are visited for herring, and occasional animals, usually sick or dead, come ashore between British Columbia and California. They do not travel in large herds, but singly, or in small groups of up to about ten animals. It is the young seals of both sexes and the adult females that range furthest. It is believed that the adult bulls spend the winter in the north, probably in the Gulf of Alaska.

The migration has started by October when some of the adult females have already left the Pribilofs. In November large numbers are leaving, and by December the majority have left the Pribilofs. Most of them head south-east through the Aleutian passes towards the coast and even in December are found as far south as San Francisco. Between January and April the seals may be found anywhere along their migration route from Sitka, Alaska to California; in April there is the beginning of the northward movement, and by May there are again large numbers in the Gulf of Alaska. From June to October most of the seals are in the vicinity of the Pribilof Islands, although younger animals that have no occasion to be on the breeding grounds are still widely scattered.

In the western Pacific the seals are first seen off Hokkaido in October, the numbers increasing until they reach a maximum in December. They then move south to Honshu and are present in large numbers in April off Miyagi Prefecture, northern Honshu. The return journey starts later than on the American side and the last seals are still in Hokkaido waters in July.

The San Miguel herd remains there all year, but most of the seals from Robben Island winter in the Sea of Japan; most of the Commander Island animals migrate towards Japan, and most of the Pribilof herd moves down the North American coast. There is, however, a certain amount of mixing between the American and Asian herds, especially of the immature animals of about three to five years old. About 30 per cent of those found off the coast of Japan in winter have been tagged on the Pribilof Islands. Some of the Commander Island males get taken in the Pribilof commercial harvest and vice versa (NOAA, 1977). There is no evidence of any specific or subspecific difference between the different populations of the Northern Fur Seal (Taylor *et al.*, 1955).

DESCRIPTION
The nose to tail length of adult males is about 2·13 m and of the females 1·5 m, and their weights are, respectively 182–272 kg and 43–50 kg. The colour of the adult males is dark rich brown, the intermingling of white hairs with the dark ones on the neck and shoulders giving a greyish tinge. The females are dark grey dorsally and lighter grey with a tinge of chestnut ventrally. In both sexes there is a characteristic light patch across the chest, and the rich underfur is chestnut colour. When born the pup has coarse black hair but at about eight weeks old this coat is shed for one that is steel grey dorsally and creamy white ventrally. After the animal is two years old the coloration becomes darker and gradually assumes the adult pattern and the male starts to develop a short bushy mane when he is about six years old. Occasional albino and partly albino animals are reported (Scheffer, 1962).

BREEDING
The pups when born are 66 cm in length and 5·4 kg (♂), and 63 cm, 4·5 kg (♀). The peak of the pupping season is between 20 June and 20 July. For about a week after birth the pup stays with its mother, but after this time the mother goes to sea except for about

Northern Fur Seal, Callorhinus ursinus *on St Paul Island, Pribilof Islands. An adult male, showing the very short snout of this animal. Photo. E. H. Miller.*

one day a week when the pup is suckled. On the intervening days the pups gather in groups or 'pods' in those areas of the rookery where they are least disturbed and, apart from the brief periods of suckling, their whole time is spent with their own age group. Pups wander about the rookery, and spend a lot of their time sleeping and playing. They are able to swim at birth but normally do not enter the water until they are about a month old. Suckling lasts for three months, and towards the end of this time a pup may take in 4·5 litres (a gallon) of milk at a time.

The breeding season starts when the adult males return to the rookery in early June, frequently returning to the same spot year after year (Gentry, 1980). They establish their territories, roaring and fighting, and by the time the females arrive, in mid-June, the system is well established, the harem bulls having the territories, while the bachelors – those not quite strong enough to compete – hang around the edges of the territories usually inland, intent on the possible waylaying of a female. The harem bulls, threatening, fighting and mating, hold their stations and an average of 50 or more cows, without returning to the sea for feeding, for up to two months (Osgood *et al.*, 1916). The harems are primarily the result of the gregariousness of the females rather than any controlling activity by the bull (Peterson, 1968).

About two days after the female comes ashore the pup is born, usually head first, delivery taking less than ten minutes. For a few days the mother is attentive to her pup, smelling it carefully, pulling it away from danger and threatening other females or pups that come too near. The bull is at all times indifferent to the pups, and even the mother's parental care diminishes rapidly and is soon restricted to suckling (Bartholomew, 1959). Mating takes place about a week after the birth of the pup and after this time only the weekly feeding visits are paid to the rookery. The female will feed only her pup and how she locates it amongst the many thousand others is still something of a mystery. The first clue is apparently geographical – she comes ashore at the right place, and normally the pup has not wandered far. A characteristic call stimulates the hungry pups to approach, but it is presumably scent that establishes which is the correct pup, as the mother sniffs at the nose of each one and will only accept her own, threatening and rebuffing all others, although a pup is willing to feed from any available female.

Towards the end of the breeding season, when the harem system is breaking up, the virgin females start to come ashore, some at least of them to be mated by the young bachelors. The females, sexually mature at three years, may bear their first pup when four years old, but are more likely to do so the following year, or even later. Males are sexually mature at five or six years, begin to breed at about eight years but do not attain harem bull status until they are twelve years old, and they may go on being harem masters until they are at least twenty. After mating the blastocyst does not become implanted for 3·5–4 months. The adult seals start to moult in the middle of August, and the new coat of guard hairs is not full grown until the end of October.

Marked animals have been seen on the Pribilof Islands after twenty-one years (Kenyon & Scheffer, 1954), and it has been suggested that the seals live at least until they are thirty years old (Scheffer & Wilke, 1953).

MORTALITY, PREDATORS AND PARASITES
Apart from man, killer whales and white sharks are probably the chief enemies of the fur seal at sea. Nearly all fur seals are infected with nematodes, and the hookworm (*Uncinaria lucasi*) is one of the major causes of pup death (Keyes, 1965). The larvae live in the blubber of the mother, and reach the pup in the milk the first time it suckles, but not subsequently (Keyes, 1965). On the change in diet from milk to fish the pup acquires many of its other parasites – the nematodes like *Phocanema* and *Contracaecum* which are attached to the lining of the stomach, causing ulcers, and whose larvae may penetrate the stomach wall and cause peritonitis (Keyes, 1965). Tapeworms, flukes and thorny headed worms also come from fish. Seven different species of helminths, one represented by over a thousand small but heavily spined flukes (*Pricetrema*) in the stomach caused the death of an eight-month-old seal (Neiland, 1961).

Sucking lice are present externally, one genus (*Proechinophthirus*) on the fur, and the other (*Antarctophthirus*) on the naked parts of the body. The pups are infected from the mother by body contact within a few hours of birth, and the greatest infection of lice is on the black pups of one to three months old which are largely terrestrial and also have a higher body temperature than the adults (Kim, 1972, 1975). Nasal mites are almost universally found, with the highest numbers in adult seals. Larvae of the mites live in the mucus-filled turbinals and attached to the mucous lining of the nasopharynx, and also in the lungs. The presence of the mites causes irritation and inflammation, impairment of respiration, and can predispose the animal to more serious respiratory diseases (Kim *et al.*, 1980).

Apart from parasites many pups suffer from malnutrition – a significant cause of death, and also from wounds by bites, crushing, and other traumatic contacts with adult seals. The predation by Stellers Sea Lions on fur seal pups is mentioned in the section on this sea lion (Gentry & Johnson, 1981). In rookery conditions the wounds may soon become infected with a variety of bacteria (Keyes, 1965). Leptospirosis also causes a high mortality (Lander, 1979). Brown and red algae, and also barnacles are known from fur seals that have spent some time at sea (Scheffer, 1962). Increasing numbers of animals are being affected, sometimes lethally, by man made debris such as portions of fishing nets and plastic wrapping bands. Over three thousand fur seals are caught and drowned annually in Japanese gill nets, and Japanese fishermen use some seven thousand animals for food (Lander, 1979).

FEEDING

The seals digest their food while at sea, so that stomachs taken from the more easily accessible animals on land are nearly always empty; and as the animals feed on the high seas their habits are not easily observable. However, it has been established that they feed during the evening, night, and early morning and usually sleep during the day (Ichihara & Yoshida, 1972). They have been recorded down to depths of 73m and the presence of 'seal fish' (*Bathylagus callorhinus*) remains indicates that some food at least is obtained in deep water, while abundant herring and squid remains show that pelagic fishing is also practised. The contents of many hundred stomachs have been examined and it appears that squid, herring, pollack and lantern fish are among the most commonly occurring food items, while seal fish, rock fish, lamprey, cod and other similar fish are frequently found. Fishermen complain that the seals attack salmon and ruin gear, but as most of the fish eaten are of only slight commercial value and salmon is found in relatively few stomachs, it is considered that the annual 5 million dollars' worth of fur and by-products from the fur seal outweighs any damage to fish and gear (Taylor *et al.*, 1955).

EXPLOITATION

At the time of their discovery in 1786–87 by the Russian navigator and fur trader Gerassim Pribilof, the fur seals of the islands named after him numbered about 2·5 million. From 1786 until 1867 the islands were under Russian management, but in 1868 they were acquired by America. Extravagant sealing operations to start with reduced the herd but occasional rest periods and killing restrictions allowed it to recuperate a little. The start of American sealing was even more extravagant, and enormous numbers of animals were taken both on land and at sea, the latter procedure being particularly wasteful as there was no means of distinguishing and protecting females. 'From 1889 to 1909, over 600 000 animals were taken – and at least that many or more were lost after being wounded and not recovered. The herd had now been reduced from

an estimated two million to probably 300 000' (NOAA, 1977).

In 1911 agreement was reached between Great Britain (including Canada), Japan, Russia and the United States to prohibit pelagic sealing in the north Pacific, except for the small numbers taken by native hunters. This agreement was broken in 1941 by the withdrawal of Japan, but a combined programme of research by Canada, Japan and United States was accomplished in 1952, and in 1957 the four original nations drew up a new convention. Pelagic sealing was again prohibited and in return for this Canada and Japan each receive 15 per cent of the Pribilof skins and, subject to some regulations, a similar percentage of the Russian skins from Robben and Commander Islands.

When the United States Government took over the control of the Pribilof herd in 1911, the seals had reached their lowest numbers, being reduced to approximately 200 000 animals. Since then, under strict control, the herd has increased to its present numbers.

The commercial sealing season starts about 20 June and lasts for a month or five weeks, depending on the movements of the seals. Only bachelors are taken, and only those whose length is between 1·04m and 1·14m. It has been shown that about two-thirds of these are three-year-olds, about a third are four-year-olds and there are a few of two years and five years. As well as having the best quality fur, and skins undamaged by scars, the killing of these bachelors does not affect the structure or breeding performance of the herd because of the animals' polygamous habits.

A watch is kept on the production of pups so that the annual commercial harvest may be estimated. On the Pribilof Islands, 350–400 000 pups born annually give the maximum sustainable commercial productivity (NOAA, 1977). The commercial kill on the Pribilof Islands averages about 30 000 young males. On the Commander Islands 2000 seals are taken annually and on Robben Islands 3000 seals. In recent years there has been a decline in the number of pups born in these two colonies so that the quota has been reduced to the above figures (Lander, 1979).

As the bachelors are separated from the breeding herd it is easy to herd them, driving them slowly away from the beach until suitable flat ground is reached, when the animals of the correct size are selected and the others allowed to go back to the beach. The measuring and skinning of each dead seal takes only two minutes, and the skins are then washed, deblubbered, salted and packed in barrels. When the Japanese and Canadian skins have been despatched, the rest are sent to a firm in Missouri for processing, in the course of which 125 operations and three months elapse between the raw skin and the prepared and finished fur. The long guard hairs are removed, the underfur straightened from its natural kinked state, dyed to black or shades of brown, and then the pelts are dressed to make them light and supple. The carcasses of the seals are converted into meal for chicken and mink food, and the blubber oil is used in soap.

WALRUS

Walrus Odobenus rosmarus

DISTRIBUTION

Walruses are inhabitants of the shallower regions of the circumpolar Arctic coasts. They have a preference for moving pack ice in areas where the sea is about 80–100 m deep. They can thus haul out to rest on these mobile platforms while battling with rough seas, and still remain close to their food supply. If there is no ice available they will haul out on small rocky islands, or on traditional sites on land called uglit in Eskimo (singular = ugli).

Two subspecies are recognized, *Odobenus rosmarus rosmarus* (Linnaeus, 1758) in the North Atlantic, and *Odobenus rosmarus divergens* (Illiger, 1815) in the North Pacific. Apart from distribution, the two groups differ mainly in size. The Pacific Walrus is the bigger animal, and has longer tusks and wider skull, and the Atlantic Walrus is the smaller, the population in the Eastern Canadian Arctic being the smallest in size. A third subspecies has been suggested, but the name is not normally used – *Odobenus rosmarus laptevi*, Chapskii, 1940, based on specimens from the Laptev Sea. The skull characters show that it is similar to the Pacific Walrus, though intermediate in size between Atlantic and Pacific forms (Fay, 1981).

The Atlantic and Pacific walruses each occur in two geographically isolated groups. For the Atlantic Walrus these are (1) eastern Canadian Arctic and western Greenland; (2) east coast of Greenland, Spitsbergen, Franz Josef Land, Barents and Kara Seas. For the Pacific Walrus the two groups are (3) Bering and Chukchi Seas, and (4) Laptev Sea (Mansfield, 1973).

1 In the Eastern Canadian Arctic the greatest numbers of walruses occur in northern Foxe Basin, and in northern Hudson Bay there are also large populations in the vicinity of Coats Island and southern Southampton Island, and there is a small population on the Belcher Islands in southern Hudson Bay. Small groups are found in Ungava Bay and along the south and east coasts of Baffin Island from Hudson Strait to Lancaster Sound where they may be found as far west as Bathurst Island. There are also walruses around the coasts of Ellesmere Island from Norwegian Bay in the west,

through Jones Sound in the south to Kane Basin in the east.

In western Greenland the main population occurs in the Thule area, from Cape York to Smith Sound. There are also walruses further south near Upernavik and in the area between Egedesminde and Holsteinborg, though rarely south of Godthaab (Mansfield, 1973).

2 On the east coast of Greenland walruses occur from Angmagssalik to as far north as 81°N. Around Spitsbergen walruses used to be abundant, but hunting has reduced their numbers to such an extent that only 42 sightings were made between 1960 and 1971. However, they are now protected and show signs of increasing, and there is a small permanent population on one of the islands (Kvitøya) of the group (Nyholm, 1975). Around Franz Josef Land too, the numbers have diminished and only a few hundred animals were believed to be there in about 1969 (Øritsland, 1973).

For the Novaya Zemlya region, and the Barents and Kara Seas there is little recent information, but it seems that during this century the numbers have declined and it is estimated that only a few thousand animals may remain (Bychkov, 1973a, Chapskii, 1936).

3 The Pacific Walrus has its headquarters in the Bering and Chukchi Seas and on Wrangel Island, its distribution extending eastwards about as far as Point Barrow and westwards approximately to Kolyma Bay. In the Bering Sea it ranges from the Gulf of Anadyr on the Siberian side to Bristol Bay on the Alaskan side (Brooks, 1954). Since a Soviet ban on walrus hunting about a thousand animals have reappeared on a small island off the Kamchatka coast (anon., *Oryx*, 1980).

4 The remaining population is in the Laptev Sea and may extend on to the New Siberian Islands (Bychkov, 1973b).

The total world population of walrus is thought to be in the region of 250 000. Most of these (200 000) are in the North Pacific, while the Canadian and West Greenland population is about 25 000. The Laptev population is thought to be about 4–5000 (Fay, 1981, Reeves, 1978).

As they prefer to remain associated with the edge of the ice, most walruses move north in summer and south in winter. Movements of the Pacific Walrus are known reasonably well. The animals spend the winter (Dec–April) in the central and southern parts of the Bering Sea, from the Gulf of Anadyr to Bristol Bay. In May and June, as the ice melts, the walruses move north past Nunivak Island and St Lawrence Island, appearing in the Bering Strait in the first half of June, the females with their newborn calves being at the head of the migration and the males following in a separate herd later. The walruses spend the summer (July–Sept–Oct) in the Chukchi Sea as far north as the ice will allow, feeding in the shallow waters. Several thousand bulls, however, remain in the south during the summer, hauling out on places such as Round Island and other islands in Bristol Bay and in the Gulf of Anadyr. In November there is a general movement south, back into the Bering Sea, though the speed of this migration depends mainly on ice and weather conditions, and the entire population may not reach its winter quarters until January. During this migration they haul out to rest at their traditional uglit on many of the islands in the Bering Sea. Some animals, mostly males, may spend the winter north of the Bering Strait (Brooks, 1954, Fay, 1981). Although living mainly in moving pack ice with plenty of natural openings, walruses are capable of breaking through 20cm of ice by banging against it with their heads.

Pacific walrus, Odobenus rosmarus divergens. *A group of males on the beach of Round Island, Bering Sea, in summer. Photo. E. H. Miller.*

The details of the movements of the Canadian population are less known, but they seem to remain in the same general vicinity all year, with local movements connected with the ice (Loughrey, 1959). Even in the high Canadian Arctic, between Ellesmere Island and Bathurst Island, there are areas where the tide currents are very strong, resulting in thinner ice, making the area available to walruses all through the winter (Kiliaan & Stirling, 1978).

Walruses used to be more common further south in the Canadian Arctic and were plentiful as far south as Sable Island and the Gulf of St Lawrence in the sixteenth century (Mansfield, 1966a, Allen, 1880). They were certainly plentiful enough to be taken commercially when the skin at least was exported to America for carriage traces, and to England for glue (Shuldham, 1775). Nowadays, Atlantic walruses only occasionally reach Newfoundland (Mercer, 1967), but a few animals may extend westward as far as Prince Patrick Island (Harington, 1966). Walruses that probably belong to the Pacific group occur occasionally along the Canadian Arctic coast as far as Banks Island and Bathurst Inlet. Between the latter animals and the Atlantic Walruses on Prince Patrick Island, Melville

Island and the Boothia Peninsula is a 480 km wide area that is covered with solid ice throughout the year. It is centred on M'Clure Strait and M'Clintoch Channel and Harington (1966) suggests that this forms a barrier between the Atlantic and Pacific walruses.

The Pacific Walrus too, used to have a greatly extended range. In the first half of the nineteenth century they used to be common as far south as the Shumagin Islands off the Alaska Peninsula, and on the Pribilof Islands, but they have never occurred in any quantity on the Aleutian Islands or the Commander Islands. There is evidence that occasional animals reached the Sea of Okhotsk and the southern tip of Kamchatka. By the second half of the nineteenth century the herds on the Pribilof Islands had been virtually wiped out, and at the present time the Pribilofs are south of the main area of distribution and only occasional animals are seen there (Fay, 1957). There are reports of less than half a dozen walruses ever having reached Japan.

Walruses still occasionally occur off Iceland, and there are some 31 records of walruses on the Norwegian coast between 1900 and 1967 (Brun *et al.*, 1968). Very occasionally walruses get as far as Germany

Atlantic walrus, Odobenus rosmarus rosmarus. *The head of an adult male, taken off Coats Island, Hudson Bay, Canada. Photo. E. H. Miller.*

(Lubeck Bay, Erhardt, 1940) and the Netherlands and Belgium (van Bree, 1977).

Early British records are those of William Caxton who records a walrus taken in the Thames in 1456 (Kiparsky, 1952), and of the historian Hector Boece who wrote in 1527 of the walrus in the Orkneys (Ritchie, 1921). Between 1815 and 1954 there are 27 records of walruses having been seen or killed, all of them off the Scottish coasts, except one shot in the Severn in 1839, and one seen in the Shannon, Ireland in 1897. Of the last two Scottish occurrences one was in 1928 when a dead animal was washed up at Gairloch, Ross-shire, and the last was in 1954 when one was seen on the shore at Collieston, Aberdeen.

DESCRIPTION

An adult male walrus is a large animal, second only in size to the elephant seals amongst the pinnipeds. The Atlantic animals are very slightly shorter and lighter than the Pacific animals.

	Pacific	Atlantic
Adult male,		
nose to tail length	3·2 m	3 m
weight	1215 kg	1200 kg
Adult female,		
nose to tail length	2·6 m	2·5 m
weight	810 kg	800 kg

The external appearance of a walrus is well known, the truncated head appearing rather small on the large body, and bearing the large array of stiff whiskers and the long conspicuous tusks that are present in both sexes. The eyes are small and frequently bloodshot; the pinna is absent but the external auditory meatus is protected by a fold of skin (Murie, 1872a). The skin is rough and wrinkled and thrown into conspicuous folds, giving the appearance of being several sizes too large for the animal inside. The thickness of the skin increases with age, reaching about 2·5cm in the adult. On the neck and shoulders of adult males the skin is much thicker (5–7cm) and may be raised up into numerous nodules and tubercles 3cm in diameter and 1cm high. These are absent in females and are thought to be a secondary sexual character, though they are sometimes regarded just as scars (Brooks, 1954, Fay, 1981). Under the skin the thickness of blubber averages 6–7 cm, though it may sometimes reach 15 cm and weigh over 410 kg.

The colour of walruses varies according to age and other conditions. Cinnamon brown is the general overall colour of skin and hair, but the younger animals have much darker skin and hair, while some of the old males are so pale they appear almost as albinos. The appearance of the animals also depends to a large extent on temperature. In cold water the blood is withdrawn from the skin, which then appears much paler. As the ambient temperature increases more blood flows through the skin and the animal appears a pinkish red colour which has often been mistaken for sunburn (see chapter on skin and temperature). Adult males moult during June and July and appear naked for the short period (*c.* one month) before the new hair grows again. The females are believed to moult a little later, over a more extended period.

The pharyngeal pouches or air sacs that have been described (Fay, 1960a) in the adult walrus are not known in any other seal. The lateral walls of the extremely elastic pharynx are expanded as a pair of large pouches just lateral to the glottis. These extend backwards between the muscles of the neck as far as the front edge of the scapula, or sometimes even to the posterior edge of the thoracic cavity. Inflation of the pouches takes place with air from the lungs, and it is then prevented from escaping by muscular constrictors round the openings. When the walrus is sleeping or wounded, the pouches can be inflated and used as buoys, but their main function is thought to be for a resonance chamber to enhance the bell-like note that the animals make while under water. Young animals appear not to have the pouches developed, and in some adults examined, while the pharyngeal walls are very pliable, the pouches were not present. Native populations of the north have long used these pouches in the construction of drums or as containers for food.

DENTITION

The most noticeable part of the walrus dentition is the pair of long tusks, present in both sexes but more slender in females. These are the upper canines, and their great size modifies the whole of the anterior end of the skull. The permanent upper canines erupt when the animal is about four months old. They have a small cap of enamel, but this is worn away in about two years so that the tooth is composed of a thin outer layer of cement, with dentine internally. The canines have persistently open pulp cavities and continue to grow throughout the life of the animal. The granular nature of the core of dentine filling the pulp cavity is characteristic of walrus ivory, and its presence in carvings is used as a method of identifying the ivory.

The tusks are about 2·5cm long when the animal is one year old, about 10 cm at two years, and 29 cm at five years. Those of males are larger and heavier than those of females, and a single tusk may reach a metre in length (Wilke, 1942) and a weight of 5·4 kg, though about 35cm is an average length. Supernumerary canines are sometimes found.

Besides the canines there are, in the adult walrus, four teeth on each side of both upper and lower jaws (Cobb, 1933). In the upper jaw these teeth are the third incisor and three postcanines, and in the lower jaw, the canine, which is molariform, and three post-canines. The milk dentition and the succession of the permanent teeth will be mentioned in the section on dentition. The cheek teeth have a small cap of enamel when they erupt, but again this becomes worn off in about two years. The pulp cavities remain open until about the fourth or fifth year, and then gradually close. There is a large deposit of cementum round the postcanine teeth which begins at the root tip and gradually builds up to the biting surface. Both upper and lower cheek teeth are roughly conical when they erupt, but the upper ones soon get worn very flat.

The tendency of captive walrus to 'dig' in their concrete pools is mentioned in the chapter on pathology. Excessive wear of the tusks this way leads to exposure and infection of the pulp cavity, which can lead to abscesses on the outside of the head where the maxilla has been damaged by the tooth. Covering the bottom of the pool with rubber is one method that has been used to alleviate this problem (Brown & Asper, 1966).

Mention was made above of the characteristic granular or marbled nature of the secondary dentine in the tusks, and the presence of this can be used to identify quite definitely the ivory of many carvings. In many museums there can be seen most delicately carved walrus ivory boxes from the early twelfth century, carved ivory panels, statues of the Virgin and Child, and draftsmen, all from the same period. Perhaps some of the best known carved walrus ivory pieces are the Lewis chessmen (photo). In 1831, on Lewis in the Outer Hebrides, a man was digging in a sandbank and unearthed some seventy carved ivory chessmen (Madden, 1832). They date from the middle of the twelfth century and are thought to have been the stock of an Icelandic merchant whose ship was probably wrecked. The chessmen were then swept on shore and buried, not to be found again for seven centuries. These chessmen, many of them in remarkably good condition, and delightful to behold, are now in the British Museum (Taylor, 1978).

Walrus tusks have always been held in high esteem and were considered as suitable gifts for kings. In about AD1050 a high official of Greenland who wished to acquire the favour of the King of Norway, sent to the King the three most precious gifts that Greenland could produce – a fully grown tame Polar bear, a set of carved chessmen and the skull of a walrus with the tusks carved and ornamented with gold. The tusks were also made into sword handles, and rings that would protect one from cramp (Madden, 1832). Walrus ivory is said to be harder than elephant ivory, and does not turn yellow so soon. For that reason, it was at one time preferred by dentists for making false teeth (Allen, 1928).

A king from the Lewis chessmen. This twelfth century figure is carved from walrus ivory. Photo. British Museum.

The correlation between long tusks and dominance is mentioned later. The tusks may be used when the animal is coming out of the water on to the ice. They are then hooked over the edge of the ice and the animal pulls itself up. From this action of the tusks in locomotion is derived the generic name *Odobenus* – the 'tooth walker'.

BREEDING

The winter season, between January and April, while the Pacific Walruses are in the central and southern parts of the Bering Sea, is also the mating season. Those females that are ready for mating gather in herds at traditional places, separate from the pregnant animals. Each herd of oestrous females on the ice is attended by a number of adult bulls, in the ratio of about one bull to fifteen cows. The bulls take up their stations in the water and display to advertise their presence, fighting should their display station be usurped by another. Much of their display is vocal and the sounds are made both in air and under water. Various growling and barking sounds are made at the surface, and a soft whistle just before submerging. The walrus seems to submerge with the pharyngeal pouches fully inflated, and these act as resonators for the bell-like sound that is usually made under water. Rasps, clicks and pulses of sound are also made under water (Ray & Watkins, 1975, Schevill *et al.*, 1966).

The traditional breeding places may cover several hundred square kilometres and the animals within them move about according to the dictates of the weather, ice or food supplies. The adult males follow the groups of females and are continually displaying to them. Subadult males remain on the fringes of the activity and do not display. Females leave the herd to meet the chosen male and copulation is believed to take place under water (Fay & Ray, 1979, Fay, 1981).

Birth of the calves takes place between mid-April and mid-June, with a peak in May, while the pregnant females are on their northward migration in spring. At birth the nose to tail length of the calf is 1·2 m, and it weighs about 63 kg. About two or three months before birth the calf sheds a fine white lanugo (soft foetal hair), and when it is born it appears greyish all over, becoming more tawny after a week of two while the flippers become black. There is another moult in July when they are about two months old and thereafter the moult is annual.

The female looks after her young calf, defending it and protecting it from the worst weather by sheltering it under her chest, between her foreflippers. The lactation period may last for up to two years, though over the last 18 months the calf will be weaned gradually. It will remain associated with its mother for at least two years, or even longer if she does not produce another calf. As the calf has yet only short tusks it probably relies to a certain extent on food material stirred up by its mother. When the calf becomes independent the young males tend to join up with herds of other males while the young females remain with the other females of assorted ages.

After a cow has mated, say in mid-February, there is the usual four-month delay before the blastocyst becomes implanted in mid-June. After this there is active gestation for 11 months. The total gestation period is therefore 15 months and obviously annual breeding is not possible. After the calf is born in about May there is a post-partum oestrous, but at that time of year there are no fertile males around, so the female has to wait until the mating season in about February. Thus only a proportion of the adult females are capable of breeding in any one year. Young adults may produce a calf every two years, but the interval gets longer as the female gets older. Thus a herd of young adults will produce considerably more calves than a herd of more mature individuals (Fay, 1981).

Most females become sexually mature between five and six years old. Males are mature at eight or nine years old but cannot compete with the older males until they are about 13–15 years old.

Walruses are gregarious animals and positively thigmotactic. Even when there is plenty of room available they crowd on to their chosen beaches or ice floes and lie in contact with each other, their long tusks resting on a neighbour's back. As they come in from the sea they attempt to work their way towards the centre of the group where bodily contact is greatest. The largest walruses, with the longest tusks are the dominant animals and have been described as bullies – aggressive and threatening. They use their weight to forge a path through the others, jabbing with their tusks until the desired position is reached. Smaller animals with smaller tusks, or those with broken tusks are at a disadvantage, and although they may attempt to force a way through the herd, the frequent jabbings draw blood and they are eventually forced to remain on the periphery of the group (Miller, 1975a, 1976, Bruemmer, 1977).

MORTALITY, PREDATORS AND PARASITES

Apart from man, killer whales and polar bears are probably the main enemies of walruses. Killer whales are credited with ramming a female walrus and knocking the calf away from her, and in such instances the female walrus could well sustain lethal injuries. On St Lawrence Island in the Bering Sea killer whales are said to have panicked a herd of walrus ashore, so that many of the walrus were crushed (Brooks, 1954). St Lawrence Island lies in the path of the autumn southward migration and there are records of occasional large numbers of dead walruses being found there. A recent occurrence was in 1978 when several thousand walruses came ashore at places where they had not been seen before in this century (Fay & Kelly, 1980). Over a thousand of these animals were later found dead, very many from severe torsion of the neck and resultant bleeding into the cranium, also intestinal prolapse and compressed collapsed lungs with internal bleeding. Gunshot wounds were identified in a few, also lacerated flippers associated with broken bones and internal bleeding which were probably the result of killer whale attack. Many of these dead animals were very lean and there was quite a high percentage of aborted foetuses and young calves. Most of the injuries were consistent with the walruses piling up on the beach and crushing each other to death, but what precipitated this is not known.

Polar bears are said to attack only the young walrus. Walrus meat is not infrequently found in polar bear stomachs but this could well come at least in part from carcasses left by hunters (Nyholm, 1975). Adult walruses do not always show fear of polar bears, and sometimes conflict between the two results in a dead bear. Bears may also stampede walrus herds and collect the separated calves (Loughrey, 1959).

FEEDING

Although there is no actual fasting period, less food is taken while on the northward migration. Foraging for the bottom dwelling forms on which they live does not take place at depths of over 73 metres (40 fathoms) and the walrus seem to prefer the shallower waters round the coasts where the sea bottom is gravelly, and supporting various kinds of molluscs. Stomachs of walruses from Greenland that have been examined show that the three bivalve molluscs that form the greater part of the food are the soft shell clam *Mya truncata*, the Arctic rock borer *Saxicava arctica* and the cockle *Serripes groenlandicus*. Stomachs that contain the cockles are regarded as a great delicacy by the Eskimos. The Pacific Walrus also eats large quantitites of the same species of *Mya* and *Saxicava* and the full stomach of a bull walrus has been recorded as weighing 49 kg, and containing about 23 kg of *Mya* siphons and 16 kg of the feet of another mollusc *Clinocardium nuttalli*. Cows and young animals like the smaller molluscs such as *Astarte sp.* and *Macoma calcarea*, as well as an annelid (*Nephtys* sp.) and a sipunculid worm. The echinoderm *Molpadia arctica* is also eaten, and small quantities of the polar cod *Boreogadus saida* are sometimes taken (Brooks, 1954).

Walrus tusks are frequently worn along the front and sides and this, together with the pattern of wear on the mystacial vibrissae suggests that the animals virtually stand on their heads under water while searching for food. They move forward, keeping the tusks and whiskers in contact with the sea bottom, disturbing the sediment and finding their prey by touch. It seems unlikely that they do any digging with their tusks (Fay, 1981).

Although the walrus feeds mainly on molluscs, only very seldom indeed are pieces of shell found in the stomach. The feet and fleshy siphons of the molluscs are either torn off or sucked off and swallowed whole, as chewed up fragments are not found. Mussels and probably other molluscs may be taken into the mouth, the soft parts sucked out and swallowed and the shell rejected. Shells that have probably been treated in this way often occur round walrus breathing holes. In captivity a young walrus used its whiskers to hold a fish down on the ground while sucking the flesh from it, and another was seen to break the mollusc shell by a sharp blow from its moustachial pad before sucking out the soft parts. Sucking undoubtedly plays a large part in feeding. Captive walruses have been known to suck up a metal plug weighing 2·3kg from the outlet of their pool, even when the pool was full of water. It has been determined experimentally that a captive walrus can draw in a vacuum of at least one negative atmosphere, and that this is quite sufficient to remove the siphon from a soft-shelled clam (Fay, 1981).

The probable correlation between sucking and the shape of the lower jaw and the concave palate in walruses and some other seals is mentioned in the chapter on the skull. Nyholm (1975) mentions that he saw three walruses in the water crush shells between their foreflippers. The soft parts of the mollusc then floated to the surface, where they were eaten. Although they do not chew their food, the cheek teeth of walruses get very worn and smooth, probably from abrasion by the sand and gravel that they inadvertently swallow with their food.

Remains of young Ringed and Bearded Seals, and even of young walruses, have also been recorded from walrus stomachs, but it is believed that these are eaten only when other food is scarce, or it may well be carrion. Habitual seal-eaters, usually males, are less frequent, and their tusks and skin get stained with grease so the individuals are easily recognizable. Walruses have been known to eat narwhal flesh, and although on at least two occasions the narwhals were only recently dead, it is not known whether the walruses actually killed them. There appears to be a certain correlation between the carnivorous walruses and the incidence of hypervitaminosis and trichinosis, neither of these two conditions being present to any great extent in normal mollusc-eating walruses (Fay, 1960b).

EXPLOITATION

At the present time it is not believed that there is any commercial exploitation of walruses, except by local populations of Eskimos, Aleuts etc. who are permitted to take them for sustenance. Before 1972, when the Marine Mammal Act came into force, there was a market for walrus hides. These are very thick, fibrous and abrasive, and were ideal for making buffing wheels for polishing silver and other metals. Synthetic materials are now used for this purpose, but are more expensive and not as good (Fay, 1979). Walrus hides have also been used for the tips of billiard cues, for ropes, harness, boot leather and machine belts. When split the hides make a tough covering for the Eskimo oomiak (open boat made of skin and wood). Many parts of the walrus are eaten by the Eskimos – stomachs containing cockles and clams are prized, hearts, flippers, intestines and lean meat are all eaten, but the use of the carcass varies very much according to the locality and much of it may be used as dog food. Rain parkas used to be made out of the intestines, though probably modern materials are now entering Eskimo culture. Carved ivory is most probably still sold to tourists.

HISTORY

One of the earliest works in which the walrus was mentioned, the *Speculum Regale* (see Allen, 1880), was

Albrecht Dürer's drawing of the head of a walrus. The original is in the British Museum. Photo. British Museum.

published in the thirteenth century, where it says, quite correctly, that walruses are related to seals, and this opinion was held by most writers on natural history until about the middle of the eighteenth century. From then until the end of the nineteenth century, when they reached their position as a family of the Pinnipedia, they were grouped with such diverse creatures as cetaceans, sirenians and even the platypus, thus indicating the confusion of the time. The first person to draw a walrus seems to have been the artist Albrecht Dürer. The drawing, in brown ink, is dated 1521, and at the side Dürer noted that the animal was captured in Dutch waters, was twelve ells long, brownish in colour, and had four feet. It is of interest to consider how Dürer came across such a relatively unknown animal. In 1520 the Bishop of Trondheim sent the head of a walrus, in a barrel of salt to Pope Leo X in Rome (Kiparsky, 1952). It seems very unlikely that there were two walruses being taken about Europe at this time, so it was probably the gift to the Pope that Dürer saw, and the fact that this was a preserved head would explain both why Dürer drew only the head, and why his drawing has a slightly distorted appearance.

The first walrus to be seen alive in England came from Bear Island in 1608. It was a young animal and they 'brought our living Morse to the Court, where the king and many honourable personages beheld it with admiration for the strangenesse of the same, the like whereof had never before been seene alive in England. Not long after it fell sicke and died' (*Purchas his Pilgrimes* 1624, ref. from Allen, 1880 p. 140).

The first to be exhibited by the Zoological Society of London arrived from Spitsbergen in 1853 but lived only a few days (Owen, 1853). Walruses have been exhibited in many zoos since these early days, particularly in America and Denmark, and this has enabled much work to be done on their growth and development. Although of relatively large size – a ten-month-old walrus will weigh about 195 kg, and will increase in weight at the rate of about 16 kg a month – they have been found to be charming, friendly and curious animals, and are great favourites with their keepers.

NB A comprehensive work on the Pacific Walrus has just been written. Unfortunately it has appeared too recently for its subject matter to be included here. It should be consulted for information resulting from nearly 30 years' observation and work on this animal (Fay, 1982).

NORTHERN PHOCIDS

Grey Seal Halichoerus grypus

DISTRIBUTION

The Grey Seal occurs in temperate and subarctic waters on both sides of the North Atlantic – in the Gulf of St Lawrence area, Iceland, British and Norwegian coasts, and in the Baltic.

Western Atlantic population In Canadian waters Grey Seals are more widespread in summer and may be found approximately from Cape Chidley on the Labrador coast to Nova Scotia. They are present on the islands in the Gulf of St Lawrence – on the Mingan Islands, Anticosti, Magdalen Islands, on islands in Northumberland Strait and on Cape Breton Island. They are in the Bay of Fundy and on the shores of Nova Scotia; on Sable Island, in the Miramichi estuary, and in the estuary of the St Lawrence River about as far as Trois Pistoles. The shores of the French Miquelon Islands support a large population, and there are Grey Seals in the Straits of Belle Isle and up the Labrador Coast as far as Cape Chidley.

Breeding colonies are less widespread. They are found in two situations – on islands in ice-free waters, and on fast ice and islands closely connected with fast ice. The single biggest island breeding colony is on Sable Island where the number of pups born has increased from 134 in 1962 to over 2000 in 1976, indicating a population of some 7500 older seals and being some 31 per cent of the total Canadian population. Other island colonies are on the Basque Islands off the

southeast coast of Cape Breton Island, and Camp and nearby islands off the eastern coast of Nova Scotia, all of which produce about 5–600 pups a year. Colonies on fast ice are more common in the eastern part of Northumberland Strait between Amet Island and Cape Breton Island, and produce about 3700 pups a year. In years when there is little ice large numbers of seals crowd on to the beaches of Amet Island to breed. There is another colony on Deadman Island, close to the Magdalen Islands. Some hundreds of seals, both Grey and Harbour, congregate in the marine waters of the Bras d'Or Lakes in autumn, and may remain there all winter. Pregnant Grey Seals usually move out to the shore ice to pup, but occasionally pups are born in the lakes.

It has been estimated that about 6400 pups are produced annually in eastern Canada, which would indicate about 24 000 older seals (Mansfield & Beck, 1977).

A small breeding colony, possibly about 20 animals exists in American waters at the west end of Nantucket, and on the small islands of Muskeget and Tuckernuck in this area. This colony of 'horseheads' has been known since the early part of the twentieth century, but it seems that until recently, only fishermen have taken an interest in them. Records of Grey Seal bones from Indian midden sites are known from various places along the coasts of Maine and Massachusetts as far as Block Island, Rhode Island. Special

legislation was passed in March 1965 giving complete protection to Grey Seals (in American waters), and it is hoped the Nantucket herd will increase (Andrews & Mott, 1967).

The great increase in the Canadian Grey Seal population that has taken place since about 1965 has caused problems with the inshore fishing. Culling operations started in 1967, concentrating on moulted pups so that their skins could provide some financial recompense. About 800 or more pups were taken every year, and as many as 2300 in 1975. This had little effect on the population as a whole, which continued increasing. Since 1976 the Canadian sealing regulations decree that Grey Seals may not be taken in January and February, partly because this is the breeding season and partly so that continued supervised culling may be carried out. A bounty system, however, allows the fishermen to kill seals to reduce local predation of fish. All age classes are killed for bounty, and it is hoped that this may eventually solve the problem of predation by seals, although there is not enough information yet (Mansfield & Beck, 1977).

Eastern Atlantic population

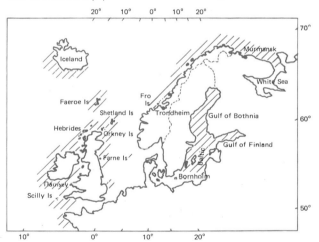

ICELAND Grey Seals are very much less abundant than Harbour Seals in Iceland. Grey Seals are found on the south and west coasts of the island, and have their headquarters in Breidafjordur, breeding on the many islands there. About 1000 Grey Seals are estimated to live in Iceland.

FAEROES The Grey Seal is the only seal that breeds on the Faeroe Islands and it is estimated that there are about 3000 seals there.

NORWAY There are small breeding colonies of Grey Seals at various places along the Norwegian coast from North Cape to Stavanger, but relatively small numbers of pups are born. The island of Halten, just outside Trondheimfjord, with about 90 pups a year has probably the largest concentration. A count of pups in the area between the Lofoten Islands and Trondheim was made in 1976 and the estimated total population for this area is about 3000 (Benjaminsen *et al.*, 1977). Occasional animals reach the west coast of Sweden and the Danish islands in the Kattegat.

NORTH EUROPE Along the Murman coast of the Kola Peninsula up to about 800 adult females have been reported on the islands in the vicinity of Murmansk, indicating a total population of about 2000 (Karpovich *et al.*, 1967). Spitsbergen is reported to have a few Grey Seals.

BALTIC The Grey Seal ranges widely over the Baltic in the Gulf of Bothnia and Finland, but not extending as far west as Bornholm and being scarcer in the northern parts of the Gulf of Bothnia. They usually keep the open sea or the outer skerries. Figures of between 5000 and 10 000 animals have been given, but the popula-

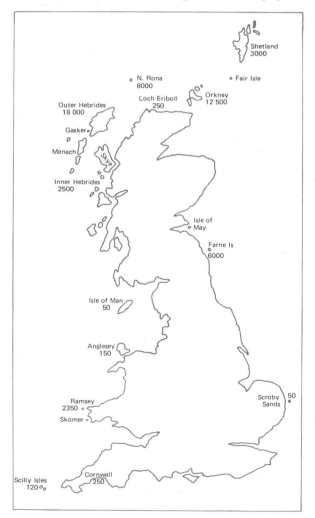

Estimated total populations of Grey Seals in breeding colonies in Great Britain.

tion is believed to be declining and the lower figure is thought to be the better estimate (Bonner, 1972, 1973).

The Baltic seals used to be hunted to a far greater extent than they are now. A certain amount of opportunistic hunting takes place in summer and autumn, and bounties on deal seals are paid by Sweden and Finland, although the claims are declining. Some protection is given to seals in Swedish waters (Bonner, 1973). The Baltic Grey Seals are remarkable for breeding on ice, in February and March. In this respect they resemble the Canadian population and differ from other European colonies which pup in the autumn. As the ice moves about the exact location of pupping groups varies, and may depend on the weather. A few days before the pup is born the female appears at the edge of the ice, apparently cruising about and seeking a suitable pupping spot, at which a group of females may eventually congregate. The pups are born, usually between the last week in February and the first two weeks of March, on drifting or fast ice, but close to open water so the female can escape if necessary. She spends much time in the water, joining the pup only so it may suckle. Lactation lasts 20 days and when the female then abandons the pup it moves around on the ice, may head towards land and may even reach the Swedish mainland. The polygamous adult bulls do not appear to join the females until the end of lactation, when copulation presumably occurs. The adults, after a brief interval at sea, return to the ice in April and May to moult (Curry-Lindahl, 1970).

IRELAND Small breeding groups are found at any suitable place around the Irish coast, with most of the pups being born on the small islands and rocky skerries of the west coast. At any single place only small numbers of pups are born, a total of 40 pups on Clare Is being about the highest. There are perhaps 2000 Grey Seals in Ireland (Lockley, 1966).

GREAT BRITAIN Grey Seals are increasing in numbers, and published figures are always likely to be out of date. The world population of Grey Seals was given as about 82 000 animals in 1976 (Bonner, 1976), but more recent figures suggest a figure closer to 101 000. Two thirds of the world population lives in Britain, and the British population is in the region of 70 000–76 000 (Summers, 1979, NERC, 1981).

Outside the breeding season Grey Seals may be dispersed over much of their range, and large numbers may haul out at places where no breeding occurs. The above figures are based on breeding groups and are usually arrived at by counting the pups and multiplying by 3·5 – a figure derived from the life table of the animal (Bonner, 1976).

HEBRIDES AND NORTH RONA Grey Seals breed on many islands of the Inner and Outer Hebrides. In the Outer Hebrides, Gasker has the largest breeding population, with about 1400 pups observed in 1974, giving an estimated total of 2100 pups. All islands have shown an increase in the population of Grey Seals, but the largest increase has occurred on the Monach Isles, where 15 pups were seen in 1966, and more than 2000 in 1975. The total pup production of the Outer Hebrides has more than doubled in recent years, from about 1600 in 1961, to 5200 in 1974, which would give a total population of about 18 000 seals. The increase continues, and 1981 figures are in the region of about 33 000 (NERC, 1981). North Rona supports about 8–9000 seals.

The Inner Hebridean Grey Seals may be a different stock from those on the Outer Isles. They also have increased in numbers, though no so markedly, and total pup production in 1974 was estimated to be 800, and the total population about 2500. By 1981 the population had doubled to 5000.

ORKNEY Grey Seals have increased enormously on the Orkney Islands. They were regarded as rare there at the beginning of this century, and in 1974 about 3600 pups were produced, indicating a total population of about 12 500 animals, and in 1981 the population had risen to 16 500 (NERC, 1981).

SHETLAND Breeding places are very inaccessible on these islands, but the population seems to be relatively stable with about 3500 seals (Anderson, 1981).

SCOTLAND A few pups, hardly more than 100, may be born at a few places on the Scottish mainland, such as in the sea caves in Loch Eriboll.

FARNE ISLANDS This is a very well documented colony that has shown a phenomenal increase in numbers. Some 120 pups were born there in the early 1930's, and this figure had increased to about 2000 pups in 1971. Overcrowding led to confusion amongst the seals with an increase in aggressiveness of the adults. More pups were separated from their mothers, leading to increased deaths from starving and also from the increased susceptibility to disease of the undernourished pups. This led to increasingly unhygienic conditions on the rookery. Too many seals on the small islands of the Farnes led to disappearance of the vegetation and to soil erosion. This state of affairs was started by the colonies of burrowing puffins, but exacerbated by the heavy seal bodies (Bonner & Hickling, 1971, 1974, Harwood, 1978). Management of the population involves culling a certain proportion of the breeding cows together with their pups, with the aim of returning to the 1960 population level of 1000 breeding females. The first official cull was in 1972 with the removal of 603 breeding cows, but by 1974 pup production had again climbed to 1655 animals, and careful culling is continuing. There is a total population of about 8000 seals on these islands.

Grey Seals also breed on the Isle of May in the Firth of Forth, and the population is increasing there, approximately 500 pups being born there in 1980 (Prime, 1981). About 12 pups a year are produced in the small colony on Scroby Sands, Norfolk. Both these groups are probably derived from the Farne Islands colony.

WALES A survey of the Welsh coast in 1974, counting pups and extrapolating from this, gave a total population of 2500 Grey Seals in this area. Of these, by far the greater number (2350) was found in South Wales, with Ramsey and Skomer Islands being the centres of the breeding population. The Welsh population has doubled since the 1950s, but since just about all the available breeding sites are now occupied, further major increase is not envisaged (Anderson, 1977).

CORNWALL and SCILLY ISLES Relatively small numbers of Grey Seals breed in these areas, the total population in Cornwall, mostly on the north coast, being 250, and 120 in the Scilly Isles. Winter haul outs in the Scilly Isles boost the numbers, but these are not part of the breeding population (Summers, 1974).

OTHER AREAS Grey Seals are frequently found on the French coast, mainly on the shores of Brittany. Tags on some of them indicate they have come from Ramsey Island. A small colony exists on the Ile d'Ouessant, Brittany, but very small numbers of pups are born there (Brien, 1974, Prieur & Duguy, 1981). Probably less than 50 seals live on the Isle of Man. There are occasional Portuguese strandings.

Summary of Grey Seal populations

World population	101 000	NERC, 1981
Total Canadian population	24 000	Mansfield & Beck, 1977
Sable Island	7500	Mansfield & Beck, 1977
Iceland	1000	Smith, 1966
Norway	3000	Benjaminsen *et al.*, 1977
Northern Europe	2000	Karpovich *et al.*, 1967
Baltic	5000	Bonner, 1972
Ireland	2000	Lockley, 1966
Great Britain	76 000	NERC, 1981
Faeroe Islands	3000	Smith, 1966
North Rona	8–9000	Bonner, 1976
Outer Hebrides and North Rona	40 000	NERC, 1981
Inner Hebrides	5000	NERC, 1981
Orkney	16 500	NERC, 1981
Shetland	3500	Anderson, 1981
Farne Islands	8000	NERC, 1981
Wales	2500	Anderson, 1977
Cornwall & Scilly Isles	370	Summers, 1974

Pups from most colonies have been marked or tagged in some way. Recoveries of tagged animals show that they range outwards from their breeding places more or less in a random fashion (Bonner, 1972). Pups marked on the Farnes for instance make their way to the shores of Norway, Denmark and the Netherlands; pups marked on North Rona reach Iceland, the Faeroes and Norway, and the Pembroke seals include Ireland and the Brittany coast in their wanderings. Allowing three weeks between birth and departure from the colony, some of the Ramsey pups have reached Brittany, some 400 km away in a maximum of 16 days swimming. A pup tagged on the Farne Islands at the beginning of November when it was one week old, was recovered from the Frø Islands just outside Trondheimfjord about 960 km away when it was nine weeks old, and another tagged on the Farne Islands was first recovered on the Isle of May when it was four weeks old, and nine days after this was found at Karmøy, Norway about 580 km away, possibly swimming about 64 km a day to do this (Hickling, 1962).

In Canadian waters too animals in their first year move widely between Labrador and Nantucket (between about lat. 55° and 42°), and to many parts in the Gulf of St Lawrence. A 'moulter', perhaps three or four weeks old was tagged on Sable Island on 5 February, and recovered at Barnegat Light, New Jersey on 2 March, having covered 1280 km in 25 days – about 50 km a day (Mansfield & Beck, 1977).

EXPLOITATION

The Conservation of Seals Act 1970 applies to Grey Seals as well as to Harbour Seals. Grey Seals in Britain are protected by a close season from 1 September to 31 December, though this does not apply to those seals caught 'red-handed' in the vicinity of fishermen's nets. Under licence, seals may be taken for scientific purposes, for culling of the population, and for the commercial use of this population surplus. Grey Seal pups are taken in Orkney and Shetland, and also in the Faeroes and the Baltic, though the skins are of less value than those of Harbour Seals. They can be used for such things as fur boots and tourist trinkets.

In Denmark, Grey Seals are totally protected, and they are protected at certain places in Norway. In Sweden, hunting is permitted only by professional fishermen (Bonner, 1972).

DESCRIPTION

Adult males reach their maximum nose to tail length at about 11 years and are then 2·2 m and weigh 220 kg. Adult females are nearly 15 years old when they reach their maximum length of about 1·8 m, with a weight of 150 kg, though weights can of course be very variable. The coat colour varies, and all shades of dark and light grey, brown and silver may be found. Both males and females have darker backs and lighter bellies, and have varying degrees of spotting, but the sexes can be identified, even when wet, by the distribution of the darker and lighter tones of colour. In bulls the darker tone, whether brown, black or grey forms a continuous background, darker above and lighter below, upon which may be greater or less amounts of irregular spotting of the lighter tone. In females it is the lighter tone that forms the background colour, upon which are spots and patches of darker colour (Hewer & Backhouse, 1959). This characteristic pigmentation is visible in the skin of the foetus as early as 115 days of active gestation, but is at first obscured by the white natal coat. An unusual rusty or reddish orange colour has been seen on the snouts, throats, flanks and undersides of flippers of some seals from the Inner Hebrides and also from the Scilly Isles.

The high, arched 'Roman' nose of the adult male Grey Seal is characteristic; the adult female has a straight profile to the dorsal surface of the head. An anterior view of the nostrils of a Grey Seal shows them to be more nearly parallel than those of a Harbour Seal.

BREEDING

Grey Seals usually produce their pups in the colder months of the year. Round the coast of Britain the pupping season is between the beginning of September and the middle of December, but the pupping time in each island group is not exactly the same. The seals of South Wales, Cornwall and the Scilly Isles, and possibly Ireland as well, form an earlier breeding group, producing their pups between early September and late October, with a peak around the latter half of September (Anderson, 1977). The colonies on North Rona, the Hebrides and possibly those from North Wales, may form a separate group, pupping between the beginning of September and the middle of November, with a peak in the first half of October (Summers, Burton & Anderson, 1975, Hewer, 1957). It is possible that the population on the Inner Hebrides may be separate from the above stock as there is some evidence that the breeding season is earlier (Bonner, 1976). The seals on the Shetlands and Orkneys may form another separate group pupping between late September and the end of October, with the greatest numbers in the last half of October (Vaughan, 1975). On the Farne Islands the pups are produced between mid-October and mid-December with the greatest numbers being born during the first fortnight of November (Hickling, 1962). Thus the peak pupping times differ round the coast and are at the end of September (Welsh), the first half of October (N. Rona), the last week of October (Orkney), and the first two weeks of November (Farnes).

The timing of the winter pupping season is continued in Canada where the pups are born between the end of the December and the beginning of February, with the peak in mid-January (Mansfield, 1966b). The Baltic Grey Seals, in spite of being geographically closer to the British colonies, pup in early spring, between the last week in February and the first two weeks of March (Curry-Lindahl, 1970).

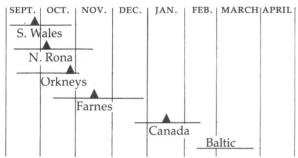

SEPT.	OCT.	NOV.	DEC.	JAN.	FEB.	MARCH	APRIL

Diagram showing pupping season and peak pupping times of some Grey Seal colonies

A young Grey Seal which has shed its first, white coat from the front end of the body. The damp patch round the eye is caused by lacrimal secretion. The flexed tip of the foreflipper is also visible. Photo. J. E. King.

In spite of these differences in timing, the main events of the reproductive cycle take place, as far as is known, at the same times relative to each other.

Although most British pups are born in the autumn, some spring pups (March–May) have been recorded from South Wales. These are believed to be the first offspring of newly mature cows (Backhouse & Hewer, 1957). The production of a pup in March presupposes fertilization in April of the preceding year, which would in turn suggest that the production of sperm by the bulls in the Welsh colonies follows a slightly different pattern from that in other colonies (Hewer & Backhouse, 1968).

At birth the pup is about 76cm in nose to tail length and weighs about 14kg. It is covered in long creamy white silky hair which is usually shed at the end of the animal's third week, the moult being complete in four to five days. The moulted pup is then clad in a coat of short hair with the patterning and sexual difference in colour of the adult. Pups are able to swim at birth, and sometimes do, but normally they do not enter the water until the moult is complete. A bond is formed between the mother and her pup at birth and the cow can recognize the pup by its call and its smell. A sleeping pup will wake and call, and the mother, recognizing the call of her pup will come to it, sniff it to establish its identity, and then allow it to suckle. Lactation lasts for about 16–21 days and during this time the cow spends much time just offshore if the geography of the site allows easy access, but probably does not feed. The pup feeds for about six minutes every five or six hours and puts on weight rapidly, increasing by about 1·5 kg daily. By the time it is weaned it weighs about 50 kg and is a fat barrel, hardly

moving, and scarcely longer than at birth. After weaning the pup may go to sea almost immediately, or may spend a month or so on land before it finally leaves. They will eventually return to the site of their birth but range widely in the adjacent seas before this. The movements of tagged pups in their first winter has already been mentioned.

About a month before the pupping time, large numbers of pregnant cows, and bulls of all ages, assemble on or near the breeding beaches, the largest bulls tending to keep away from the rest. Frequently, on landing, a bull will challenge others, and it is from the results of these challenges, the vanquished animals retiring to the sea, that the younger breeding bulls of the season are selected. The birth of the first pups seems to be the signal for the beginning of territorial activities. The older bulls take up their territories, usually starting inland, the best territories – those offering the greatest opportunities for future matings – being occupied first. These are large areas, and it is only later, when the attentions of the bull are diverted by mating, that successful occupation of the fringes of this territory can be achieved by bulls of lesser status. Before this all trespassers are challenged, and, if necessary, fought, to preserve territorial rights. On their journeys to and from the sea the cows may pass through several territories, and are frequently accosted by bulls.

This is the general pattern of events, but many of the

Grey Seal colonies have been studied in much detail and for precise details of behaviour these original accounts should be consulted, and each one of these has a relevant list of earlier references. The Canadian colonies have been studied by Cameron (1967, 1969, 1970), and Mansfield & Beck (1977). Papers on the British colonies include Anderson *et al.* (1975 – Rona), Hewer (1957 – Hebrides), Hickling (1962 – Farne Islands), and Bonner (1972) and Hewer (1974) give general summaries of Grey Seal biology. The Baltic animals have been observed by Curry-Lindahl (1970).

Birth of the pup is rapid, the umbilical cord being severed by the sudden movement of the mother's hindquarters. The placenta may be shed within a few minutes or there may be a delay of nearly an hour before it appears. It is completely ignored by the cow and is eaten by gulls. At the end of the lactation period the female comes into oestrous, is receptive to the attentions of the bull, and mates. Mating may take place either on land or in the water depending on the territory, though the latter seems to be most usual. The number of cows covered by a single bull is probably six or seven, and an oestrous cow may copulate with more than one bull. There are no non-pregnant cows on the breeding beaches, and the newly mature virgin cows are probably mated by the younger bulls away from the breeding area. The males are monogamous if there is ample breeding space, as on Sable Island, but on most colonies there is a certain degree of crowding, and the males are then polygynous (Mansfield, 1966b).

After mating, the fertilized egg takes about eight to ten days to develop into a blastocyst. After this it remains dormant, lying in the uterine horn for 100 days (*c.* 3·5 months) before recommencing development. It becomes attached in the normal mammalian way, via the placenta, to the wall of the uterus, and develops for a further period of 240 days, when it is

Grey Seal, Halichoerus grypus. *On this female the darker spotting on the lighter background colour can be seen.* Photo. Bill Vaughan.

born. This, as mentioned in the chapter on reproduction, gives eight months of active gestation, and a total gestation period of *c*. 350 days (*c*. 11·5 months).

Taking the Orkney and Shetland colonies as examples of this timing, the middle of October (the sixteenth) is the mean time of pupping, lactation lasts for about 14 days, the mean date of mating is 30 October, and after the period of suspended development the blastocyst recommences development on 16 February and becomes attached to the uterine wall on 21 February. The period of active gestation then ends on 16 October when the pup is born. Descriptions of a number of stages in development, from blastocyst to small embryo to full-term foetus is given in Hewer & Backhouse (1968). Cows usually produce their first pup when four or five years old. Bulls are sexually mature at six years but do not normally appear on the breeding beaches till they are eight years old. Most of the breeding bulls are between 12 and 18 years old.

Neither bulls nor cows feed during the pupping and mating season and both sexes lose a lot of weight during this period. A captive female lost 43kg (25·6 per cent of her weight) in the 18 days between the birth and weaning of her pup (Matthews, 1950). During the period of feeding at sea between about late November to mid-January, the seals regain condition and then undergo their annual moult. Most females are moulting in early February, and the moult of the male animals is later, the peak being in mid-March. At this time they spend long periods hauled out, though not at the breeding sites.

MORTALITY, PREDATORS AND PARASITES

A failure of the mother–pup bond leading to starvation is one of the main causes of death of pups. Dead pups are usually underweight, and an undernourished pup will easily fall sick from some of the many pathogenic organisms around. They are also less likely to withstand the traumatic effects of bites, tossings or tramplings by adult seals. Infections which enter via the unhealed umbilicus may set up peritonitis, and the unhygienic conditions of the rookery frequently lead to septic eyes. Injuries from adults, particularly in the more crowded rookeries, or from accident, also cause death. Many pups may fall off cliffs, or be washed away by strong seas (Anderson *et al.*, 1979, Bonner, 1972).

Adults may suffer from skin lesions from which the organism *Corynebacterium* has been isolated (see also Harbour Seal), and they may have pneumonia, sometimes caused by lungworms. Uterine tumours have been described in a seal estimated to be at least 46 years old (Mawdesley-Thomas & Bonner, 1971). A very young animal that came ashore in the Netherlands was found to have the atlas fused to the skull – possibly the result of a dislocation at birth (van Bree, 1972).

The nematode *Terranova* is more abundant in Grey Seals than in Harbour Seals and cod infected with the larvae (codworm) are thus more plentiful in areas where Grey Seals are increasing (see chapter on Harbour Seal). Twelve Norwegian Grey Seals had an average of 2005 nematodes in each stomach, 94·5 per cent of which were *Terranova* (Benjaminsen *et al.*, 1977). Other nematodes such as *Contracaecum* and *Anisakis* are found in the stomach. *Dioctophyme* in the kidney, and *Otostrongylus* in the lungs where they may be associated with pneumonia. Acanthocephalans, cestodes and trematodes are found in the gut, mites in the nasal cavities and lice on the skin (Bonner, 1972).

FEEDING

The Grey Seal is a fish eater, showing little preference for any particular species, and at least 29 different sorts of fish have been recorded as being eaten. Although pelagic and midwater fish are eaten, bottom living fish from depths of 70 metres (40 fathoms) or more are also taken. Some crustaceans and molluscs are also included in the diet. In salmon fishing areas the seals do not avoid the salmon and the taking of these fish, and the damage to nets and lines earns the fishermen's wrath (Bonner, 1972, Mansfield & Beck, 1977, Rae, 1973).

Harbour Seals

Harbour Seals – also known as Common Seals or Spotted Seals are shore living animals found principally, although not exclusively in estuaries and in areas where sandbanks are uncovered at low tide. They are also found on shingle beaches and on rocky shores that shelve gradually into the water and are easy of access. Not infrequently seals will position themselves at their chosen hauling out place, to which they can be very faithful, long before low tide. In this way they may become perched on some apparently inaccessible rock, and then remain there until the tide reaches them again and they are able to swim off. Harbour Seals occur on the coasts on either side of both North Atlantic and North Pacific Oceans.

Atlantic Harbour Seals

In the eastern North Atlantic, round the shores of Britain and Europe the Harbour Seal is *Phoca vitulina vitulina*, Linnaeus, 1758. In the Western North Atlantic it is *P. vitulina concolor* DeKay, 1842. Landlocked seals, previously regarded as subspecifically distinct are discussed with the latter animal.

Eastern Atlantic Harbour Seal
Phoca vitulina vitulina

DISTRIBUTION
The largest single group of these Harbour Seals in Britain, and possibly in the world, is in the Wash. This shallow indentation in the east coast of England, bordered by the counties of Norfolk and Lincolnshire, receives four main rivers, and has extensive sandbanks, which are exposed at low tide. There are also smaller breeding colonies on the English east coast – at Donna Nook, Lincs; Blakeney, Norfolk; Scroby Sands, Norfolk, and in the Thames Estuary. A total of about 100 pups a year is produced in these smaller colonies, but there is believed to be considerable intermingling of adults with the Wash colony. The estimated population in the Wash area is about 5–7000 seals, producing nearly 1500 pups a year.

On the eastern Scottish coast Harbour Seals are found in all the main firths and estuaries, except for the Firth of Forth. The Moray-Dornoch Firth area has about 700 seals, while the Firth of Tay has about 500. On the Orkney and Shetland Islands, Harbour Seals are widely distributed with about 4000 on Shetland and 3500 on Orkney. On the Outer Hebrides there are perhaps 1300 seals, about 4500 on the Inner Hebrides and along the west coast of Scotland down to the Firth of Clyde, though less is known about the animals in this area.

The total population of Harbour Seals in British waters is estimated to be about 20000 (Bonner, 1976, Anderson, 1981, NERC, 1981).

The shallow waters of the eastern and north eastern coasts of Ireland support the bulk of the Irish seal population which is about 1500–2000 animals. In Northern Ireland, Co. Down has the sheltered coast with sandbanks exposed at low tide that the Harbour Seal prefers. Strangford Lough, with about 260 adults, has the bulk of the Northern Ireland population which is estimated to be about 600 animals. Legislation to protect the Harbour Seal in Northern Ireland is being considered, while the Republic of Ireland's Wildlife Act 1976 controls hunting (Nairn, 1979). In the Republic of Ireland, Galway Bay has probably the largest breeding group with about 110 animals. South of this there are barely 50 animals, but between Galway Bay

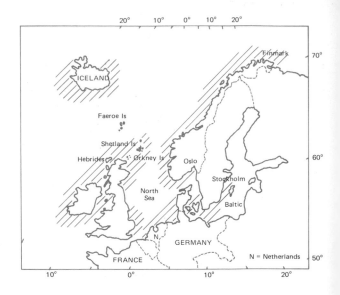

and Donegal Bay there are eight breeding sites with a total of some 370 seals (Summers *et al.*, 1980). The sandy shores of the Netherlands and the adjacent parts of the German and Danish coasts have 3–4000 Harbour Seals, though pollution of this area in the last 15 years has reduced the stock from over 5000 (Summers *et al.*, 1978). Small numbers of seals extend into the Baltic as far as Stockholm.

Harbour Seals are present along all the Norwegian coast, but are less abundant than they used to be. They are most common in the southern part of the country, in the Alesund–Bergen–Stavanger area, but also extend to Oslofjord, and may be found far up some of the fjords (Øynes, 1966). There are perhaps 1000 Harbour Seals in Norway. The more southerly shores of Iceland support a considerably larger population, and as many as 28000 have been suggested (Arnlaugsson in Bonner, 1976). The total eastern Atlantic population of Harbour Seals is thought to be in the region of 50000 animals (Bonner, 1976).

Harbour Seals, of course, often range outside their main areas of distribution, and stragglers may not infrequently be found at many places round the coasts. They often ascend rivers and have been recorded as far up the River Thames as Teddington. Several strandings occur almost every year on the French coast, the greater number of them being in the Boulogne area, and there are a few Portuguese records.

DESCRIPTION
In size, adult males are 1·5–1·8m nose to tail length, and adult females 1·2–1·5m and the maximum weight is about 113 kg. The colour is very variable, the ground colour being any shade of grey or brownish grey, and the body liberally covered with small black spots

which rarely fuse into patches. There is no obvious difference in colour between males and females, though skins from more northern areas, such as Shetland, are more conspicuously marked (Bonner, 1972). Harbour Seals have a shorter, rounder head than Grey Seals. Both animals may occur in the same areas, and another character which will distinguish them is the position of the nostrils. In Harbour Seals the nostrils are set in a wide V, with the lower ends almost touching, while in the Grey Seal the nostrils are more nearly parallel.

BREEDING

There is no obvious social organization at the breeding season, and the seals appear to be promiscuous. Most births take place at the end of June or the beginning of July. The pups are usually born on the shore, between tide marks, or on a sandbank between one high tide and the next. They are about 85 cm long and weigh about 11–12 kg. Their embryonal coat of long white hair is normally shed in utero, and may be found on the sand at the birth site. The pups are born with the short adult-type coat and are ready to swim almost immediately.

A considerable amount of attention is shown by the mother to her pup. She will conduct it into the sea if the occasion demands, and will even dive, holding the pup to her if danger threatens. Lactation lasts four to six weeks, and may take place on land or in the water.

From about the middle of August to the middle of September the adult seals moult, though there is no change in their behaviour at this time. After moulting, the mating season starts, with a peak in copulations at the end of September. An increase in excitement is seen in the Shetland seals at this time. There is much splashing and energetic rolling in pairs in the water, and this normally rather silent seal snarls and yelps. The males take great runs through the water, 'porpois-

Harbour or Common Seal, Phoca vitulina vitulina, *the Eastern Atlantic subspecies. Photo. Bill Vaughan.*

ing' so their bodies are clear of the water (Hewer, 1964). The fertilized egg develops into a blastocyst which remains unattached for two months, attachment to the uterine wall taking place at the end of November. Sexual maturity is probably reached at the same age as in the Western Atlantic Harbour Seal – males at six years and females at three to four years. This seems to be the general pattern of the life history for most Harbour Seals, though slight variations in timing may occur (Bonner, 1972, 1976).

MORTALITY, PREDATORS AND PARASITES

According to the distribution of the seals their enemies may be polar bears, killer whales, sharks or even sea eagles which may attack the pups. Not much is known about diseases in the wild though there are old accounts of epizootics that killed many Harbour Seals in Shetland (in Bonner, 1972). Lesions of the skin particularly of the ventral surface, with associated ulcers, have been found in both Harbour and Grey Seals. These raw patches, sometimes quite large, may be due to secondary infections of minor wounds by *Corynebacterium* (Bonner, 1972, van Haaften, 1967, Anderson *et al.*, 1974). Such conditions as tuberculosis, pneumonia, septicaemia following the rupture of the pancreatic duct, and malignant leukaemic lymphoma have been recorded in captive Harbour Seals from all parts of their range.

Harbour Seals carry a large population of parasites, as do most pinnipeds. Nematodes seem to cause the most trouble and high numbers of such forms as *Parafilaroides* and *Otostrongylus* in the bronchi and lungs may cause catarrhal pneumonia and associated diseases. The heartworm *Skrjabinaria* may also be found, and another heartworm *Dipetalonema* has been known to block the pulmonary artery (see chapter on parasites). The nematode *Terranova* (codworm) is also found in Harbour Seals, the larvae are present in the muscles of cod, and although harmless to man, do not enhance the appearance of the fish from a commercial point of view. *Anisakis* is another seal nematode whose larvae are found in herring. Again, these would be destroyed by cooking, but might cause problems in places where people eat herring virtually raw (Bonner, 1972). Cestodes, trematodes and acathocephalans are also found in the guts of Harbour Seals, and lice on their skin.

FEEDING

The Harbour Seal is a fish eater, some 29 different sorts of fish having been recorded as being taken. Usually the commoner species of fish in the area are taken and the diet varies according to the time of the year and the fish available. After weaning the pups will eat shrimps and gradually progress to adult food. Round the British coasts flounder, sole, herring, eel, goby, cod and whiting are amongst the fish eaten, and squids, whelks, crabs and mussels are also taken. Taking of fish from nets, and damage to nets and gear bring the seals into conflict with fishermen (Bonner, 1972, Rae, 1973).

EXPLOITATION

In Britain the Conservation of Seals Act 1970 provides a close season for Harbour Seals from 1 June to 31 August, during which time it is unlawful to injure, kill or take a seal. Exceptions to this are those seals killed in the vicinity of nets etc., to prevent them damaging fishing equipment, and those seals killed under licence for scientific or management purposes. This gives scope for the taking of a certain number of animals for commercial purposes and in Britain about 2000 young Harbour Seals are taken each year for their fur. However, most such furs – known as 'rangers' on the London fur market – come from Canada, with some from Iceland (Bonner, 1972). Bonner (1972) also notes the Legislation on Harbour Seals in some other European countries. In Denmark hunting is allowed between 1 September and 31 May; in Norway they are protected at certain specified places; in Sweden hunting is restricted, and in the Netherlands the seals are totally protected.

Western Atlantic Harbour Seal
Phoca vitulina concolor

DISTRIBUTION

This Harbour Seal occurs on the Greenland coast from about Upernavik on the west coast to Angmagssalik on the east coast, and across to the southern and western shores of Baffin Island, the shores of Hudson Bay and the coast of the Ungava Peninsula. Although capable of remaining in these arctic areas all the year round, the seals are less frequently found north of Southampton Island (Mansfield, 1967b). It is more common, however, further south, and its main area of distribution is from Labrador to New Brunswick. It is reasonably abundant on the coasts of Newfoundland and Nova Scotia, at some places in the Gulf of St

Lawrence, and on many of the islands in this area such as Sable Island, Prince Edward Island and the Magdalen Islands. It is occasionally reported from the coasts of Virginia and Carolina, and a straggler has even reached Florida.

This seal is not an animal of the ice, and in arctic localities it tends to remain in areas where swift flowing river outlets and tide rips keep areas free from ice in the winter. It has a liking for fresh water and will often ascend rivers at some distance from the sea. This habit has apparently culminated in some seals spending their entire lives in landlocked lakes. Apparent differences in jaw morphology and coat colour led Doutt (1942) to describe the landlocked seals of Upper and Lower Seal Lakes, Ungava, as a separate subspecies *Phoca vitulina mellonae*. However, the separation of this population is not regarded as conclusive on anatomical grounds, and the dark coat colour may be merely an expression of its northerly distribution (Mansfield, 1967b).

It has since been found that these are not the only seals in Canadian lakes, and seals have been found in Edehon and Ennedai Lakes which drain into the west side of Hudson Bay. Although these animals may remain in the lakes all year, they are not physically isolated from the sea (Mansfield, 1967b, Beck *et al.*, 1970).

These Canadian seals are found in small, rather isolated populations and there is some evidence to suggest that the Sable Island population of some 1500 animals is distinct from the mainland population (Boulva & McLaren, 1979). The probable lack of mixing of these populations is also suggested by the tendency of the Bay of Fundy group to have more cheek teeth than other groups of Harbour Seals.

The estimated population of Harbour Seals in eastern Canada, south of Labrador is about 12700 (Boulva & McLaren, 1979).

DESCRIPTION

Size, weight and colour of this Canadian Harbour Seal are much the same as the Eastern Atlantic animals, though Mansfield (1967b) also notes a whitish network that is superimposed on the blue-grey ground colour with black spots. The belly is lighter, and as with the British Harbour Seal, there is a tendency for the animals from the more northerly parts of the range to be a bit darker.

BREEDING

In the southern part of its range, in the Newfoundland–Nova Scotia area, the pups are born between mid-May and mid-June, but in more arctic regions the season is later, between late June and early July. The pups are much the same size as those on the other side of the Atlantic, being about 75 cm in length and about

Distribution of P.v.mellonae *and* P.v.concolor.

10 kg in weight. About 75 per cent of the pups are born in a smooth hair coat, having shed their white lanugo in utero, but the remainder of the pups are born in the white coat and may keep it for up to nine days after birth (Boulva & McLaren, 1979). The pups are weaned when about three or four weeks old, and shortly after this gather in small groups. They tend to feed on small invertebrates such as shrimps before becoming competent fish catchers.

On Sable Island signs of sexual excitement such as chasing in the water have been seen from April to late July but most breeding behaviour seems to be aquatic and difficult to observe. Sperm contents of the testis, however, indicate that this is the breeding season. This is confirmed by observations of the female reproductive systems which suggest that most ovulations take place in mid-June, near the end of lactation. There is a delay of about three months before the blastocyst is attached in about mid-September. Normally, sexual maturity is attained in males by six years and in females between three and four years (Boulva & McLaren, 1979). Moulting seems to take place in July and would thus appear to be after mating, instead of before it as in the British animals.

FEEDING

Predictably, the Harbour Seal feeds mainly on easily available fish in the locality, and herring and flounder figure largely in its diet. They seem to eat one meal daily and have no fasting season (Boulva & McLaren, 1979).

PREDATORS AND PARASITES
Parasites are much as in the Eastern Atlantic Harbour Seal. Sharks are said to be the most important predators in eastern Canada.

EXPLOITATION
Because of a certain amount of damage to fisheries, a bounty system was in operation from about 1927 until it was discontinued in 1976. A certain number of seals are shot, but the animal does not have a great economic value.

Pacific Harbour Seals

Three forms of Harbour Seal live in the North Pacific. Although their distribution overlaps to a certain extent, their breeding areas are separate, they show preferences for different types of terrain, and there are characters of colour, skull, dentition and body size that separate them.

Phoca largha associates with pack ice during the breeding season and the pups are born with a white woolly natal coat. This, and other characters suggest that there are greater differences between *P. largha* and the other two forms, and full specific rank has been suggested. *Phoca vitulina stejnegeri* (Allen, 1902) and *P. vitulina richardsi* (Gray, 1864) occur mainly in ice-free areas and breed on rocky islets, sand bars etc. A full account of the taxonomic history, synonymy and distribution of these three forms is given by Shaughnessy & Fay (1977), and they will be discussed separately here.

Insular Seal *Phoca vitulina stejnegeri*

DISTRIBUTION
This seal occurs from the Nemuro Peninsula on Hokkaido, northward along the Kuril Islands and eastern Kamchatka as far north as the Commander Islands. It breeds in these same areas. It occurs mainly in ice-free areas, and is not associated with the pack ice at breeding time as is the Larga Seal.

DESCRIPTION
Of the three North Pacific Harbour seals, *P.v. stejnegeri* is the largest, and shows most sexual dimorphism. Adult males have a nose to tail length of 1·7–1·85m, and adult females 1·6–1·7m. In colour the Insular Seal seems to be indistinguishable from the Pacific Harbour Seal (see below), though with a predominance of darker animals.

BREEDING
Breeding occurs on rocky islets at any suitable place in the range and seems to take place between April and June. At this time of year the adults congregate in large

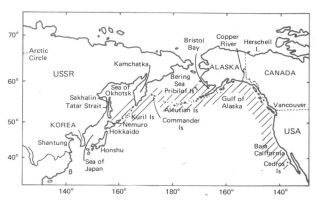

Distribution of the Insular seal P.v.stejnegeri *in the western Pacific, and the Pacific Harbour seal* P.v.richardsi *in the eastern Pacific.*

groups that may number several hundred individuals. Males may have scars or wounds about the throat, suggesting competition, but little is as yet known about this. The pups are about 98cm long at birth, and shed their white natal coat before they are born, emerging into the world in a coat not unlike that of the adult. Occasionally however, a few pups still retain the white coat for a while. The pups are weaned four weeks after birth.

Pacific Harbour Seal *Phoca vitulina richardsi*

DISTRIBUTION
This seal occurs in the eastern North Pacific. Its distribution ranges from the coast of Baja California, about as far south as Cedros Island, northwards along the coast and islands of the United States and Canada to the Gulf of Alaska, and along the Alaska Peninsula and Aleutian Islands. In the eastern Bering Sea it reaches the Pribilof Islands and Bristol Bay. It has been estimated that there are about 35 000 seals in British Columbian waters just prior to the pupping season (Bigg, 1969) and the total population from Alaska to California is in the region of 300 000 animals (Bonner, 1979a).

DESCRIPTION

In size, the Pacific Harbour Seal is approximately midway between the two others just mentioned. The average nose to tail length of adult males is 1·62 m, of females 1·5 m. Adult males weigh about 73 kg, and adult females 59 kg. The colour of the adult coat is rather variable and there are light and dark individuals. The light coloured animals have a dark dorsal surface, and the pale sides and belly have dark spots. There is a superficial resemblance to *P. largha*, but the dark mid-dorsal area is made up of closely packed dark spots, some of which have obvious white rings round them. The spots are larger than in *P. largha* and tend to coalesce. The pale areas round the eyes and snout are present in all ages. The dark coloured animals have over all the body, a black or nearly black background made up of closely packed black spots. Some of the spots are encircled by light rings which are more numerous ventrally.

BREEDING

At the breeding season the adults gather in large aggregations on such places as rocky islets, sand bars and mud flats exposed at low tide. The season of birth varies according to the latitude. In the area of Cedros Island they are born in early February; in March and April on the Channel Islands of Southern California; during May along the Washington coast, and between late June and September, with a peak in late July in southern British Columbia and Puget Sound. In the Gulf of Alaska pups are born between mid-May and late June, and between June and mid-July on the Pribilof Islands and in Bristol Bay. Thus in the Bristol Bay area the pups of *P. largha* are produced early, in about April on the ice floes, and those of *P.v. richardsi* some two months later when the ice has gone.

Pups of the Pacific Harbour Seal have the adult-type pelage at birth, nearly all of them having previously shed their white coat. At birth they are about 82cm in nose to tail length, and weigh about 10kg. They suckle for about five or six weeks, reaching a weight of about 24 kg by then (Bigg, 1969).

Adult females ovulate and mate at the end of weaning and there is a two-month delay before the blastocyst becomes attached. Most females become sexually mature and mate for the first time when they are three or four years old, and most males by five years old (Bigg, 1969).

FEEDING

As with other Harbour Seals, fish form the main items of diet, flounders, herring, tomcod and hake being commonly taken, but other fish, and squid and lampreys are also eaten (Scheffer & Slipp, 1944).

MORTALITY, PREDATORS AND PARASITES

Parasites are as usual common in these seals. Stranded animals show many causes of death from ruptured arteries from a possible fall to bacterial lesions and tumours. Although protected, many animals are shot, presumably because they have been thought to be taking salmon (Stroud & Roffe, 1979).

Larga Seal Phoca largha

DISTRIBUTION

This seal breeds in association with the pack ice, but at other times of the year is found on all shores of Japan, Okhotsk, Bering and Chukchi Seas, on the Pacific coast of Kamchatka, and on the Commander Islands. On the Russian Arctic coast it may occur as far west as Chaun Bay (70°N 170°E). On the North American side it occurs from Herschell Island and Point Barrow to the Pribilof Islands, Bristol Bay, and the eastern Aleutian Islands. It occurs as far south as the Yellow Sea, Korea, and possibly as far as the Yangtse River.

There are eight major areas where breeding takes place, three of these in the Bering Sea. The breeding areas are: (1) in southeastern Bering Sea, from Bristol Bay westward to long. 170°W (2) in northwest Bering Sea in the Gulf of Anadyr (3) in southwestern Bering Sea, in the vicinity of Likte Strait and Karaginski Island (4) in the northern Okhotsk Sea, north of lat 55°N (5) in

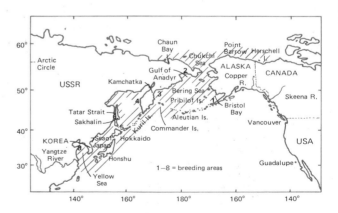

the southern Okhotsk Sea from eastern Sakhalin to northern Hokkaido (6) Tatar Strait (7) Peter the Great Bay, near Vladivostok (8) Po Hai Sea in the north western Yellow Sea (see map).

DESCRIPTION

Larga Seals are the smallest of the North Pacific Harbour Seals, adult males having a nose to tail length of 1·6–1·7 m, while adult females are 1·4–1·6 m. The pelage has a pale silver background with a darker grey area along the dorsal surface which may be broken by fine rings of the pale background colour. Superimposed on both pale and dark areas is a scattering of brown-black oval spots. In the youngest individuals the areas round the eyes and snout are pale.

BREEDING

In the breeding season males and females occur in pairs, remaining together for about two months, and widely separated from other such couples. The pups are born on the ice floes at the southern fringe of the seasonal pack ice in the Bering Sea, and on ice of similar quality in other areas. The time of their birth varies according to the latitude. In the Bering Sea most pups are born between late March and mid-May; in the northern Okhotsk Sea the pupping peak is in mid-April; in the southern Okhotsk Sea and Tatar Strait it is in March, and in the most southerly breeding areas pups are born from early February to mid-March.

The pups at birth are about 85 cm in length and have a natal coat of long white woolly hair that is shed two to four weeks after birth. They are fed by their mothers for four weeks, during which time they triple their birth weight. A two-week-old pup weighs about 18 kg, and a four-week-old pup 28 kg (Burns *et al.*, 1972).

An estimate of the population suggests that there are about 200 000 animals in the Bering Sea, and a similar number in the Okhotsk Sea. Some 15 000 animals are caught annually for their skins (Bonner, 1979b).

FURTHER CHARACTERS OF THE NORTH PACIFIC HARBOUR SEALS

As well as the difference in their breeding terrain and colour of the newborn pup, some of the other differences between the three North Pacific Harbour Seals may be summarized. *P. largha* has the smallest and most fragile skull (condylobasal length of adult male skull 218 mm), the premolar teeth are usually set straight along the long axis of the jaw, and in the hyoid apparatus the tympanohyals are well developed.

In *P. vitulina stejnegeri* the skull is larger and more massive (condylobasal length of adult male skull 236 mm), and in *P.v. richardsi* the size is more variable (condylobasal length adult male *c.* 225 mm). In both these two animals the premolars tend to be set obliquely in the jaw, and the hyoid has abbreviated tympanohyals and stylohyals. No difference has yet been found between the blood proteins of typical *P. largha* and typical *P.v. richardsi*.

Much more remains to be done to establish their relationships, but research to date would indicate that *P. largha* appears to differ more from both *P.v. stejnegeri* and *P.v. richardsi* than these two do from each other, but the taxonomic status of *P.v. stejnegeri* has not yet been finally settled (Shaughnessy & Fay, 1977).

Larga Seal, Phoca largha. *Photo. F. H. Fay.*

Ringed Seal *Phoca hispida*

DISTRIBUTION

The Ringed Seal is an animal of the circumpolar Arctic coasts. It is normally fairly solitary, but is the commonest seal of the Arctic and is found wherever there is open water in the fast ice, even as far as the North Pole. It occurs rarely in the open sea or on floating pack ice, but is common in fjords and bays where the ice is firm (see map). The wide distribution of this seal has led to its being divided into a large number of subspecies, some of which are given below. This may, or may not be justified, as the differences seem to be mainly geographical. As used here the subspecific names give mainly an indication of distribution.

P. hispida hispida Schreber, 1775 is known from all the Arctic coasts of Russia, Europe, Canada and Alaska, including the islands north of the mainland coasts such as Novaya Zemlya, Spitsbergen and Baffin Island. It is found in Nettilling Lake in the centre of Baffin Island, in Hudson Bay and on all the coasts of Greenland. It is seen on the northerly coast of Iceland during the winter, but does not breed there, and its range may just extend down the Labrador coast as far as Newfoundland. Occasional stragglers may move further south, and have been recorded from the Pribilof Islands, the Atlantic coast of France, Germany, Netherlands, Portugal, Norway, Ireland, Isle of Man and Norfolk.

P. hispida krascheninikovi Naumov & Smirnov, 1936 occurs in the northern parts of the Bering Sea, not on the Aleutian Islands, occasionally on the Commander Islands in spring and summer, and once from the Pribilof Islands. To the west this animal merges with *P. hispida ochotensis* Pallas, 1811 of the Okhotsk Sea. The latter ranges south from the Okhotsk Sea and the western coast of Kamchatka, along the shores of Sakhalin, the Tatar Strait, and along the Pacific Japanese coasts as far south as about 35°N. It is the most common Japanese seal, and is also said to occur along the eastern coast of Korea.

The Ringed Seals of the Baltic, including the Gulfs of Finland and Bothnia are called *P. hispida botnica* Gmelin, 1785, those from the Finnish Lake Saimaa and the Russian Lake Ladoga are, respectively, *P. hispida saimensis* Nordquist, 1899 and *P. hispida ladogensis* Nordquist, 1899. These two freshwater lakes are not entirely cut off from the sea, as the River Neva connects Lake Ladoga to the sea in the Gulf of Finland at Leningrad, and Lake Saimaa is connected to Lake Ladoga by a swift stream, and also has the Saimaa Canal into the Gulf of Finland. There is, however, no evidence that the lake seals travel to the sea, and in fact the Saimaa seals avoid the southern parts of the lake because of its pollution. The Ringed Seal does not migrate, and its movements are only local.

Distribution of Ringed, Baikal and Caspian Seals.

DESCRIPTION

Both male and female animals grow to about 1·4–1·5m nose to tail length, and 68kg in weight, though females tend to be slightly smaller. Small Ringed Seals, often reported as dwarfs in the literature are sometimes given distinct specific or subspecific epithets. It would seem though, that these smaller animals are from areas of drift ice where because of unstable conditions, parental care may be curtailed, and the animals may have been thus deprived of food. Certainly, starving animals are more common in such areas. The largest and heaviest seals are found on the stable fast ice (McLaren, 1958b).

The colour is very variable, but is usually a light grey background spotted, particularly on the back, with black. Many of the spots are surrounded by ring shaped lighter marks, giving the seal its popular name, but down the centre of the back the dark spots may be so close together as to be confluent. The belly is a silver grey. There seems to be a tendency for the Gulf of Bothnia and associated lake seals to be slightly darker in colour.

ECOLOGY

In general, Ringed Seals are animals of inshore waters. In winter, adults and some of the immature seals stay under the ice in bays and fjords, while the younger seals are further out, at the edge of the fast ice. After the April pupping season, and with the coming of warmer weather in June, most seals lie out on the ice, basking in the sun, moulting and fasting. The seals surface to breathe at open cracks in the ice, and during the winter breathing holes are kept open by the seals who abrade the ice with the claws of their foreflippers. Holes may be maintained in ice over 2m thick, and are cone shaped with the smaller part towards the ice surface. A small ice dome, only about 4 cm high, with a small hole in the top, is all that is usually visible of a breathing hole in ice. When slabs of ice are cracked and piled up by pressure, snow accumulates to the windward and leeward sides of the pile, and if the snow is more than 20cm deep it may be hollowed out and used as a lair by the seals and has of course a breathing hole opening into it from the water below. Such lairs, usually oval and single chambered, may be used for hauling out by a number of seals, or by single animals such as a breeding male. Birth lairs are larger, and are distinguished by having a series of tunnels into the nearby snow. From the diameter of these tunnels it is evident they are made by the pup. These lairs would be useful in giving some protection to the seals from such land predators as polar bears and arctic foxes, and also in protecting the seals from the cold and wind. A young, blubberless pup would derive considerable thermal benefit from the warmth generated by its mother in the narrow confines of the lair (Smith & Stirling, 1975).

BREEDING

The pups are born, in the snow lair between about the middle of March and the middle of April, with the peak at the beginning of April. They are about 65cm in nose to tail length and weigh about 4·5 kg. At birth the pups have long creamy white fur which they begin to shed after two weeks, completing this moult within about six weeks. This freshly acquired adult-type coat is of fine texture, dark grey dorsally and silver grey ventrally. It is of greater commercial value than the adult skin, and is known as a 'silver jar', whereas skins of other immature and adult Ringed Seals are known as 'common jars'. The very young pups are of course 'whitecoats'.

The pup is suckled for nearly two months, a rather long period of parental care which probably has some correlation with being born on fast, rather than on moving ice. The adult males are ready to mate from March until mid-May, and at this time have a strong offensive odour – from which an early specific epithet – *foetida* – was derived. The peak of mating activity probably occurs in mid-April, shortly after parturition and while the female is still lactating. The blastocyst is implanted at the beginning of July, after a delay of 81 days, so that there is an active gestation period of 270 days (Smith, 1973a). The male is sexually mature when seven years old and most females are mature when five years old.

PREDATORS AND PARASITES

Enemies, apart from man, include polar bears, killer whales and arctic foxes, and young pups are sometimes eaten by walruses. Polar bears pose a constant threat to Ringed Seals – the bears remain on the sea ice as long as possible, and hunt seals all the year. When hauled out on the ice the seals are constantly alert, and they are well dispersed over the area so that there are plenty of breathing holes available for escape. Bears take large numbers of the immature seals and show a preference for eating the blubber, often leaving the rest of the carcass. In this way the bears are serving their own needs for a high energy diet. Bears may dig into lairs, but often the newborn pups are killed but not eaten (Stirling & McEwan, 1975, Smith, 1980). Arctic foxes are an important predator of Ringed Seals, but they tend to eat pups only, digging down into the lairs to get them (Smith, 1976).

Parasites include the usual gut infestations of the round, thorny headed and tapeworms, and roundworms have also been found in the lungs and venous sinuses of the liver. Ringed Seals also get infected with the nematode *Trichinella* which they acquire through eating infected amphipods (see chapter on parasites). Externally, lice have been recorded.

FEEDING

The seals feed on a wide variety of small pelagic amphipods, euphausians and other crustaceans, and also on small fish. Seventy-two food species were identified in the stomachs of seals from the eastern Canadian Arctic. Here it was noticed that in shallow, inshore waters the seals were feeding near the bottom, chiefly on polar cod (*Boreogadus saida*), of average length 18cm, and on the small crustacean *Mysis* while those in the deeper offshore waters were catching the planktonic amphipod *Parathemisto*. There is little evidence on the depths to which they can go while feeding, but several have been killed in over 37 metres (20 fathoms) of water, and they may be able to go as deep as 91 metres (50 fathoms). Although feeding largely on planktonic organisms, the Ringed Seal does not have greatly subdivided teeth like the Crabeater Seal, and there is some evidence that it may pick out the larger animals in a shoal of *Parathemisto*, whose size ranges from 19–63 mm, and catch them individually, probably by suction. However, such individual selection of food items is unlikely to be practised when small amphipods of about 10mm in length, and occurring in thick swarms, are eaten (McLaren, 1958b).

EXPLOITATION

Many thousands of Ringed Seals are caught annually, from all areas where they occur, mostly for their skins, which are used for leather or for their decorative fur, and also for blubber and meat. They have always been useful to the Polar Eskimos – meat and blubber, liver and intestines to be eaten by men and dogs, blubber oil for lamps, white pup skins for use as underclothes – fur side inside, adult skins for clothes, bags, dogharness and tents, and many other things. Livers of large Ringed Seals have sufficient vitamin A in them to be toxic (see hypervitaminosis in chapter on viscera). Nowadays, perhaps the seals are less useful to the Eskimos for basic living as modern fuel and clothes will have replaced seal products to a certain extent. But taking the pelts for sale to the general market has become more profitable and is much more of a big business.

Early spring – March and April – when the seals are in good condition is the best time for hunting. A hunter and his dog team will wait at the edge of the fast ice, and the seal, who is curious and with not too good eyesight will approach the hunter, only to be shot with a high-powered rifle. The curiosity of the seals can be aroused by making scratching noises on the ice. By July most of the adults are moulting, so their skins are not good, and they have little blubber after fasting, so tend to sink if shot in the water. At this season the first year 'silver jars' are taken. Canoes with

outboard motors are used in summer. Little hunting is done in winter. Eskimos used to spend hours waiting at a breathing hole in order to harpoon a seal, but this slow method of hunting is not used much now. The stalking of seals on the ice, by an Eskimo crawling along, pushing in front of him a small sledge carrying a gun and a white screen which hides him, has been at least partly superseded now by mechanized snow-mobiles, partly camouflaged by white canvas. These machines are partly status symbols, and allow hunting to take place at greater distances from base, but they are noisy, expensive, break down, and produce clouds of polluting gas. There is still a lot to be said for a dog team (Smith, 1973b).

DEMOGRAPHY

The total population of the Ringed Seal is impossible to estimate, though a figure of 6–7 million has been suggested (Stirling & Calvert, 1979). Work was carried out in Baffin Island, particularly in the Cumberland Sound, Hoare Bay and Home Bay regions on the east coast of the island, between 1966 and 1969 to get some idea of the population of seals and the number taken in these areas (Smith, 1973a). Aerial surveys were made, and the figures corrected for a certain proportion of

Ringed Seal, Phoca hispida, *the commonest seal of the Arctic. Photo. Thomas G. Smith.*

seals under the ice and not visible. The corrected estimates were 70684 in Home Bay, 36376 in Hoare Bay, and 58782 in Cumberland Sound. This gives almost 166000 seals in a relatively small part of the total range occupied by these seals. The total annual catch in Home Bay (corrected for animals lost etc.) is just over 5000, and just over 9000 in Cumberland Sound. The latter place is not good Ringed Seal habitat as the ice is unstable, and it is thought that the population is being supplemented from neighbouring areas. The possibility of future increased hunting using more sophisticated techniques makes the institution of a management programme necessary.

Caspian Seal *Phoca caspica*

DISTRIBUTION

The Caspian Seal is an animal of very restricted distribution, and is found only in the Caspian Sea (see map). In winter the seals stay in the northern parts of the sea, while in summer they are to be found in the southern regions, moving back to the north in late autumn. The reason for this move across the lake is the effect of its differing depths, and the preference the seal has for the ice. The northern part is shallower and freezes in winter, ice beginning to form in November and December and lasting until April, but the water becomes too warm for the seals in summer and they then move south into the deeper, cooler water (Naumov, 1933).

Caspian Seal, Phoca caspica, *found only in the Caspian Sea. Photo. Novosti Press Agency.*

The stock of Ringed Seals in the Okhotsk Sea was estimated to be about 800000 in 1966 (Fedoseev, 1968). The seals in Lake Saimaa were persecuted and fell to the very low figure of 40 in 1958. They have been totally protected since 1958 and the numbers increased to 200–250 by 1966. Even so, the fishermen still complained and licences were issued to shoot some seals (*IUCN Red Data Book*). The Ladoga Ringed Seal population is said to be 5000 animals, and special reserves for it, protected from hunters and tourists are recommended (Bychkov & Antoniuk, 1975). However, another author (Popov, 1979a) gives a figure of 10000–12000 from aerial surveys made in winter.

DESCRIPTION

Adult male animals may reach 1·5 m in nose to tail length, and weigh up to 86 kg, while females are slightly smaller, only reaching about 1·4 m. Both sexes are a deep grey dorsally, and greyish white ventrally, the back liberally spotted, though there is considerable variation. Males tend to be spotted all over, whereas the spots are lighter and mainly on the back in females.

BREEDING

At the end of the year there is a great migration of seals to the northeastern corner of the Caspian Sea. The ice there is in the form of big floes, that under the influence of strong winds become tilted and thrust up against each other forming rough hills. In the quiet seclusion of this rough ice terrain the seals keep open smaller breathing holes and larger exit holes, and the pups are born safely away from the ice edges. The pupping season is from late January to the beginning of February, with most pups being born at the end of January. They are about 64–79 cm in length, about 5 kg in weight at birth, and are clothed in long white fur that is shed after about three weeks, giving place to a coat of short dark grey hair. They are fed by their mothers for about a month (Ognev, 1935).

Moulting of the adult animals takes place at different times by different groups. The pupping females are first and start during lactation at the beginning of February. Later, the groups of moulting females are joined by the adult males who start to moult at the end of March, and have usually finished by the beginning of May. The immature animals moult a little earlier and have finished by the middle of April. After moulting most of the seals move to the southern part of the lake. Mating takes place between the end of February and the middle of March and as the animals usually occur in pairs they are possibly monogamous. Sexual maturity is attained at five years in the female, but most do not reproduce until they are six or seven years

Caspian Seal pup still clad in its long white fur. Photo. Novosti Press Agency.

old. The males are sexually mature at six to seven years old (Popov, 1979a).

FEEDING

During the autumn and winter the seals eat gobies, sculpins and small crustaceans, and in the spring sprats, herring, and any available fish such as sand smelts, roach and carp are added to the diet. Sculpins are particularly favoured and the seals will move to parts of the lake where these fish are most abundant, sometimes even gorging themselves to death on the fish.

PREDATORS

There is a certain amount of predation on the seals by wolves and eagles.

EXPLOITATION

The total population of the Caspian Seal is estimated to be about 500 000–600 000 animals, and the numbers are believed to be reasonably stable (Popov, 1979a). About 60–65 000 pups are taken annually for their skins. Regulations forbid the killing of adult females during the breeding season, and it is not permitted to take immature animals and pregnant females. For the origin of this seal in the Caspian Sea, see the chapter on fossils.

Baikal Seal *Phoca sibirica*

DISTRIBUTION

The Baikal Seal, as its name suggests, is found only in Lake Baikal, a large freshwater lake in eastern Russia (see map). This is the deepest lake in the world, its average depth being about 700m, and its greatest depth about 1·7 km. The seals are more numerous in the northern parts of the lake, particularly in winter and the beginning of spring, when most of the seals are on the northeast shore. In early summer they are much more evenly distributed, but at the end of June they begin to come on shore in herds, especially in the north, in the less inhabited regions. When the ice begins to form at the end of October, the herds break up and the seals disappear, spending the winter in the water which of course is considerably less cold than the air. The seals breathe at holes in the ice that they keep open by constant use, and reappear on the ice again in February (Naumov, 1933, Ognev, 1935).

DESCRIPTION

Baikal Seals are small animals, growing to a nose to tail length of about 1·3m, and a weight of about 80–90 kg. The colour is a dark brownish grey, shading to a lighter yellowish grey ventrally.

BREEDING

Only the pregnant females do not spend the winter in the water. They come on to the ice and make a lair under the snow where most of the pups are born in the middle of March. The pups are about 70cm in length and 3 kg in weight, and are covered with a long white woolly coat that they lose after about a fortnight for a slightly paler version of the adult coat. They are fed by their mothers for about 2–2·5 months (Popov, 1979b). Both mating and moulting take place in May and June, and the seals are thought to be polygamous. The females are mature at two to five years old but do not breed until they are about six years old, while the males are not mature until they are eight years old (Popov, 1979b).

FEEDING

The seals are fish eaters, gobies and the deep water fish *Comephorus* having been recorded as part of their diet, and as is usual in these instances, they are said to do a certain amount of harm to the fishery business.

EXPLOITATION

The total population of Baikal Seals is estimated to be

40–50000 animals, and the population has obviously recovered from the over fishing that seriously reduced the numbers in the 1930s. About 2–3000 pups are taken commercially each year for their skins (Popov, 1979b). In earlier times the skins were sometimes used as clothing, the flesh eaten, and the blubber treated as a delicacy, being eaten in long strips without salt or bread (Radde, 1862). For the origin of this seal in Lake Baikal see the chapter on fossils.

HISTORY

The Baikal Seal was apparently first mentioned in print in 1763, and although named and described in German and Russian papers during the nineteenth century, no specimens reached Britain until 1909, and no live animals until 1959 when Moscow Zoo sent a pair of these seals to London Zoo.

The specimens that arrived in 1909 were collected by Dr Charles Hose who was at that time making a journey to Sarawak on the Trans-Siberian Railway. During a two-day stop at Irkutsk near Lake Baikal, Dr Hose managed to persuade, with roubles, some fierce looking black bearded fishermen, clad in smelly sheepskins to catch some seals for him. He resumed his train journey with three seals on the luggage rack of his compartment, where they dripped water and were fed on herrings from the refreshment car. Unfortunately two of the seals died after a couple of days and rather than waste his rare material, Dr Hose dissected them there and then, flinging the more perishable parts out of the train window, to the consternation of fellow passengers. The third seal also died while in the ship for Shanghai, but luckily the three skulls were saved, and they, together with two of the skins are still kept in the British Museum (Natural History) (Hose, 1927).

Baikal Seal, Phoca sibirica, *found only in Lake Baikal. Photo. Bill Vaughan.*

Harp Seal *Phoca groenlandica*

DISTRIBUTION

The Greenland, Saddleback or Harp Seal is found in the open sea of the Arctic Atlantic (see map). It is known off the coasts of Europe and Asia from Severnaya Zemlya and Cape Chelyuskin to northern Norway, including the Kara Sea, Novaya Zemlya, Franz Josef Land, White Sea, Spitsbergen (Svalbard) and Jan Mayen. It occurs on all coasts of Greenland except the extreme north, Baffin Island, Southampton Island, Labrador, the east coast of Newfoundland and the Gulf of St Lawrence. It is not normally found far into Hudson Bay, and only rarely visits the northern coast of Iceland. Stragglers are occasionally reported from Norway, Scotland and the Shetland Islands, and even from as far south as the Bristol Channel and River Teign in Great Britain. There is a possible record from Co. Galway, Ireland; one from the River Mulde, near Dessau, Germany of an animal believed to have swum up the River Elbe; a jaw of Magladenian age was found in the Grotte de Raymonden, Perigueux, France, and an animal was once seen in the English Channel. On the western side of the Atlantic an adult Harp Seal was caught near Cape Henry, Virginia (lat. 37°N) and one far across on the Arctic coast of Canada at Aklavik in the Mackenzie River delta.

The Harp Seal is a gregarious migratory animal, its range extending north to the far open waters of the Arctic in summer and early autumn, and the animals coming south in late autumn and winter in time for the spring breeding. While on migration the seals leap and jump out of the water like dolphins.

DESCRIPTION

Both adult males and females are about 1·6 m in nose to tail length, and weigh about 136 kg. There is considerable change in coat colour as the animal grows older, and a certain amount of variation even in the adult coat. The adult male is a light silvery grey over most of the body. The head, to just behind the eyes is black and there is a characteristic horseshoe-shaped black band running along the flanks and across the back. In the female the face and 'harp' colour is paler and may be broken up into spots.

The newborn pups are about 90 cm long and weigh about 6–10 kg, and are covered in a coat of white silky fur. When about a week old the moult starts, and is complete in three or four weeks. The young animal then has a short-haired coat of silvery grey irregularly spotted with darker grey and black. There is enormous variation in the shade of ground colour and degree of spotting, but all immature seals, of both sexes, show the same general pattern.

BREEDING

Harp Seals breed at some distance in from the margins

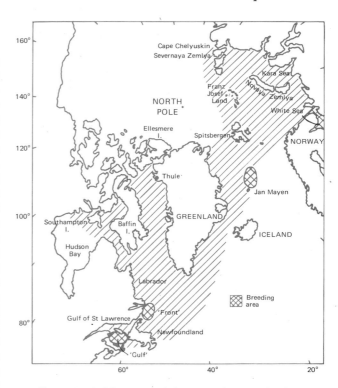

of large ice fields, in rough hummocky ice which gives a certain amount of shelter for the pups, and through which holes are kept open for access to the sea. Large pupping rookeries are formed, but each male mates with only one female. There are three main breeding grounds – in the White Sea, in the Greenland Sea north of Jan Mayen, and in the Newfoundland area. In this latter area there are two centres, one in the Gulf of St Lawrence, usually to the north or west of the Magdalen Islands (known as the 'Gulf'), and the other off the coast of Labrador between Belle Isle and Hamilton Inlet (known as the 'Front') (Mansfield, 1967). After the breeding season the seals disperse from their breeding concentrations, but tagging has shown that the Newfoundland animals do not, in general, move in an easterly direction past the southern tip of Greenland, (a single animal tagged in the Gulf of St Lawrence was recovered off northern Norway – Sergeant, 1973b), while the occasional Jan Mayen seal may be found in the White Sea.

Analysis of skulls from the three centres has shown that there are statistically significant differences between the populations, but the difference is greatest between the Newfoundland herd and the rest (Yablokov & Sergeant, 1963). Analysis by electrophoresis of the haemoglobins and serum proteins indicate that Harp Seals can be divided into a western stock (Newfoundland) and an eastern stock (Jan Mayen and White Sea) (Naevdal, 1965), but there is no difference between the blood proteins of 'Gulf' and 'Front' seals

(Naevdal, 1969). Thus the three populations are distinct, but that of Newfoundland is the most isolated. This difference is reflected also in the slightly different breeding seasons of the three centres, which is related to the difference in the onset of warmer spring weather.

From their summer feeding grounds at the edge of the ice near Spitsbergen and Franz Josef Land, the seals move into the White Sea from October onwards, the adults first, the immature animals not until April. They stay in the water till pupping begins. The pupping season may last from the end of January to the beginning of April, but most of the pups are born between 20 February and 5 March, and during this time the female feeds very little. Suckling lasts for ten to twelve days, and about two weeks after parturition the adults mate and it is only at this time that the males are seen on the ice. When the moult of the pups white coat is complete at about four weeks, they then take to the water and are independent. From mid-April to the end of May, first the immature seals, then the adults, collect in rookeries on the ice for moulting. These moulting rookeries are much larger than the pupping ones, and tens of thousands of animals may collect together. At the end of moulting the seals move north out of the White Sea (Sivertsen, 1941).

In the Jan Mayen area the movements are similar, but the pupping season is slightly later, most of the pups being born between 18 and 20 March. The ice drifts southwards, and after mating the seals swim again to the ice north of Jan Mayen for moulting in April, and by the middle of May they have moved to their northerly feeding grounds between Spitsbergen and Greenland.

The Newfoundland population have the same general movements. The seals spend the summer and early autumn feeding intensively in the waters of west Greenland and the eastern Canadian Arctic. They may extend as far north as Thule and the southern end of Ellesmere Island, and may just enter Hudson Bay. In September they start moving south. In January part of the population enters the Gulf of St Lawrence and the rest remain in the waters round Newfoundland. In February the adults congregate in the pupping grounds of the 'Gulf' and the 'Front'. The pups are born from late February to mid-March and are suckled for eight to twelve days. At the end of this period the females mate, usually in the water, and then leave the area for some concentrated feeding. During April and May the seals gather in aggregations for moulting, adult males gather separately, immatures of both sexes together, and the adult females return from feeding to be the last to join the herds. After this the summer northward migration starts again and by mid-June most of the animals have reached the waters of

south-west Greenland again (Mansfield, 1967a, Sergeant, 1976, Lavigne, 1979).

Both males and females are sexually mature at four to five years old, the females bearing their first pup when five years old. This is the age of maturity in a heavily exploited population, crowding resulting in a reduction of the age of maturity (Sergeant, 1973a). It is believed that females can bear pups at least until their sixteenth year, and possibly up to twenty years or more. A delay of 4·5 months in the implantation of the blastocyst means that gestation proper – the *active* gestation period – does not begin until July and lasts for about 7 months.

Work on the Gulf of St Lawrence animals shows that pups grow at a rate of 2·5 kg a day during the lactation period, 1·9 kg of this is laid down as blubber, the rest going to form the body in general. Five days after birth 50 per cent of the weight of the pup is the skin and blubber, and by weaning this can be as much as 60 per cent. After weaning the weight of the body core decreases by some 35 per cent, but the blubber and skin weight decreases by only 10 per cent. At the time of weaning, not only is the milk supply stopped, but the pup moults, losing the insulation of the thick fur. It is therefore important that the blubber layer be kept as intact as possible for insulation until the pup has mastered the art of getting its own food (Stewart & Lavigne, 1980).

FEEDING

For their first year of life the pups are solitary and feed primarily on pelagic crustaceans such as the euphausid *Thysanoessa*, the amphipod *Anonyx*, and a little fish. Young animals watched while feeding in captivity, took small crustaceans individually, largely by suction. Older seals are more gregarious and although still including the smaller custaceans in their diet, they now include pelagic schooling fish as well. Capelin, herring, redfish, and polar cod are among the fish eaten.

EXPLOITATION

The white coated pup is one of the main objectives of the commercial exploitation, though current market fluctuations determine which of the age groups produces skin of the most value. Pups when just born are of little value, and the 'whitecoat' skin must be taken when the pup is between two or three and ten days old, as after this the long white hairs start to fall out, and the animal is known as a 'ragged-jacket'. The 'ragged-jacket' stage starts about 12 days after birth, immediately after weaning, and lasts for about six days. When this moult is complete and the pup has the grey spotted coat, it is known as a 'beater' and this coat being softer and thicker is more valuable than that of later age groups. Immature seals or 'bedlamers' and adults from the moulting assemblies are taken for blubber oil and leather, and the skin may be used for making slippers.

In the Jan Mayen area the first recorded sealing expedition is from 1720, many nations taking part until about the 1880s after which time the Norwegians did most of the hunting. By the early twentieth century the seal stocks were depleted and the Norwegians turned their attention to the White Sea. In turn, these stocks became low and the Norwegians then explored the possibilities of the Newfoundland waters. About half the Norwegian sealing income is now derived from Newfoundland.

Soviet sealing from ships, as distinct from shore-based, developed in the 1920s in the White Sea, spread to Jan Mayen and then to Newfoundland in the 1960s, but is now mainly back in the White Sea again.

Newfoundland sealing developed in the nineteenth century, but has now been taken over by Norwegian companies based in Canada and with Canadian crews.

The number of Harp Seals taken commercially is set by a quota system agreed by the sealing nations under the International Commission for the northwest Atlantic Fisheries (ICNAF) treaty, and this body also sets the opening and closing dates for the sealing. In the White Sea the season is between 23 March and 30 April and a quota of 27 000 animals is set. At Jan Mayen the season is 23 March to 5 May, with a quota of 15 000. At the 'Front' the season is 12 March to 24 April, the Norwegians taking 60 000, the Canadians 60 000, and 30 000 is the target for the catch by Canadian lands-men. The closing date is arranged so that it is *before* the late arrival of the females at the moulting sites, and is to prevent their capture. The 'Gulf' has been closed to sealing ships since 1972 (Sergeant, 1976).

DEMOGRAPHY

Estimation of the total population of such a widely distributed animal, even with the incentive of commercial use, is difficult. It is not easy to distinguish and count the white coated pups lying on a white background when seen by aerial photography, but the use of ultra-violet photography makes the white pups appear black and greatly facilitates the counting. Some 250 000 pups are born in the western North Atlantic, and this is equivalent to a population of about a million animals before the pups are produced. The total world population is believed to be about 2·5 million animals. The 'primeval population' has been estimated to be some 9 or 10 million animals (Sergeant, 1976). The western herds of Harp Seals were estimated to be about 3–4 million in the early 1950s, and had probably almost reached their maximum levels with the absence of sealing during the war. During the next twenty years or so the number of seals was halved, but observations showed that the general condition of the herds had improved, and the reproductive rate had increased. Danger to the seal herds is not so much from the carefully controlled sealing, but from increased fishing by man. Capelin is now fished by man in ever increasing tonnages, thus removing a large part of the food supply of the Harp Seal (Sergeant, 1976).

Harp Seal, Phoca groenlandica. *An adult showing the characteristic horseshoe-shaped black mark on the coat. Two whitecoat pups are present in the foreground. Photo. J. Boulva.*

Ribbon Seal *Phoca fasciata*

DISTRIBUTION

The Ribbon Seal is a rather solitary animal of the seasonal ice floes of the North Pacific and its distribution varies according to the ice formation (see map). It occurs in the Sea of Okhotsk and probably also in the Tatar Strait, and in the region of Kamchatka and the Kuril Islands, and may reach the northern coast of Hokkaido at times when there is much ice. It is also found in the Chukchi Sea and in the Bering Sea, including the Aleutian Islands, and ranges from Point Barrow to Unalaska Island (Burns, 1981). Stragglers have been reported from Cordova in the Gulf of Alaska, and from Morro Bay, south of San Francisco where a curiously hairless male was captured and kept on exhibition for a month before it died.

During the late winter, spring and early summer the seals are associated with the sea ice, far from land, where the pups are born and the adults moult, but their distribution during the rest of the year is hardly known. As the ice melts the seals disappear, and its is assumed they then live at sea as they are rarely found on land (Burns, 1981).

The average annual Soviet catch of Ribbon Seals from the Bering Sea for the years between 1961 and 1967 was about 13 000 animals, and in the early 1960s the total Bering Sea population was estimated to be 80–90 000. Sealing at this level reduced the population and the quota was reduced to 6290 in 1968 and to 3000 in 1969 and has remained approximately at this level since. Only about 100 seals are taken annually in Alaska.

The total population of Ribbon Seals is now about 240 000, with 100 000 in the Bering Sea, and 140 000 in the Sea of Okhotsk (Burns, 1981).

DESCRIPTION

Both male and female Ribbon Seals are approximately 1·5 m in nose to tail length, and weigh about 90 kg, though females may weigh a little less than males of similar size. The nose to tail length of an adult male from the Okhotsk Sea has been given as 1·6 m, weight 94·8 kg, while an adult female 2·5 cm longer weighed 78·9 kg. Burns (1981) notes that the largest Ribbon Seal that he had seen was a pregnant female 1·8 m nose to tail length and weighing 148 kg, 7·3 kg of this being the foetus she was carrying.

Ribbon Seals, as their name implies, have a very distinctly patterned coat as illustrated. Adult males are dark chocolate brown with wide, white or yellowish-white ribbon-like bands round the neck, the hind end of the body, and forming a large circle round each foreflipper, although there can be variation in detail. Females are paler and the bands are less distinctly marked. After the moult of the white natal coat the

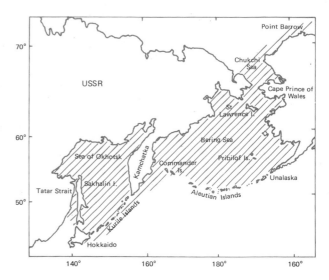

young seals are blue-grey dorsally and silver-grey ventrally, rather like a young Hooded Seal. The characteristic ribbon pattern is first seen in young males after the moult at two years old, and the full adult pattern is reached at three years old (Naito & Oshima, 1976).

Ribbon Seals are apparently more streamlined than other phocids of the Bering Sea. They have a long flexible neck, and when disturbed on the ice hold their heads up high looking for the intruder, though in air their sight appears poor and they can be approached with ease. Their fast sinuous movement over the ice is mentioned in the chapter on locomotion, and the curious tracheal air sac of males is mentioned in the chapter on the respiratory system.

BREEDING

The pups are born on the ice between the beginning of April and early May, though most births are between 5–15 April, with, of course, some variation due to weather conditions (Burns, 1981). At birth the pups are about 90cm long and weigh approximately 10·5kg. They are born in a coat of long white hair which is shed by the time they are five weeks old. They are suckled for 3–4 weeks and by the time they are weaned they have more than doubled their birth weight to about 27–30kg. During this period the pups are usually left unattended on the ice, though the mothers may be close by in the sea. They are weaned abruptly when the mother leaves them. The mother continues to feed throughout lactation. After weaning the pups spend two or three weeks learning to dive and feed themselves, and they become reasonably proficient at this by mid-June when the sea ice of the Bering Sea is disappearing.

Mating of the adults occurs shortly after the pups are weaned, with a peak during the first week of May.

The males are probably polygamous, but only approach the females who are in oestrous, and are not around in the earlier parts of the breeding season. Nearly all females are sexually mature when four years old, though some may be mature at two or three years. Males are sexually mature when five years old, though again some mature earlier. They are in breeding condition from March to mid-June (Burns, 1981). Ribbon Seals moult between late March and the end of July, subadults having finished moulting by mid-May, at which time the adults are just starting to moult. The normal life span is thought to be about 20 years, though the maximum may be in the order of 30 years.

FEEDING
There is little information on feeding habits, especially from seals taken in midsummer when the Okhotsk and Bering Seas are free of ice, and as the adults sink when killed it makes collection of material difficult. Seals collected in May and June, probably mostly from the ice, had less food in their stomachs than those taken in the water. In the Bering Sea in spring, shrimps and crabs are taken, followed in order of abundance by fish and cephalopods (Burns, 1981).

Other collections of seals made in spring have shown that animals in the southerly parts of the Bering Sea eat large numbers of pollock (*Theragra chalcogramma*) and eelpout (*Lycodes* sp), while those in the north consume the locally occurring arctic cod (*Gadus macrocephalus*) and appear to select the larger individuals. Most of the stomachs of the seals collected were empty, but the food eaten could be identified from the otoliths and other hard parts that remained (Frost & Lowry, 1980).

Ribbon Seal, Phoca fasciata. *Adult male, clearly showing the ribbon markings. Photo John Burns.*

Hooded Seal *Cystophora cristata*

DISTRIBUTION

The Hooded or Crested Seal is an animal of the North Atlantic and Arctic seas. It prefers deep water and thick, drifting ice floes, and is not often to be found on firm ice. Hooded Seals occur mainly from Bear Island and Spitsbergen to Jan Mayen, Iceland, Denmark Strait, Greenland, the east coast of Baffin Island and Labrador (see map). They are for the most part, solitary animals, but during the breeding season, in March, they collect in widely scattered groups of families. These breeding concentrations occur in the region of Jan Mayen (the 'West Ice'), to the north of Newfoundland (the 'Front') with other smaller groups in the Gulf of St Lawrence, and in Davis Strait at 64°N latitude. The herds on the West Ice and Front have been known for some 150 years, but although the Davis Strait group was known to nineteenth century whalers, it has only been rediscovered as recently as 1974.

Commercial sealing takes place on the West Ice and the Front and it is believed that it is only through replenishment of the Front herd from the Davis Strait group that has prevented the Newfoundland herd being exterminated (Sergeant, 1974, 1976). Recovery of tagged specimens and the similarity of the dates of the breeding season in all the breeding concentrations suggest that there is much mixing between the herds. A second concentration of Hooded Seals takes place in July and August when the animals moult. The only known moulting grounds are in the Denmark Strait, and further north off the Greenland coast between 72–74°N. It has been shown by tagging that the Newfoundland animals frequently reach the southern Greenland coasts and it seems that the Denmark Strait moulting group is formed mostly from the western North Atlantic animals. Presumably the Jan Mayen animals form the more northerly moulting patch. At the end of the summer the seals disperse again to their solitary winter habitats, about which little is known.

Although primarily Arctic animals, stragglers have been recorded from further afield – some half dozen or so down the Atlantic American coast, including one as far south as Cape Canaveral, Florida; and along the Canadian Arctic coast as far west as Herschell Island. On the other side of the Atlantic they are not uncommon off the Norwegian coast and a single pup has been born there (Øritsland & Bondø, 1980); there are about eight occurrences in the British Isles, three from the French coast of the Bay of Biscay, and two from Portugal. Some at least of these are very young animals. One from Hendaye, near the French–Spanish border, captured alive in July 1978, was a young female 128cm in nose to tail length, which died after

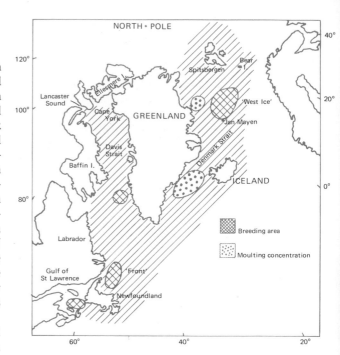

two days in captivity, probably from a nematode (*D. spirocauda*) infection of its lungs. The most recent Portuguese specimen, stranded in June 1979 was from the south of Portugal, on the Algarve coast. It was a male, only 104cm in length (?nose to tail), weighed only 27kg and was still in its first 'blue-back' coat (Pouvreau *et al.*, 1980).

DESCRIPTION

Adult males reach a nose to tail length of about 2·5–2·7m, and a weight of about 408kg, the females being slightly smaller (2·2m, 350kg). The colour of the adult is grey, covered with black patches of irregular size and shape, from 50–80cm^2 on the back, to much smaller spots on the neck and abdomen.

The most striking feature of the Hooded Seal is the presence in the adult male of an enlargement of the nasal cavity to form an inflatable crest or hood on the top of the head. When not inflated, the hood, which starts just behind the level of the eyes, is slack and wrinkled, and the tip hangs down in front of the mouth, but when blown up it forms a high cushion on top of the head, and may be twice the size of a football. The hood starts developing when the young male is about four years old, and continues increasing in size with age and increasing body size. After the age of thirteen years, and with a body size of more than 220cm, the males possess large hoods, though there is always variation.

The hood is inflated with air, the nostrils being closed before inflation, and in this respect it is different from the method of inflation of the Elephant Seal proboscis where the nostrils remain open. The hood may be inflated when the animal is disturbed, or during the pupping and mating season. But inflation frequently occurs when the animal is lying quietly on the ice, and the seal appears to 'play' with its hood, moving the air between its anterior and posterior parts.

The hood is an enlargement of the nasal cavity, it is divided into two by the nasal septum, and lined by a continuation of the mucous membrane of the nose. The skin of the hood is very elastic, and the hairs are shorter and grow less thickly than on other parts of the body. It has no blubber, this being replaced by elastic tissue, but there are muscle fibres, particularly round the nostrils, where they form an annular constrictor muscle.

Hooded Seals also possess the ability to blow from one nostril a curious, red balloon-shaped structure. This balloon usually protrudes from the left nostril, is formed from the very extensible membranous part of

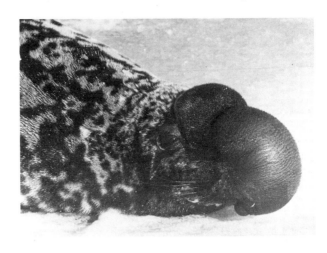

An adult male Hooded Seal showing the inflatable hood. Photo. Fred Bruemmer.

Adult Hooded Seal, Cystophora cristata. *Part of the hood can be seen, but is mostly obscured by the curious red 'balloon' blown from the left nostril. Photo. B. Bergflødt.*

the internasal septum, and may reach at least the size of an ostrich egg. The animal causes this balloon to appear by closing one nostril, and blowing air into the hood. Only the side of the hood with the closed nostril will start to inflate, and if the hood is kept low by muscular action, the pressure causes the elastic internasal septum to bulge, and eventually to appear as a balloon through the open nostril of the other side.

Both the inflation of the hood, and the extrusion of the balloon have been seen when the animals are disturbed by humans, or excited during the mating season. But this activity of the hood and balloon also occurs when the animals are completely undisturbed (Berland, 1966).

BREEDING

The pups are born in the second half of March. They are about 1m long and weigh about 15kg. They shed their light grey embryonal hair before birth, and this is swallowed by the foetus and excreted into the amniotic fluid as greyish felted discs. When born the pups have a very beautiful coat which is silvery blue-grey dorsally, darker on the face, and separated by a sharp line of demarcation from the creamy white ventral surface. The hairs do not lie quite so flat as in other seals, and this gives a furry appearance to the coat. It is these young animals, or 'blue-backs' as they are called in the fur trade, that are hunted for their skins. Pups are suckled for only ten to twelve days, during which

time they grow rapidly. At this time the adult male lies out on the ice near the female. Mating occurs at the end of lactation, the adults then returning to the sea to feed, while the pups remain on the drifting ice floes for about another fortnight, before they too take to the sea. Most females reach sexual maturity at three years, and produce their first pup when four years old. There is a delay of 3·5–4 months in the implantation of the blastocyst (Øritsland, 1964).

Although solitary for most of the year, family groups of a bull, cow and pup are formed during the breeding period, and the adults will fiercely defend their pup. The next gathering of seals is in July and August when they assemble for moulting, as previously mentioned. After this they leave the area, and their movements for the rest of the year are uncertain.

MORTALITY, PREDATORS AND PARASITES

Apart from predation by man, ice accidents and attack by polar bears cause death. The usual internal parasites are found, though it is very unlikely that these are fatal. Little is known of the diseases that attack the animals in the wild, though tuberculous lungs and an infection of the cranial cavity have caused death in captive animals. An adult male Hooded Seal, known as 'Hansi', was captured off the Faeroes in 1954 and held in captivity at Bremerhaven for just over a year. The animal was healthy until it got the cranial infection that caused its death, but dissection revealed that the

Young Hooded Seal in its 'blue-back' coat. Photo. J. Boulva.

atlas and axis were firmly fused to each other, and the atlas to the skull. Also the ninth and tenth, and the eleventh and twelfth thoracic vertebrae were fused together, so probably the seal had recovered successfully from some accident (Ehlers *et al.*, 1958). The increase in the fishing industry, man taking large amounts of small pelagic fish in the northwest Atlantic, may well be depriving the seals of their food and thus affecting their numbers.

FEEDING

The seals do not feed during the breeding season, so that when stomachs are easily available they are empty. At other times of the year it is less easy to see what the seals are eating. Greenland halibut, redfish, capelin, cod and squid have been found in stomachs and it has been suggested that the seals dive to about 180m (*c*. 100 fathoms) to get their food.

EXPLOITATION

Commercial sealing of Hooded Seals is based mainly on the newborn pups, or blue-backs, that are killed as they drift south on the ice floes. The pups are hunted for their beautiful coats, and oil and skin are also taken from the adults. The leather is not so tough as that from the Bearded and Harp Seals, and also, being porous it is less popular for kayak construction.

Hooded Seals occur further out to sea than Harp Seals, but pup at approximately the same time, so that the same sealing expedition can be used for catching both kinds of animal. Consequently it is almost impossible to distinguish between them in the earlier literature on sealing.

Within the International Committee for Northwest Atlantic Fisheries (ICNAF) there is a special committee on seals that recommends the quotas. It is prohibited to catch Hooded Seals in the Gulf of St Lawrence, and also during the moulting haul out in Denmark Strait. At Jan Mayen sealing takes place between 20 March and 5 May and a quota of 30 000 was set in 1972. Off Newfoundland 6000 seals may be taken by Norway and 6000 by Canada between 12 March and 29 March, and from 30 March to 24 April a further 3000 seals may be taken by both nations (Sergeant, 1976, Mercer, 1976). Frequently, however, the animals disperse and the full quota is not taken. Canada and Norway have agreed to restrict the kill of adult females to less than 10 per cent of the total catch.

DEMOGRAPHY

In the Newfoundland region about 30 000 pups are produced annually – which indicates a total population in this area of about 100 000 animals. The Davis Strait group has about 40 000 adults and produces about 11 300 pups annually. The Jan Mayen group produces about 50 000 pups and there are about 225 000 subadults and adults there before the breeding season (Øritsland in litt.).

Bearded Seal Erignathus barbatus

DISTRIBUTION

Bearded Seals are circumpolar in their distribution, and are to be found all along the European, Asiatic and North American coasts of the Arctic Ocean, and on all the associated islands (see map). Their northernmost limit seems to be about latitude 80–85°N, and they occur as far south as the Sea of Okhotsk, Tatar Strait, Sakhalin, Hokkaido and Hudson Bay. A few animals reach the Gulf of St Lawrence and sometimes even as far south as Cape Cod, and isolated stragglers occasionally appear on the northern Norwegian coasts, and from Scotland, Norfolk and Normandy.

Two subspecies of *Erignathus* have been proposed. *E.b. barbatus* (Erxleben, 1777) which occurs from the Laptev Sea westward to the Hudson Bay area, and *E.b. nauticus* (Pallas, 1811) from the Laptev Sea eastward to the Canadian Arctic, although the exact limitations of the two forms have never been clearly defined. Two separate examinations of groups of over two hundred skulls have not solved the problem of subspecies, as work by Manning (1974) suggests the sub-

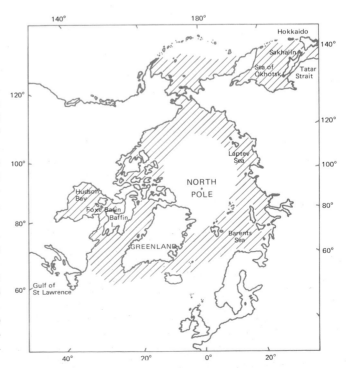

species are valid, whereas the opinion of Kosygin & Potelov (1971) is in favour of a single widely spread species.

These seals are not gregarious and are not found in very large numbers in any one locality, although they become more obvious in summer when they haul out on the beaches. Even on land or ice they always lie with their heads very close to the water so they can escape quickly. They prefer shallow waters near to coasts that are free of fast ice in winter, gravel beaches, and ice floes that are not too far out to sea. They are associated with moving ice floes with open leads between them, but they also maintain breathing holes in thin ice when necessary. This association with drifting ice floes for much of the year means that the seals may move considerable distances north and south with the ice.

DESCRIPTION

Bearded Seals are the largest of the northern phocids. Adult males and females reach approximately the same nose to tail length of about 2·25m, and weigh about 250kg, though for their size they have a disproportionately small head. The colour is also the same in the two sexes, and is unremarkable, being grey dorsally, slightly darker down the midline, with a brownish or reddish tinge on the head, and slightly lighter grey ventrally. The most characteristic feature of the Bearded Seal, and that which gives the animal its name, is the great profusion of long, very sensitive, glistening white moustachial whiskers which make the animal recognizable from some distance away. These differ from the whiskers of other phocids, except the Monk Seals, in being straight and not beaded. They are also curious in curling, sometimes forming tight spirals at the tips, particularly when they get dry. Another character that the Bearded Seal shares with *Monachus* is the possession of four mammary teats, other phocids having only two. The Norwegian sealers' name of 'Square flipper' which they used for the Bearded Seal is due to the square shape of the foreflippers, as the third digit is slightly longer than the others. The nails on the foreflippers are strong, those on the hind flippers slender and pointed.

Bearded Seal, Erignathus barbatus, *showing the fine profusion of whiskers, and also the rather square foreflipper. Photo. Ivor Christens.*

BREEDING

No particular concentration of animals takes place at the breeding season, and the pups are born in the open on ice floes. In Alaska the average pupping date is about 20 April (Burns & Frost, 1979), in the Canadian Arctic it is about 1 May, and in the Barents Sea the pups are born between 25 March and 12 May (Mansfield, 1967a, McLaren, 1958a, Potelov, 1975). At birth the pups are 1·2–1·3m and weigh about 30–40kg. They are covered with a short, dense woolly coat of greyish brown hairs with, quite frequently, a lighter region down the centre of the back and some white on the snout and crown. After a short time, probably at the end of lactation, the pups moult and exchange this woolly coat for one of stiff hairs like that of the adult, and of similar colour, although sometimes slightly spotted.

Lactation lasts for 12–18 days, but the pups may start to take solid food such as shrimps before the end of this period. During lactation the female remains close to her pup and may defend it if necessary.

Most matings take place during May both in Alaska and in the Canadian Arctic, and from the end of March to the end of May in the Barents Sea, the females ovulating at the end of lactation. The blastocyst becomes implanted after a delay of about two months, most becoming implanted in July. The total gestation period is therefore eleven months, and the active gestation period nine months (Burns & Frost, 1979). It is probable that most males mate for the first time when six or seven years old, and most six-year-old females are bearing their first pups. Times given for the adult moult vary between March and June, but as it will undoubtedly take place after mating, the time will vary accordingly.

Males in the breeding season make a long drawn out warbling note which ends in a low moan or sigh. This sound is apparently made only during courtship and may indicate a proclamation of territory or of breeding condition (Ray *et al.*, 1969).

MORTALITY, PREDATORS AND PARASITES

Bearded Seals are said to be very curious animals, but if suddenly frightened they seem to become paralyzed with fear. When diving they roll forwards, exposing first the back and then the hind flippers. Man and polar bears are the only real enemies of the Bearded Seal (Smith, 1980), although they become heavily parasitized.

FEEDING

Food consists almost entirely of bottom-living animals taken in shallow water of up to 130m or less. Animals eaten include shrimps, crabs, holothurians, clams, whelks, snails, octopus, and bottom fish such as sculpin, flounder and polar cod. They appear to have a special liking for whelks, although the shells are never found in the stomach. In Alaskan waters crabs form an important food source. Although the exact details of the feeding habits of Bearded Seals have not been recorded, the animals show a number of similarities with the walrus and may use suction to a certain extent. Both are bottom feeders with a large array of moustachial whiskers that could be used to help sort out the smaller food animals; both eat molluscs and it is very seldom that remains of shells are found in the stomach, although opercula are found in Bearded Seal stomachs; and both get extremely worn cheek teeth.

EXPLOITATION

Bearded Seals are of great importance to the coastal native in Alaska, and round the shores of the Bering, Chukchi and Beaufort Seas. The meat is taken for food for men and dogs, and some Eskimos are said to dislike it fresh or boiled, and prefer it frozen and rather high. Of the carcass, after the skin and blubber have been removed, 70 per cent is usable meat. The liver, however, which is said to be very lobate and rather like a kidney in appearance, is seldom eaten even by dogs – it frequently contains sufficient concentrations of vitamin A to be poisonous, so that quite severe illness can result from its consumption. Dogs will not touch the liver when fresh, but may eat it when frozen (see Hypervitaminosis, p. 171).

The seal body provides other useful by-products. The hide is strong, durable and elastic, and is used for footwear, heavy ropes, dog harness and kayak covering, but is too thick to be used for clothing. Seal products were put to considerably more use in times past. The intestine was used for windows and waterproof clothing, the blubber was used as fuel for lamps, and dyes and waterproofing compounds were made from the blood (Burns & Frost, 1979).

In the 18 months between 1 Jan 1977 and 30 June 1978 slightly over 6000 Bearded Seals were taken by Alaskan natives. In the Bering and Chukchi Sea areas, as well as use by the natives, the seals are also caught commercially by Soviet sealing vessels (Burns & Frost, 1979).

DEMOGRAPHY

There are no reliable estimates of the total population of this widely distributed seal. The Bering and Chukchi Seas support reasonable numbers, and they are also plentiful where conditions are suitable in the Canadian Arctic, with the greatest concentrations in eastern Hudson Bay and Foxe Basin. In one area off Baffin Island there is thought to be about one Bearded Seal to every 13 Ringed Seals (T. G. Smith in litt.), and the world population has been suggested to be in excess of 500000 individuals (Stirling & Archibald, 1979).

SOUTHERN PHOCIDS

Monk seals

There is only one genus of Monk seals, and this is divided into three geographically widely separated species – the Mediterranean Monk Seal *Monachus monachus* from the Mediterranean and nearby areas, the West Indian Monk Seal *M. tropicalis*, and the Hawaiian or Laysan Monk Seal *M. schauinslandi* from the Hawaiian islands.

Mediterranean Monk Seal
Monachus monachus

The presence of seals in the Mediterranean has probably always been known to the inhabitants of the area, and many myths and stories have grown up about the animals. The Greek writers Homer, Pliny and Plutarch knew and wrote of seals, and it is presumed that it was to the Mediterranean Monk Seal that they referred. Aristotle also, examined one with care, and gives quite an accurate account of its anatomy. Seal skins were used as clothing by the poor fisher folk, but the skins were also believed to have other less tangible uses. As seals were not struck by lightning, it followed that a seal skin tent would be a protection against lightning; a seal skin pulled around a field and then hung on a door would protect that field from hail stones; shoes of seal skin would drive away gout; and the right flipper of a seal under the head at night would be a cure for insomnia (Keller, 1887). Coins from about 500 BC show a picture of a seal and the previous abundance of the animals is shown by the various districts and towns that have taken their names from the seal, such as the ancient Greek district of Phocis, and the present towns of Foca in Turkey, and also in Yugoslavia. The Greek word for 'phoca' is derived from the Sanscrit for a swollen or plump animal, and the same origin can be seen in *Phocoena*, the porpoise.

DISTRIBUTION

There has been an upsurge of interest in the Mediterranean Monk Seal recently, resulting in several surveys of its distribution and numbers, and culminating in newsletters put out by IUCN and the First International Conference on this seal, held in Greece in May 1978. Although there are still gaps in our knowledge, so much information has been amassed that the picture of Monk Seal distribution has changed radically over the last ten years. Much of the recent information is summarized by Sergeant *et al.* (1978) who have also included a comprehensive list of references.

The main centre of this Monk Seal is still in the area of classical writing. The greatest numbers are to be

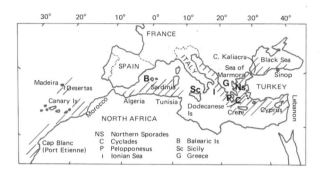

found in the eastern Aegean Sea – in the area of the Greek Dodecanese Islands and the adjacent coasts of Turkey, with numbers decreasing as one moves outwards from this centre. Lesser populations are found in the Sea of Marmara, on the islands of the Northern Sporades and the Cyclades; Crete, Pelopponesus and the Ionian Sea, and along the southern coast of Turkey, and the coasts of Lebanon and Cyprus. Further small groups occur along the Mediterranean coasts of Morocco, Algeria and Tunisia. There are still a few animals on the Balearic Islands, Sardinia and Sicily, but they have just about vanished from Corsica. Relatively few animals appear to remain in the Black Sea, and then only in the western parts. There are perhaps half a dozen seals at the Bulgarian Cap Kaliacra, and some animals are known along the Turkish coast of the Black Sea as far east as Sinop.

Outside the Mediterranean there is a population on the Desertas Islands, and another just north of Port Etienne in Spanish Sahara. This latter colony may also be referred to in the literature as Rio de Oro, Cap Blanc or La Guera. Some seals have been seen on the Canary Islands and on the Saharan coast opposite them, and there have also been a few recent records from the Azores, though an expedition in 1979 saw no Monk Seals (Ronald, 1980).

Pressure from increasing human numbers and traffic in the Mediterranean has resulted in the relatively recent extinction of this seal from the mainland coasts of Spain and France, the Crimea, and the coasts of Palestine and Egypt.

It was mentioned above that the main centre of this seal is in the Greek islands. There are, however, hardly 200 animals in this area, perhaps 50 at Port Etienne, 50 in the Black Sea, 50 in the Cyprus–Lebanon area, 50 in the Desertas, 100 in the Morocco–Algeria area, and pitiful numbers of 20 or so at various other places like the Yugoslavian coast, Tunisia etc. Sergeant *et al.* (1978) suggest that the total population of this seal lies between 500 and 1000, and is almost certainly declining.

This seal is very sensitive to disturbance by humans, and of course to their inevitable pollution. It seems likely that its original habitat was sandy beaches, but the popularity of such beaches to humans has now effectively restricted the Monk Seal to small islands, uninhabitable by man because of lack of water, and to cliff-bound rocky coasts. The Monk Seal seems to be protected, at least in theory, in most, if not all of the countries within whose limits it occurs, though it is still regarded as a pest by the Greek fishermen. The current interest in this Monk Seal has resulted at the moment in the planning of a series of reserves in the Dodecanese Islands, and the establishment of a reserve in Turkey, on the Kapidaz Peninsula, in the Sea of Marmara.

DESCRIPTION

The nose to tail length of adult animals is about 280cm and they may weigh up to 350–400kg. They are generally dark brown to black in colour, a little lighter ventrally, or with a variably occurring whitish patch on the belly.

BREEDING

Pups are usually born in the shelter of caves or grottoes. Newborn pups may be found between May and November, but the peak of the pupping season is in September and October. At birth the pups weigh about 20kg and have a nose to tail length of *c.* 80cm. They have a black woolly coat that is moulted at weaning time, which is believed to take place when the pup is about six weeks old. Little is known about the details of the life history, but mating has been seen once on the Desertas Islands at the beginning of August. The female concerned presumably had her pup at the end of May.

PARASITES AND PATHOLOGY

A diverse fauna of roundworms and tapeworms has been recorded from the stomach and intestine, and the condition of two skeletons shows that these animals suffered from ankylosing spondylitis and osteo-arthritis of the lumbar-sacral joint respectively (Zorab, 1961).

FEEDING

Local fish and octopus are eaten. This seal is possibly not a very deep diver as fishermen report that few fish are taken out of nets set deeper than 30m.

Mediterranean Monk Seal, Monachus monachus, *a drawing by Helmut Diller. Bruce Coleman Ltd.*

West Indian Monk Seal *Monachus tropicalis*

Although the West Indian Monk Seal was plentiful enough to have been captured for oil from at least the seventeenth century to about the end of the nineteenth, their existence now is doubtful. The first reference to this seal is that given in the second voyage of Columbus when, in August 1494, he anchored by a rocky island south of Haiti and killed eight 'sea wolves' that were asleep on the sand (King, 1956), but it was more than 350 years after this that the animal received a scientific name.

DISTRIBUTION
The range of this seal was from the Bahama Islands and the Yucatan Peninsula into the Caribbean Sea where it must have lived on any suitably remote sandy beach or island, as the many Seal and Lobos Cays indicate. Four seals were reported at Arrecife Triangulos, Campeche in 1948, and the last remaining colony was known on Serranilla Bank, about midway between Jamaica and Nicaragua, but there have been no reports of these animals since 1952 (Rice, 1973). In 1973 an aerial survey was made of islands and atolls from Campeche to Nicaragua and Jamaica but no seals were seen (Kenyon, 1977). All island groups showed evidence of fishermen, and as Monk Seals and fishermen are not compatible, it is not believed that any Monk Seals still exist here. However, further records from the 1970s of animals that could possibly have been Monk Seals are known from the East Bahamas Islands.

A search in 1979 did not reveal any animals, nor any trace of them, but there still seems to be hope that they are not totally extinct (Ronald, 1980).

DESCRIPTION
Details of the appearance and habits are very incomplete. The adults are about 2–2·4m in length, greyish brown on the back, shading to yellowish white ventrally (Ward, 1887, King, 1956). It is interesting to note that green algae have been found growing on the fur as in the Hawaiian Monk Seal.

BREEDING
The pups are born about the beginning of December, in soft black woolly coats, but there is no information on mating or lactation.

FEEDING
The food is believed to be fish.

Hawaiian Monk Seal
Monachus schauinslandi

DISTRIBUTION
The Hawaiian Monk Seal is confined to the atolls and islands of the Leeward Chain which stretches northwest of the main Hawaiian islands. The seals breed on four of the western atolls – Pearl and Hermes Reef, Lisianski Island, Laysan Island and French Frigates Shoal. Twenty years ago they bred on Midway and Kure Atoll as well, but constant disturbance has resulted in these areas being abandoned (Kenyon, 1973a, 1980).

   Although the seals spend much time on their breeding islands, and travel between one island and another, they also range widely in the Hawaiian chain, and even outside it. They reach other islands in the chain, as far as the main Hawaiian islands, some 1165 km from the nearest breeding place, and a pup tagged on Laysan in March 1968 was seen in July 1968 on Johnston Island, 1013 km southsouthwest of Laysan.

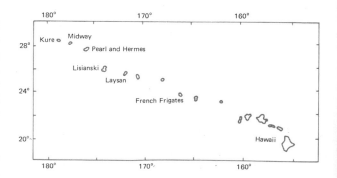

This seal was seen at intervals until December (Schreiber & Kridler, 1969). Occasional individuals presumably spend a considerable time at sea, as is shown by the growth of green algae on their fur, which disappears again after some time on the beach.

DEMOGRAPHY

Total numbers of this seal are difficult to know accurately as so many small islands are involved, and at any time an unknown number of animals may be at sea. It has been estimated, however, that over the last twenty years the population has declined by nearly 50 per cent. At present there are not thought to be more than 700 Hawaiian Monk Seals left (Kenyon, 1980). The status of the whole population is uncertain; on some of the islands there seems to be a decrease, though only French Frigates Shoal shows an increase. The seal is totally protected, and in addition, all the Leeward Islands of the Hawaiian chain (except Kure and Midway) are included in the Hawaiian Islands National Wildlife Refuge.

DESCRIPTION

Females grow to a slightly larger adult size than males, and have considerably more variation in weight. An adult male has a nose to tail length of about 214 cm, and weighs 173kg, while a female may surpass him in length by about 20 cm, and in weight by as much as 100 kg, especially just before and just after the birth of the pup. Most of this excess weight is lost by the end of the lactation period. Adult animals are light silvery grey ventrally and slate grey dorsally, males tending to be a little darker than females.

BREEDING

Pups are born over an extended period from late December to mid-August, but most of them between the middle of March and the end of May (Kenyon & Rice, 1959). Birth takes place high up the sandy beaches, near the shrubs and away from the tide line. Females congregate at this time on sheltered beaches, tolerating each other, but repelling courting males. The pups are about 16kg in weight and 100 cm in nose to tail length at birth. Their coat is of soft black hair which is moulted at approximately three to six weeks to one that is grey or blackish dorsally and silvery grey ventrally. Pups will swim when they are about four days old, but venture further afield, sometimes with their mothers, as they get older. At about six weeks old the pup has its complete dentition, and lactation is believed to continue until this time, the pup doubling its birth weight by 17 days, and quadrupling it by about six weeks.

The female apparently fasts during lactation, remaining with her pup for the entire period. During this time she loses about 2 kg for every kilogram gain by the pup – losing some 90 kg. When the female can no longer feed her pup she goes off to sea to feed, and the pup is left to fend for itself, relying on its blubber until it can catch its own food, and gradually moving about the islets.

Mating has not been observed, but possible courtship behaviour has been seen between April and November. Males were seen attempting to mount females in November (Wirtz, 1968). Tagging records over nearly two years showed that some females produced pups in both the years involved, whereas other females reproduced only once in that period (Wirtz, 1968).

Hawaiian Monk Seal, Monachus schauinslandi. A female with a newborn pup lying out on the white coral sand of French Frigates Shoals. Photo. C. Whittow.

The adult animals moult between May and November, chiefly in June, although the females do not moult until after the pup is weaned. Large patches of epidermis, with hairs attached, are shed as in *Mirounga*.

MORTALITY AND PREDATORS

These seals suffer very much from human disturbance, and many pups die before weaning, probably from malnutrition due to constant disturbance of their suckling time. Sharks are a significant cause of death, and entanglement in scraps of discarded fishing nets leaves lacerations and scars and may cause death. Many sick and dying seals were observed on Laysan Island during the first half of 1978. Liver samples from some of them were positive for ciguatoxin – a toxic substance produced by a protozoan that is concentrated through the food chain so that the flesh of affected fish is highly toxic to mammals. Ciguatoxin outbreaks occur after coral reefs have been disturbed, and such activity at Midway may well have been responsible (Kenyon, 1980).

FEEDING

Food consists principally of eels and other reef fish, and cephalopods, probably octopuses. Feeding is at night as the prey remain hidden during the day. The food animals are shallow water bottom-living forms.

ECOLOGY

In spite of being one of the most tropical of pinnipeds, living on sunbaked sandy beaches, the Hawaiian Monk Seal shows no obvious adaptations to this habitat. Its body temperature is about 36·9°C, within the temperature range of other seals; the dermis is highly vascularized and the hairs of its coat are short, so that heat may be lost this way, although this is hardly a particular adaptation. It appears to cope with its environment by behavioural means, lying quietly in shade or on the wet sand during the day, being extremely inactive, and feeding at night (see chapter on temperature). The primitive nature of this seal is mentioned in the chapter on classification.

Antarctic seals

The four genera of truly Antarctic seals might be said to share the Antarctic between them, and they manage this without serious interference with one another, because in their distribution and food habits they hardly overlap at all. The Leopard and Ross Seals differ widely in their diets, the Leopard Seal ranges further to the north, while the Ross Seal stays in the southerly, heavy pack ice. The two most abundant seals, the Weddell and Crabeater, have been likened in their way of life to the two most common penguins of the region, the Emperor and the Adelie. The Weddell Seal and the Emperor penguin are both fish eaters, of littoral distribution, non-migratory, and mostly remaining as far south all the year as open water will allow. The Crabeater Seal and the Adelie penguin feed on crustaceans, are pelagic, the penguin definitely migrates, but the seal less definitely, and both keep in touch with the pack ice.

The Antarctic Treaty which came into force in June 1961 provides protection for seals south of 60°S. The Ross Seal and all fur seals are protected, and particular areas have been set aside where all seals are protected. Seals may be taken in limited quantities, under permit. As the conservation of seals at sea does not come under the Treaty, special arrangements have been made under a Convention for the Conservation of Antarctic Seals, signed in 1972, which is complementary to the Antarctic Treaty. The Convention

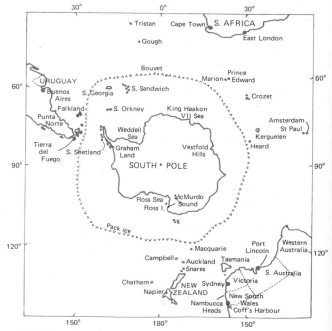

The Antarctic and neighbouring regions. The outer (spring) limit of the pack ice is indicated. The map shows the area inhabited by the four Antarctic seals with their different circumpolar distributions.

applies to the sea areas south of 60°S, and was initiated with the idea of possible future sealing in mind. In this instance, however, the regulations have been brought in before the animals have been made virtually extinct. These regulations set permissible catch limits of 175 000 Crabeaters, 1200 Leopard Seals and 5000 Weddell Seals in any one year, but allow for an annual review of the situation once an industry begins. Ross Seals, Elephant Seals and all fur seals are completely protected, as is also the stock of adult Weddell Seals when they are on the fast ice. 'There is a closed season between 1 March and 31 August, and a series of six sealing zones, each of which is to be closed to sealing, from year to year, in rotation. Three Seal Reserves are listed in which it is forbidden to kill or capture seals' (Laws, 1973). The Convention came into force in 1978.

Weddell Seal *Leptonychotes weddelli*

DISTRIBUTION

The Weddell Seal is the most southerly of the Antarctic seals. It is circumpolar and coastal, being normally found on fast ice within sight of land. It is not an animal of the pack ice. These seals do not undertake major migrations, but they do move with the ice front. In this way they move north before the onset of winter so that the more limited number of breathing holes in the more rigorous southerly parts of the range do not get too crowded. A reverse movement occurs in the spring and at any one place the number of animals hauled out varies as there is a spring increase of pregnant females, an increase of males at the end of the breeding season, and aggregations of young animals at the end of summer (Kooyman, 1968). Some Weddell Seals are to be found in the most southerly parts of their range throughout the year, though they are less obvious in winter when they presumably stay in the water where the temperature is more constant. The winter air temperature may be −20°C, but the water is about −1°C and the surface of the sea may be frozen for seven or eight months to a depth of several metres.

As well as breeding on the Antarctic mainland, there are colonies on the South Shetlands and South Orkneys. A small breeding colony in Larsen Harbour, South Georgia forms the northern breeding limit. Animals are occasionally seen at Macquarie, Heard, Falkland and Kerguelen Islands, and as far away as Uruguay, New Zealand and South Australia. Immature Weddell Seals not infrequently wander into areas from which they cannot escape as the winter weather descends. Weddell Seal bodies, mummified by the cold wind and dry conditions have been found in dry valleys in the McMurdo Sound area. Some of these remains have been dated as about 1400 years old, though seals are still getting stranded in such situations today.

DESCRIPTION

The adult male seal has a nose to tail length of about 2·5m. Adult females tend to be slightly longer, about 2·6m, and a large female measuring 3·2m nose to tail and weighing 411kg has been recorded. The head appears small in relation to the size of the animal and has been described as rather cat-like. In colour there is, as usual, a certain amount of variation, but a typical animal is black dorsally, white streaks and splashes increasing in number and extent from the dorso-lateral area to the ventro-lateral area where white is predominant. Ventrally the coat is grey with white streaks. During the summer the colour fades to a rusty greyish brown, but never bleaches to the whiteness attained by the Crabeater.

BREEDING

The adult animals may moult any time during the summer from December to March, and both feeding and swimming are continued. A few days before the pups are born, groups of pregnant cows haul out on the sea ice, usually choosing areas where existing cracks in the ice make exit holes easy to maintain. Females tend to return to the same area each year and loose aggregations composed entirely of pregnant females are formed, the individuals being very widely spaced out on the ice. Pups are born between mid-October and mid-November, with a peak in the last half of October. They are about 1·5m nose to tail length at birth, 29kg in weight, and clothed in a greyish woolly coat with a darker stripe along the spine. This natal coat is moulted when the pup is about a fortnight old, the moult itself lasting for four weeks at the end of which time the young animal has a coat similar to that of an adult. Pups may enter the water early, at about eight to ten days, while still partly at least in the woolly coat, but presumably with sufficient blubber for insulation. They have also been recorded as entering the water at various times up to about four weeks old, the variation possibly being due to local conditions, or to the effect on the pups of the human observers (Tedman & Bryden, 1979).

Lactation continues for six to seven weeks, the milk gradually being augmented by small crustaceans. The pup doubles its birth weight in ten days, and at weaning weighs about 110kg. For about the first two weeks after birth the cow and pup remain closely associated. The pup will accompany its mother to cracks in the ice and remain in the vicinity while she spends several hours each day in the water. There is, however, no evidence that she is feeding at this time, and probably does not feed until after the pup is weaned. For the rest of the lactation period, at least at one site observed, most of the cows and pups entered the water at least once a day, although not always together. Most animals took to the water during the evening and night (Tedman & Bryden, 1979). The cows mate again towards the end of the lactation period. Only a single copulation has so far been observed, and in this instance it took place under water in McMurdo Sound on 7 December, and was seen via a television camera. This particular female was with a 43-day-old pup that was not yet weaned. Virgin cows mate for the first time at the beginning of their third year. Weddell Seals are possibly polygynous, and a sex ratio of about one male to ten females has been observed. The females appear to be concerned about their pups and are aggressive towards intruders and may snap at other females. At the breeding season the adult males are also aggressive, competing to be near the females, and many show wounds on their bodies and flippers. After the breeding season, when the females have mated and the pups are weaned, subadult seals are admitted into the rookery areas and the distance between the seals becomes slightly less.

PREDATORS, PARASITES AND PATHOLOGY
Weddell Seals seem to be relatively free of natural enemies, though some bear scars from wounds thought to have been inflicted by killer whales. There is the usual mortality amongst the pups due to being crushed in the ice or by an adult, and adult females in particular bear many scars on the ventral surfaces, probably due to coming out on to sharp ice in the early spring for pupping. Cracking of the skin round the anus has been observed, probably due to the action of the cold dry climate on skin that is frequently wet. A

Weddell Seal, Leptonychotes weddelli. *The champion diver, seen here on land. Photo. Nigel Bonner.*

certain amount is known about the internal troubles of the seal. Kidney stones have been found, uterine fibroids, and fatty streaking in the aorta; the gut harbours high infestations of tapeworms and roundworms, and lice have been recorded from the skin, though it does not seem that these parasites cause the host much trouble.

During the winter, when the seals spend most of their time in the water, and can be heard calling to each other below the ice, they keep breathing holes open in the ice by sawing and biting at it with their canines. The outer upper incisors and upper canines of Weddell Seals project more horizontally than in other seals, and are admirably positioned for abrading the ice. They do, however, get very worn, sometimes right down to the pulp cavity, when infection, leading to abscesses may occur. These abscesses will not heal, and while they may not lead directly to the death of the seal, the infection may spread to other tissues, including the brain, or excessive wear may impair the function of the tooth. In the latter instance, if the breathing hole cannot be kept open then the animal may not be able to reach the water to obtain food, or may even be trapped under the ice. After the seals are about eight or nine years old large numbers of them seem to show such dental problems and these could well contribute significantly to their mortality.

FEEDING
Fish makes up more than half of the total food eaten by Weddell Seals, the remainder being made up of cephalopods and various crustaceans such as some of the larger shrimps and amphipods. Both sluggish bottom fish and active midwater fish are eaten, and sometimes the fish are of considerable size. Where it has been possible to observe feeding habits – as at the artificial holes created, a Weddell Seal was seen to bring in a large notothenid fish, *Dissostichus mawsoni* that had a length of 1·5m and an estimated weight of 31kg, and to consume it entirely within about three hours. The same seal brought in another fish of similar dimensions the next night, but ate only half of it (Kooyman, 1968).

BEHAVIOUR
Although it lives in such a remote and hostile environment, probably more is known about some aspects of the Weddell Seal than about any other seal. The seal has no fear of man, rarely tries to escape, and its reactions are those of curiosity rather than fear. When disturbed on the ice its characteristic action is to roll on its side, with its belly facing the intruder and to hold its foreflipper up in the air. The methods of studying these animals on and under the Antarctic ice, and the great depths to which they dive, are given in the chapter on diving. Weddell Seals are not gregarious animals, and when hauled out on the ice they stay at reasonable distances from each other. They tend to come up singly at holes for breathing, and will usually leave if another seal comes to the same hole, or a fight may ensue. At larger pools many seals may surface at once. While being aggressive at breathing holes, seals make an assortment of noises – amongst which are trills and chirrups, and a rapid tooth clattering which they seem to do when they pass one another when entering or leaving a breathing hole. The trill has been interpreted as occurring when emotions are tense, and the chirrups possibly as a challenge or warning.

DEMOGRAPHY
Accurate population estimates are very difficult with an animal spread over such a vast area of remote and rigorous terrain. Estimates are frequently made by counting the seals, from the air, in a relatively small area, and then extrapolating to get the population of the whole region. This of course has obvious drawbacks, but it is difficult to know how it could reasonably be improved upon. Laws (1973) has reviewed recent estimates of populations of Antarctic seals and says that about 50000 is a reliable estimate for the number of Weddell Seals in the western Ross Sea and that a conservative estimate for the whole population would be 250–500000 and higher estimates giving a minimum of 750000 have been suggested (De Master, 1979).

Ross Seal Ommatophoca rossi

DISTRIBUTION
The Ross Seal is still one of the least known of the Antarctic seals, though records of its occurrence are becoming slightly more frequent. It is usually a solitary animal, with apparently local areas of abundance in its wide distribution. As with all the Antarctic seals it is circumpolar, and seems to occur most often in

pack ice of medium to high concentration, preferring the smaller, smooth surfaced floes, and often being found some distance from the coastline. Although there are records of Ross Seals occurring all round the Antarctic, there are slightly more in the Ross Sea area, and distinctly more in the King Haakon VII Sea. Recent censuses in this latter area (approx 0°–6°W, and up to about 55km (35 miles) off the coast) in January 1974 and 1975 have recorded the greatest densities of Ross Seals known so far. In ten days at the end of January 1975, 132 Ross Seals were seen, forming 32 per cent of the total seals counted in the area and giving a density of 0·55 seals per km² (Hall-Martin, 1974, Wilson, 1975c). The Ross Seal is virtually unknown outside the Antarctic, but a young female, about 1·6m long was seen at Heard Island in September 1953 (Ingham, 1960), and a young male 1·8m long came ashore at Beachport, South Australia in January 1978. A single fossil lower jaw ramus has been found in Early Pleistocene levels from Napier, New Zealand (King, 1973).

DESCRIPTION
Females with a nose to tail length up to 2·5m, and males up to 3m have been recorded. Weights range between 200–210kg. In colour the Ross Seal is a dark silver-grey dorsally and silver-white ventrally. Laterally, at the line of demarcation of the two colours there

are spots of grey and white, and between the face and foreflippers the light colour of the throat is marked by parallel streaks of grey. The body is plump, the head short and wide, and the neck is particularly thick and appears swollen under the chin. The mouth is rather small for the size of the animal, and the protruding eyes are the only external indication of the large size of the eyeballs, as the openings of the eyelids are unremarkable in length. The shape of the flippers with their very reduced nails is mentioned later.

When disturbed, the Ross Seal lifts its anterior end, opens its mouth widely and gives forth a series of trills and thumping noises. While doing this the particularly long soft palate of this animal is inflated forwards in the mouth and meets the back of the raised tongue, so that it looks as if there are two pink tennis balls at the back of the mouth (Ray, 1966). The Ross Seal can produce a range of loud bird-like chirps, trills, clucking and cooing sounds, and before making the sounds the throat is expanded and bulges outwards, the sounds appearing to be made with an almost completely closed glottis (Bryden & Felts, 1974). Some of

Ross Seal, Ommatophoca rossi *on the Antarctic ice. The photo shows the streaky colouring of the throat, and tears running from the large eye. The raised head and open mouth suggest that the seal is making its characteristic trilling sounds. Photo. Thomas G. Smith.*

the very strong pharyngeal muscles may be involved in making these sounds.

DEMOGRAPHY

Estimating the total population of Ross Seals is of necessity a very difficult task. Reviewing all the information available a figure of 100–150 000 has been reached (Hofman, Erickson & Siniff, 1973), but as mentioned above, there are local areas of abundance and other areas where the seals are very thinly scattered.

BREEDING

Nothing is known directly about the breeding habits, but the generally solitary way of life would suggest that probably association is in pairs rather than in larger breeding groups. A foetus 101cm long has been found at the end of September, and other records suggest that the foetus grows to about 100 cm during October (Øritsland, 1970). The lack of sperm in mature males captured throughout October, and the presence of recent corpora lutea and implanted blastocysts in females collected at the end of January, suggest that mating takes place possibly at the end of December, so that by extrapolation, the pups could be born perhaps in mid-November. The only well-documented record of a new born Ross Seal was made on 14 November 1978 when a mother and pup were seen on the ice in Dallman Bay, near the Antarctic Peninsula. The pup, a male, had a nose to tail length of 96·5cm and a live weight of 16·8kg. The presence of a soft, bleeding umbilical stump, and the unfrozen afterbirth indicated very recent birth. The pup was dark in colour dorsally, light ventrally and had striping on the throat similar to that of the adult. During the 30 minutes that the pup and its mother were observed, the pup was seen to vocalize, to crawl round the floe, and to move to an adjacent floe with total submergence of the body as it swam (Thomas *et al.*, 1980).

Other young pups estimated to be between 5 and 15 days old were collected between 18–21 November 1968 south of the Scott Islands. These pups, one of which still had an umbilical cord, had nose to tail lengths of 109–138cm, and weights of 40–75kg. The peak pupping season is believed to be between 3–18 November (Tikhomirov, 1975). This latter author gives early to mid-December as the mating period, and implantation takes place in late February to early March after a delay of 2·5–3 months. Moulting of the adults is in January, and they do not apparently go to sea to feed at this time.

PREDATORS AND PARASITES

They seem to have few natural enemies and are not heavily scarred, though some scars are attributed to attacks by Leopard Seals. Roundworms and tapeworms have been recorded from the gut, and lice from the skin.

FEEDING

Cephalopods, sometimes large ones, form the main food of the Ross Seal, although fish and krill have also been found in the stomach. For catching its swift moving and slippery prey the seal would have to be a strong swimmer, capable of rapid manoevres under water. The front limbs are more specialized and flipper-like in shape than those of most phocids, and yet compared with other Antarctic seals the bones and muscles are relatively small (Bryden & Felts, 1974). The large eyes are for the perception of movement in the dimly lit waters under heavy ice. The canines and incisors have delicate, sharply recurved points that would help secure the slippery prey, and all the muscles in the throat that are concerned with gripping and swallowing are very well developed (King, 1969b, Bryden & Felts, 1974).

Crabeater Seal Lobodon carcinophagus

DISTRIBUTION

The Crabeater Seal is the most abundant seal in the world. It is gregarious, circumpolar and pelagic, being found on the drifting pack ice. Its northward movements coincide approximately with those of the pack ice in winter, but in summer they are able to move further south as the ice breaks up. These seals are seen in great numbers in the summer months of January to March, and are particularly numerous in the coastal waters west of Graham Land, and in the southern part of the Ross Sea, where this sudden influx is said to be semi-migratory in character. Stragglers have been recorded from Australia, New Zealand, Heard Island, South Africa and the Atlantic coast of South America about as far as Rio de Janeiro.

Between 1968 and 1975 there were nine records of Crabeaters from South Africa, but only one record prior to this (Ross *et al.*, 1976). This probably reflects a greater number of interested biologists rather than a substantial increase in Crabeaters reaching South Africa. The length of most of these stragglers (1·6–1·9m) indicates that they were probably pups of the previous year, and about four to six months old. Even with the help of the West Wind Drift these young seals would have had to swim strongly to reach South Africa in the time. Most of these young seals stranded in summer (December–March) at a time when the bulk of the Crabeater population would be moving south with the break-up of the winter ice. It has been suggested that these younger animals possibly get into the West Wind Drift and get carried away from the Antarctic Continent. Of the seven Australian records, four have been from Victoria, one from Tasmania, one from Sydney and one from Nambucca Heads further north in New South Wales (lat. 30°39'). These strandings have occurred between January and September, so that they are not so markedly seasonal, although half of them were in winter.

DESCRIPTION
Adult animals have a nose to tail length of up to about 2·6m; a weight of about 225kg, and the largest females may slightly exceed the largest males in length. They are slim, lithe long-snouted animals, and can move swiftly and sinuously over the ice. In colour, the young adult seals are silvery brownish-grey dorsally with very variable chocolate brown ring markings on the sides and shoulders, shading to a pale ventral surface. The coat colour fades throughout the year, particularly quickly in summer, to a creamy white, from which derives the other common name of this animal – the White Seal. In January this white coat is moulted and replaced by a new, darker one, but older animals become progressively paler, even when freshly moulted.

BREEDING
In spite of the abundance of this seal there is still remarkably little detailed information about its breeding habits. The pups are born in spring, about September to the middle of October, and are about 150cm in length. The natal coat is soft and woolly and of a greyish-brown colour; its moult starts when the pup is about a fortnight old. The length of the lactation period has not been certainly recorded, but it is thought to last four weeks. It seems that the pregnant female hauls out on a suitable ice floe just before the birth of her pup, and the male joins her just before or just after parturition. Family groups consisting of male, female and pup are frequently seen on the ice, though the male is most probably not the father of the pup but is waiting for the female to come into oestrous. He remains at a respectful distance from the female, whereas the pup stays in close bodily contact with her. The male occasionally moves closer to test the receptiveness of the female, but if she is not ready she bites him about the head and neck leaving many small scars in this region. He is, however, prepared to defend the female and pup against other male Crabeaters, Leopard Seals and humans, and is presumably defending his intention to mate with this female.

After the pup is weaned and leaves the group the male assumes a more aggressive role and will attempt to mate, though no actual mating has yet been seen (Siniff et al., 1979). Peak numbers of these male–female pairs are seen towards the end of October, and as there is plenty of sperm in the testes in October and November this is almost certainly when they mate on the ice floe. Females are believed to be sexually mature when about 3·5 years old, and males at three to six years old (Øritsland, 1970). As noted above the adults moult in January and continue feeding and swimming during this period. When disturbed they make an angry hissing noise with the mouth wide open, but do not roll on their sides as Weddell Seals do.

PREDATORS AND PARASITES
The parallel scars seen on a large proportion of Crabeater Seals are now believed to be due to attacks by Leopard Seals rather than by killer whales (Siniff & Bengtson, 1977). Certainly the distance between Leopard Seal canines, and the general appearance of the scars is very different from those of known killer whale attacks. It has been suggested that immature Crabeater Seals, up to the end of their first year, are most likely to be attacked.

Crabeater seals seem to be relatively free of disease. The usual range of gut parasites has been recorded, though the seals are not infested with tapeworms to anywhere near the same degree as in the Weddell and Leopard Seals for example. In 1955 a particular large concentration (c. 3000) of Crabeater Seals was noticed wintering on the ice in Crown Prince Gustav Channel, off the east coast of Graham Land. This was unusual as the Crabeater usually winters away from the coast. In the spring most of these seals were found to have died from what was probably a contagious virus disease specific to the Crabeater, and not caught by the Weddell Seals in the same area (Laws & Taylor, 1957).

FEEDING
The Crabeater Seal feeds almost exclusively on the small shrimp-like animals known as krill and it is probable that most feeding takes place at night as the krill are nearer the surface at dusk and dawn. It catches these animals in quantity by swimming into a shoal of them with open mouth, sucking them in and then sieving the krill from the water through the spaces between the cusps of the complicated cheek teeth. The cusps of the upper and lower teeth closely interdigitate when the mouth is shut and this sieving system is

completed posteriorly by bony protuberances from the upper and lower jaws covered in life by soft tissue, which prevent the escape of krill behind the last tooth. Two young Crabeater Seals that were kept for a short while in captivity were seen to suck small fish into their mouths with considerable force, and water was ejected from the sides of the jaws (Ross *et al.*, 1976).

DEMOGRAPHY

As with the Weddell Seal, the size of the Crabeater Seal population can only be estimated. There is no doubt that the population is very large and figures of between 8 and 50 million seals have been suggested (Laws, 1973).

Crabeater Seal, Lobodon carcinophagus *on the Antarctic ice. The scars just visible on the back may be the result of attack by Leopard Seals. Photo. P. D. Shaughnessy.*

Leopard Seal *Hydrurga leptonyx*

DISTRIBUTION

The Leopard Seal is a solitary animal, frequently seen, though nowhere common. Animals of different ages may often be found in different areas. Normally the mature animals occur in the outer fringes of the pack ice. Young animals disperse to most of the subantarctic islands in winter, and animals of about three to nine years old may reach the coast of the Antarctic Continent in summer. There are, very infrequently, records of animals found far south, even in winter. One of these was a first year animal found in McMurdo Sound, Ross Sea in June. The islands of South Georgia, Kerguelen, Heard and St Paul are the recipients of many of the winter and early spring (June–October) visitors, and they more usually come ashore during the night. On Heard Island about 750 has been suggested as the lower limit for the winter population. Any pregnant females that winter on Heard Island leave for the pack ice in September and October.

The Auckland Islands have a few Leopard Seals every year, as does Campbell Island. Only two Leopard Seals have been recorded from South Africa,

one from near Cape Town and one from near East London. South American records range from Tierra del Fuego to approximately 36°S, a little south of Buenos Aires. The most northerly record is that of two animals at Rarotonga in the Cook Islands (lat. 20°45'S).

Leopard Seals occur regularly at Macquarie Island, usually appearing during the southern winter. The first ones are usually seen in June and the last in December, with most occurring in August. The numbers seen vary from very few in some years to over 200 (max. 283) in others. As at all the subantarctic islands the number of Leopard Seals varies widely. At Macquarie Island the abundance seems to go in cycles, with major peaks of over 200 seals followed by minor peaks of over 100 seals after four years, and then another major peak after five years (Rounsevell & Eberhard, 1980). In years such as 1977 when 227 Leopard Seals were recorded at Macquarie, they were also more plentiful than usual at other places such as Australia. Most of the animals seen on Macquarie are young animals, up to three years old.

A tagging programme on Macquarie has shown that most Leopard Seals that visit the island stay only for one or two days, though some may remain for about

Leopard Seal, Hydrurga leptonyx, *in a group of Chinstrap Penguins*, Pygoscelis antarctica. *Photo. Bill Vaughan.*

three months, and a few seals tagged in one year have returned the following year. Two animals tagged on Macquarie have been re-sighted at other places. One animal swam to Campbell Island in a maximum of 14 days, so that it must have travelled at a minimum speed of 54·4km per day. The other was found dead at Bicheno, Tasmania about two months after it was last seen on Macquarie. Some of the young animals that haul out on the subantarctic island appear to be healthy; others are wounded or sick, and some are thin and tired.

The cyclic nature of the Macquarie occurrences suggests that these are periodic dispersals, rather than a migration (Rounsevell & Eberhard, 1980). These authors suggest that in winter, krill is less available because of ice cover, and young Leopard Seals, depending heavily on krill because of lack of experience in catching other foods such as Crabeater Seals, disperse northwards in search of food.

Leopard Seals are fairly regular visitors to New Zealand shores, and there are a few records from the Snares. Tasmania too, occasionally has a stranding. The southern coasts of Australia are visited regularly by Leopard Seals, and they might almost be said to be annual visitors in the Sydney region. The southeast corner of Australia, south of 30°S, between about Coffs Harbour NSW and Port Lincoln SA receives most of the Leopard Seal strandings, the most northerly being a young animal from Heron Island, Queensland who visited the beach briefly in July, 1980, and there are also strandings known from Lord Howe Island. There are a few recorded from the southwest corner of Western Australia, but none in between, probably due to the nature of the coast and the lack of observers. It may also be due to the preponderance of observers that leads to the recording of nearly as many strandings in the Sydney area as in the rest of Australia. Forty-seven of the available Australian records have the month of stranding known, and most are in August, September, and October (late winter and early spring) the numbers falling off on either side of this maximum, but all strandings occurring between July and January (winter, spring, summer). Records for other places north of the pack ice, from South America to the Snares also tend to show greater numbers in winter and spring as the animals range about the southern hemisphere. No attempt has been made to record all sightings and strandings of Leopard Seals, but it does seem that 1977 could be called the year of the Leopard Seal. There were 18 strandings recorded in Australia (seven in Sydney) that year, as already mentioned there are records of many more seals than usual being seen at Macquarie Island, Campbell Island and round the New Zealand coast. It is not always clear from stranding records whether measurements

of length are taken to the tip of the tail or the end of the hind flippers, so it is difficult to correlate the size of the animal with the month of stranding. The Australian strandings between July and December include juveniles of *c.* 2m 'length', and old adults of nose to tail length 3m, and animals of all sizes between.

DESCRIPTION

In nose to tail length adult males are about 3m and weigh about 270kg. Adult females are larger and may reach 3·6m in nose to tail length. In colour the adults are dark grey dorsally and light grey ventrally, spotted on the throat, shoulders and sides with black and dark and light grey, but there is considerable variation in the amount of spottedness present. The long slim body, disproportionately large head separated by a distinctly marked constriction at the neck, wide gape, and the curiously reptilian appearance of the head make the Leopard Seal easy to recognize. At close quarters the large cheek teeth with their three long and distinct cusps make identification certain.

BREEDING

Very little is known about the breeding habits of the Leopard Seal. It is believed that the pups are born some time between September or November and January and the wide variation in the lengths of foetuses taken at about the same time suggests an extended pupping time. Hamilton (1939a) notes newborn pups seen in September, November and December. The first one of these being about 107cm long, and Paulian (1953) also notes a full-term foetus of 102cm collected in September. On Heard Island a pregnant female was prevented from leaving at the usual time, and she gave birth to a dead pup during the night of 14–15 November. This pup was 1·6m nose to tail and weighed 29·5kg (Brown, 1952). A young female pup that stranded in Sydney in January 1976 had a nose to tail length of 1·73m. This pup was obviously very young, but it is doubtful whether it could have swum from the pack ice even if it had been born in September. Maybe it was born at some unusual place.

The natal coat is soft and thick, dark grey dorsally, with a darker central stripe down the middle of the back; the sides and ventral surface are almost white, irregularly spotted with black. Pups with the natal coat incompletely moulted have been seen in January, but the pup that stranded in Sydney in January was fully moulted. The beginning of the moult of the natal coat starts when the pup is about a fortnight old in Weddell and Crabeater Seals, and it is possible that the timing is similar in Leopard Seals. The period of lactation is probably about four weeks. A lactating female has been captured in January. Mating in captivity was observed in November, again in January, and at intervals until the beginning of February, when the female was found dead with severe lacerations, presumably inflicted by the male (Marlow, 1967). This female showed signs of having ovulated at some time before death. Tikhomirov (1975) notes that mating in the wild follows the four week lactation period, and that there is a 2·5–3 month delay before the blastocyst is implanted. Free epididymal sperm was found in wild males at the end of October. Judging from the accelerated growth of baculum and testes, sexual maturity of males is attained between three and six years. Females may ovulate for the first time between two and seven years, probably about five years. Moulting of the adult takes place at any time between January and June.

PATHOLOGY, PREDATORS AND PARASITES
Probably the killer whale is the only natural enemy of this seal, although in the more northerly parts of its range it may get attacked by sharks. Gut tapeworms and nematodes are well represented in the Leopard Seals internal fauna. A tumour of the bronchus and carcinoma of the stomach have been described, and curious bony nodules are frequently found in the nasal passages. Many of these animals recovered in Sydney have wounds and abscesses due to stingray spines (*Urolophus paucimaculatus*) that have become embedded in the flesh. Surface wounds of this nature will heal, but if the stingray spine has penetrated very deeply it remains in the seal, and a more serious and possibly lethal infection will ensue. Wounds, possibly due to the cookie-cutter shark have also been found.

FEEDING
The rather sinister and reptilian appearance of the head, and the obviously carnivorous habits of the Leopard Seal have led to its being credited, most unjustly, with an extremely fierce disposition, and unprovoked attacks on man have even been mentioned. With regard to the latter it seems most likely that the seal was making a genuine mistake or was first assaulted or annoyed, and it is known to be quicker in responding to molestation than the other Antarctic seals. It has been known for a Leopard Seal to emerge at speed through a crack in the ice, make a snap at a scientist's foot and chase the man over the ice for 100m. On another occasion a Leopard Seal, on hearing a piece of ice that had been thrown into the water, came to investigate and made a lunge at the man on the ice edge (Penney, 1969). On both these occasions it seems most likely that a hungry Leopard Seal mistook the dark vertical shape of the man for the not dissimilar shape of an Emperor penguin. A slightly more aggressive confrontation took place between two scuba divers and three Leopard Seals. The attentions of the seals led the divers to hold lengths of angle iron

between themselves and the seals. The seals dived repeatedly and struck at the ends of the metal bars until the men could retreat enough to escape (Lipps, 1980). The seals could probably be called aggressive on this occasion, but from the seals' point of view, maybe they reckoned that the presence of the men was sufficient provocation.

Certainly Leopard Seals make great use of such penguins that are abundant in any area. The seal will cruise up and down an area of coast and catch a penguin under water, emerge to the surface with it, shake it about until the skin is almost peeled away and then bite the meat off it, leaving a relatively clean skeleton. A seal will demolish a penguin in about five minutes, and on a beach that was being observed at Cape Crozier, Ross Island, about 19 Adelie penguins were being caught each day during the 15 weeks of the penguin season (Penney & Lowry, 1967, Penney, 1969). Up to 73kg of penguin remains have been taken from a single Leopard Seal stomach. Understandably the penguins are very wary about entering the water when they see a Leopard Seal there, and they tend to gather on the ice edge and go to sea in large groups, though penguins have been seen to behave in an equally indecisive manner when no Leopard Seals were around.

As well as eating penguins of many species, and other birds such as Giant petrels, Leopard Seals will attack young Crabeater, Weddell and Elephant seals, and will feed on carrion from any carcasses of seals and whales that are available. Fish and cephalopods are also eaten. The amount of krill eaten is large, and is about the same proportion of the total food as the penguin–seal part of the diet: the three-cusped cheek teeth of *Hydrurga* make as good a sieve for obtaining krill as do those of *Lobodon* (Øritsland, 1977).

VOCALIZATION
A number of different types of sounds are made by Leopard Seals – a gargling noise, grunting, and higher pitched sounds from bird-like chirps to a musical sighing, crooning or whistling. A singing sound is sometimes made when the animal is asleep, and a throaty alarm note is made by vibrating the tongue as air is expelled through the mouth.

DEMOGRAPHY
As with the other Antarctic seals that spend most of the year ranging over a wide area of ice and sea accurate and detailed population numbers are not to be expected. Laws (1973) reviewing such works suggests that a conservative estimate of the number of Leopard Seals would be between about 250 000 and 800 000.

Elephant seals

Their huge size, and the short mobile proboscis of the adult male make Elephant Seals unmistakable. There are two species of Elephant Seals – the Northern, *Mirounga angustirostris* that lives off California, and the Southern, *M. leonina* that lives on subantarctic islands. As mentioned in the chapter on fossils, the ancestor of Elephant Seals probably came from the same stock as the Monk Seals, moving from the Caribbean into the Pacific, and then down into the subantarctic regions of the southern hemisphere. It seems unlikely that they evolved from the lobodontine stock that gave rise to the Antarctic phocids, and then spread northwards, up the Pacific coast of America, ending in *M. angustir-*

ostris. Likely as their common origin with the Monk Seals may be, there are as yet no good fossils to demonstrate this, although the very few Elephant Seal fossils known, from the Pleistocene of California, are similar to modern *M. angustirostris* and also suggest the northern origin of the group. Also, many characters of the dentition, skull morphology, and growth, that are regarded as primitive, are retained by the Northern Elephant Seal (Briggs & Morejohn, 1976). This does not imply that *M. angustirostris* gave rise directly to *M. leonina* – probably a form earlier than the modern *M. angustirostris* gave rise to both.

Southern Elephant Seal Mirounga leonina

DISTRIBUTION

The Southern Elephant Seal, one of the best known of the phocids, is circumpolar and is found on most of the subantarctic islands. Breeding colonies are to be found near Punta Norte, Argentina, Tierra del Fuego, and on the islands of Falklands, South Shetlands, South Orkneys, South Georgia, South Sandwich, Gough, Marion, Crozet, Kerguelen, Heard, Macquarie and Campbell (see map). Non-breeding groups of animals are found mostly during the December–February moulting season on St Paul and Amsterdam Islands and on Bouvet, where they could have come from Kerguelen and South Georgia respectively.

A large summer hauling ground is known from the Antarctic Continent at the Vestfold Hills. Here, about 700 seals, all immature males, arrive in February for resting and moulting. They are believed to come from the Kerguelen–Heard group, and some of them have been known to return annually. Individual animals have been recorded from various points round the Antarctic Continent and in the pack ice, and are occasional visitors at any islands near major breeding sites, for example on the Auckland Islands where they presumably come from Campbell Island.

Some of the most northerly records of elephant seals are those from St Helena (lat. 16°S). These have been called manatees and sea cows, but a drawing made in 1655, and a later reference to 'eyes as big as saucers' point undoubtedly to the Elephant Seal (Fraser, 1935). They seem to have been more frequently seen there in the late seventeenth and early eighteenth centuries, but there is a record from as late as 1819. Recent northerly records from the Indian Ocean are those of two young animals caught at Rodriquez (lat. 19°48'S)

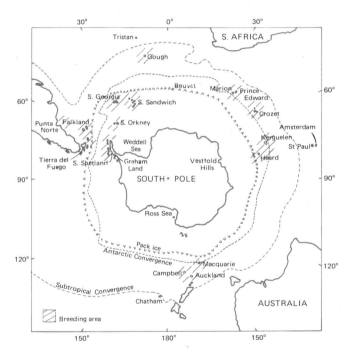

in 1942 and 1954, and another from Mauritius in 1955 (Vinson, 1956).

Navigators making hydrographic surveys in the Indian Ocean in the eighteenth and early nineteenth centuries recorded 'seals' sleeping on the beaches in the Seychelles, one part of an island was named 'Sea Monster Point', and the large size and menacing teeth of these animals are commented on. 'Seals of a large size' were seen on beaches in the Amirantes in the 1820s, and the nearby island of Coetivy was even suggested as a source of supply of oil from the 'sea lions'. In the Chagos Archipelago in 1786 the paucity of fish round one of the islands was attributed to the number of seals there (Stoddart, 1972). The fact that these animals were sleeping on the beach would suggest pinnipeds rather than dugongs. These records were made by zoologically unbiased observers, and do seem to suggest the presence of large seals in the northern Indian Ocean. Their occurrence may have been only seasonal, but at least they were in some places in numbers sufficient to consider a commercial enterprise to obtain oil. There is, of course, nothing positive from which to identify the animals, but their large size, and the fact that Elephant Seals are known to have reached Rodriquez recently, makes it very likely that these were the animals involved.

Presumably straying from the breeding colony at Punta Norte, Argentina, Elephant Seals occasionally reach the islands off the Uruguayan coast north of Montevideo up to about 33°S, and although not recorded from the coasts of Chile, there is a single record of a straggler reaching 13°50'S on the Peruvian coast near Lima.

At the time of the voyages of William Dampier in 1683 Elephant Seals were abundant on Juan Fernandez (Hubbs & Norris, 1971), though he and other travellers of those times frequently referred to them as 'Sea Lions' or 'Sea Lyons'. The skeletal remains of the anterior ends of the upper and lower jaws of a large male animal were collected on Juan Fernandez by Lord George Anson in 1744, and stored at various times in the collections of the Royal College of Surgeons and the British Museum (Natural History) where it is today. This specimen is the type, and it was on a description and figure of it that Linnaeus founded *Phoca leonina* (Hamilton, 1940, Laws, 1953).

Some 25 Elephant Seals have come ashore on the South African coast (Best, 1971), most of them round the southern tip of the continent. Most of the stranded animals are males, and although they have come ashore in most months of the year, about half the strandings have occurred in summer (December–February) this coinciding with the moulting period. Three Elephant Seals seen at Baia das Luciras (13°52'S) on the Angolan coast were probably assisted in getting so far north by the cold waters of the Benguela Current. The origin of these African animals might be from Marion Island, 1600km south, but against the current, or from Gough Island 3200km to the west where movement towards Africa would be assisted by the West Wind Drift. A single pup has been recorded as being born in South Africa, in October.

There are records of about 20 Elephant Seals coming ashore in New Zealand, all except one between September and January. These probably came from the Campbell Island colony, and most of them came ashore in the vicinity of Kaikoura, on the east coast of South Island. In contrast to the South African strandings there is a preponderance of females, about 14 of them being females, and of these six gave birth to pups in September, October and November.

On Tasmania it is reported that individual adult males come ashore at intervals between January and August. Adult females also occur, and there are records of two births, both in October, one at Strahan on the west coast, and one near St Helens on the east coast (Tyson, 1977). This is a great change from the times when the northwest corner of Tasmania formed one of the breeding grounds of the Elephant Seal. Bones found in middens there show that 2000 years ago and even earlier, the aborigines were using the pups as food, but there is no evidence of breeding there in historical times (Jones, 1966). There was still a large breeding colony on King Island in Bass Strait when the island was discovered in 1799, but this soon succumbed to the sealing fever of those days. Sea Elephant Bay and Elephant River on this island are among the reminders of the animals' abundance there.

Occasional Elephant Seals reach the rest of Australia, (i.e. excluding Tasmania) there being about six strandings in the last ten years. Both males and females are represented, and all except one of these strandings have taken place, not unreasonably, in the southeast corner of Australia, the remaining one being at Coffs Harbour NSW. One of these animals, a male approximately 3·6m long came ashore at Geelong, Victoria, in March 1977, when it lay on the beach moulting. It was tagged, and turned up again at Geelong in March 1979. Australian and Tasmanian strandings probably come from Macquarie or Campbell Islands.

It is obvious that Elephant Seals move widely about the oceans, particularly in their first few years, but usually, unless they reach land, we have no record of how far they have wandered. Probable movements between subantarctic islands, such as between Kerguelen and St Paul, Kerguelen and Vestfold Hills, Antarctica, South Georgia and Bouvet etc. have been noted, also movements between the breeding grounds and such places as South Africa, Australia, etc. Branded animals from Heard Island have been recovered both from Marion Island and from the Vestfold Hills on the Antarctic Continent. Seals branded on Macquarie have been seen on Campbell Island, Chatham Island, Swan Island off the northeast tip of Tasmania, and at two places on the southwest coast of South Island, New Zealand. The longest known migration by an Elephant Seal was made by an animal that was tagged when three weeks old at South Georgia, and found again 14 months later on the south African coast – a journey of some 4800km in its first year of life.

DESCRIPTION

The largest of the pinnipeds, an adult male Elephant Seal may be 4–5m in nose to tail length, and 3·6 tonnes in weight. The females are smaller, 2–3m in length, and about 900kg. In colour the males are dark grey, a little lighter ventrally just after the moult, but through the year this fades to a rusty greyish brown. Fighting between harem bulls leads to intensive scarring of the neck region, and the scar tissue makes the skin of the chest extremely thick, tough and cracked, looking rather like the rough and deeply fissured bark of a tree. The females are browner, and generally darker than the males, with, in adult animals a lighter coloured yoke round the neck from the many small scars resulting from bites during mating. The hair is short, stiff and harsh, with no underfur (see chapter on skin).

The inflatable proboscis is the most outstanding feature of the Elephant Seal, but its full development is only seen in adult males. In females there is no enlargement, though occasional individuals may be able to pucker the snout a little. The development of the proboscis has started with the puckering of the dorsal surface of the snout by the time the young male is about two years old, and its full size is attained at about eight years old when the status of harem bull is reached (Laws, 1953). Its tip overhangs the mouth in front, so that the nostrils open downwards, and the cushion-like shape is marked by two transverse grooves. As in the Hooded Seal, it is an enlargement of the nasal cavity, and is internally divided into two by the nasal septum. Even in the harem bulls the proboscis is flattened and less obvious in the non-breeding season, but, when breeding, it can be erected partly by muscular action, partly by blood pressure, assisted by inflation, to form a high, bolster-shaped cushion on top of the snout, the tip with its open nostrils hanging down in front of the open mouth. It may act as a resonating chamber, and certainly the roar of a big bull may carry for several kilometres. Comparison with the proboscis of the Northern Elephant Seal is made in the section on that animal, and comparison of the skull and flexibility of the back is also made.

BREEDING

The breeding season starts at the beginning of September when the bulls come ashore. They are joined, from about the middle of the month, by the first females, who at first wander about and then congregate in groups. By the end of September large numbers of cows are hauling out and harems begin to form, first only about five females to each group, later increasing to 20–40, though larger harems of up to 100 or more are known. The pups are born about a week after the cows haul out, most of them in October. Each harem is presided over by one dominant male whose main function is to mate with the females of his harem, and prevent other males from doing so. Harem bulls are older and larger than non-harem bulls, and large size gives them an advantage in fighting and retaining their dominance (McCann, 1981). The mature, but non-breeding, subordinate bulls, remain round the edge of the harems and attempt to mate with any females they can. A challenger to the reigning harem bull roars and postures for some time, to make clear his intentions, and sometimes this alone results in the withdrawal of the harem bull. Otherwise he roars back at the challenger, who may retreat or roar again. Only then does a fight for supremacy occur. Too large harems may split up. The possible correlation between length of day, the pineal gland and the breeding season is noted in the section on the pineal gland.

The pups are fed by their mothers for an average time of 23 days, the females not feeding for the time they are on land. During lactation the mother may lose about 320kg in weight, and towards the end of this period the pup is gaining 9kg a day (Laws, 1953). The pup may suckle for the first time about an hour after birth, or sometimes even three hours after, but spends a considerable amount of time at first searching for the nipples. The female lies on her side, but does not help the pup. Sometimes a pup will suckle from a female that is not its mother, particularly towards the end of the lactation period. The cows are ready to mate again about 18 days after the birth of the pup, and they leave the harems and return to sea when the pup is weaned. The blastocyst does not become attached until the beginning of March, so there is a delay of four months (Laws, 1956).

After the cows have left, the aggressiveness of the bulls diminishes, and by about November, they have lost interest in the females, and return to sea to feed, having fasted since the beginning of the breeding season. The non-breeding females are rarely seen on land, but those mating for the first time probably do so at sea in the normal breeding season. It appears that the female gives birth about a week later each year, since the gestation plus the lactation time is over a year. Eventually, a cow will pup too late in the season to mate again, and then probably mates at the beginning of the following season. Each female is estimated to produce seven pups in her lifetime, and to live about 12 years. Bulls may reach an age of 20 years, but in a commercially exploited population they reach only about 12 years. In South Georgia the females become sexually mature at two years and bear their first pup when three years old. Bulls are sexually mature at four years but are not sufficiently powerful to hold a harem until about five to seven years. In an unexploited population, such as that at Macquarie Island, the females mate for the first time when three to six years old, and while the bulls may be sexually mature at five years, they may not reach harem bull status until 12 years or more. A single adult male branded at Campbell Island in 1945 was seen in 1959 on Macquarie Island. Even at 14 years old this animal was not yet a harem master. The slower rate of growth of the Macquarie animals is attributed to disturbance in the harems, probably by the large numbers of adult males, resulting in injuries to the pups, and hence slow growth (Bryden, 1968). There seems to be no interchange between the various stocks of Elephant Seals, and the different growth rates would suggest the discreteness of the stocks (Laws, 1979).

At birth the pups have a nose to tail length of about 1·2m and weigh 35–45kg. They are clothed in a coat of black woolly hair. They start to moult this coat when about ten days old, and have finished by the time they are 34 days old, and usually wait till the moult is finished before they enter the sea. After weaning when they weigh about 140–180 kg, the pups gather in groups at the back of the harems for about two weeks, and then move towards the shore. In the Falkland Islands and Kerguelen, and other places where the soil is not frozen, they make muddy wallows in which they spend a lot of their time and this probably helps to allay the irritation of moulting.

The adult animals return to land again for their annual moult, the greatest numbers being on shore in December, January and February. The average time taken to moult is about 3–40 days, during which time no feeding takes place, but the time taken seems to increase with the age and size of the individual. The younger seals moult first and then successive age groups haul out in relays through the summer and autumn, ending with the breeding bulls in March–April. The nature of the moult, with sheets of shed hairs held together by the epidermis, is mentioned in the chapter on skin. The increased blood supply to the skin at this time leads to increased heat loss which may

Southern Elephant Seal, Mirounga leonina. *The larger male with the proboscis is in the initial stages of copulation with the smaller female. Photo. R. Lewis Smith.*

be partly compensated for by the animals lying together in groups or piles. After the moult, most seals return to sea to feed, and there are no large congregations on shore until the next breeding season. During the winter months however, immature seals not infrequently come ashore for a few days, and occasionally pregnant cows are also seen, but the adult bulls remain at sea.

MORTALITY, PREDATORS AND PARASITES

The only natural enemies of the Elephant Seal are the Leopard Seal which has been known to attack pups, and the killer whale, but neither causes any serious predation. Mortality is particularly high amongst the pups when they have left the rookery, and may be as much as 50 per cent. On land the young pups in their black coats may melt the snow under them to such an extent that they get trapped in a pit and starve, as they may also do if they lose their mothers. A certain number also get squashed by the movements of the bull. Adults also sometimes get trapped in the muddy wallows. Fighting rarely leads to death. Serious wounds occur however, large portions of the blubber may be torn off, the proboscis may get severely torn, and eyes lost, the last appearing to have little effect on the general health of the animal. Some animals have orange fungal spots on them when they come from the sea but these soon dry up.

Internally the Elephant Seal harbours tapeworms, roundworms, and thorny headed worms. The tapeworms are in the rectum where they produce nodules in the wall, the roundworms mostly in the stomach, but also in the heart, bronchi and intestine, and thorny headed worms in the stomach where their presence may cause tumours. Large numbers of ticks infest the nasal passages, and the seals of Macquarie Island harbour lice that burrow under the skin, particularly of the hind flippers, pups being most heavily infested as they spend most time out of the water. Stalked barnacles (*Conchoderma auritum*) have been recorded from an Elephant Seal that came ashore in South Africa (Best, 1971). Wounds, possibly due to the cookie-cutter shark have been found on a stranded animal.

Southern Elephant Seal, Mirounga leonina. *This roaring bull shows the inflated proboscis with the nostrils at its tip directed into the mouth. The proboscis acts as a resonator to the sound coming from the throat. The thick fissured skin of the neck is clearly visible. Photo. R. Lewis Smith.*

FEEDING

When at sea the Elephant Seal is assumed to feed mainly on fish and cephalopods. The pups after they are weaned live for about a month on their reserves of blubber, and then they start nosing about under stones and eating the small crustaceans there, before gradually passing on to the adult diet.

DEMOGRAPHY

South Georgia has the largest breeding population of Elephant Seals (310 000), followed by Kerguelen (*c.* 100 000), and Macquarie Island (95 000). Approximately another 100 000 animals are distributed among the other breeding colonies and the total mid-year population has been estimated as being between 600 000–700 000 seals (Laws, 1973).

EXPLOITATION

The story of Elephant sealing on South Georgia is well known by now. For about 45 years, from the discovery of the island by Captain Cook in 1775, it was the Fur seals that were exploited. As their numbers diminished so the attention of the sealers turned to the Elephant Seal, and by applying the same indiscriminate slaughter they soon achieved the same end of almost exterminating the animal, so that by the end of the nineteenth century even the Elephant Seal was no longer profitable. By 1910 the herd had recovered sufficiently to be considered commercially again, and licensing was introduced, to be continued until 1964, since when there has been no commercial sealing (Laws, 1973). Between 1910 and 1964 259 076 bulls were taken.

The coast of South Georgia is divided into four parts which were worked in rotation, and each year one division remained untouched. Annual quotas, in more recent years up to 6000 animals, were set, divided proportionally among the three divisions. The highest annual catch since 1910 was in 1951 when 7877 seals were taken out of a permitted quota of 8000, yielding 14 608 barrels of oil. Only bulls of above 3·5 m nose to tail length were permitted to be killed, this being the average length attained at about 4·5–5 years, and an average of 1·97 barrels of oil (409 litres, or 90 gallons) per seal was obtained. On each beach an adequate proportion of bulls had to be left, and it was not permitted to kill cows or pups.

During the sealing the chosen animals were driven towards the water's edge, shot and skinned. The flensers made cuts round the foreflippers, behind the eyes and in front of the tail and down the middle of the back. The skin with the underlying blubber, varying in thickness between 2–15 cm was then pulled off in one piece with hooks, the process taking three to four minutes for each seal. Several skins were threaded together on a rope and towed out to the waiting motor boat, which delivered them to the catcher, and so in turn to the whaling station. At the station the skins were minced up mechanically and rendered down under pressure to remove the oil. Elephant Seal oil has very much the same properties and uses as whale oil, but was not mixed with it. It was hardened and used, together with vegetable oils in the manufacture of edible fats.

The Southern Elephant Seal is fully protected, and since sealing stopped there may have been an increase in numbers. However, the fish of the Antarctic are being taken by fleets of trawlers, and the effect of the depletion of the fish stock on the seals is yet to be assessed. This is the reverse of the usual situation where fishermen complain that seals are interfering with the fishing.

Northern Elephant Seal
Mirounga angustirostris

DISTRIBUTION

The Northern Elephant Seal has shown a remarkable recovery. Before commercial exploitation of them started, their range extended from about Point Reyes (lat. 38°N), near San Francisco, along the coast to Magdalena Bay, Baja California, and they were presumably plentiful. Commercial use for their blubber began about 1818, and by 1860 they were so scarce that they were no longer important as a source of oil and writers noted that they were nearly extinct. A small herd of less than 400 animals found on the mainland coast south of Cedros Island in 1880 was soon cleared up, with the last few being taken as scientific specimens. In fact, the Northern Elephant seal was not recognized as distinct from the southern species until there were practically no northern animals left (Bartholomew & Hubbs, 1960). By 1890 there was only a single herd, perhaps as few as 20, and certainly not more than 100 animals on Guadalupe Island, and it is from this herd that the numbers have built up to their present level, and from Guadalupe that the Elephant

Seal has spread so that it now occupies much of its former range.

At the present time 90 per cent of the breeding population is present on Guadalupe and San Benito. Breeding colonies are also present on many of the Californian Channel Islands – on San Miguel, Santa Cruz, San Nicholas, Santa Barbara and San Clemente, and seals are present though not breeding on Santa Rosa and Anacapa. To the south, animals are present, possibly breeding on San Geronimo, and there is breeding on Los Coronados, Isla Cedros, Isla San Martin and Isla Natividad. On the last two islands breeding has only recently re-started, pups being seen in 1977 and 1975 respectively (Le Bouef & Mate, 1978). To the north of the Channel Islands, Elephant Seals reappeared on Ano Nuevo in 1955 for the first time for about 100 years, and started breeding again in 1961. They were seen on the Farallones in 1959, though there was no breeding until 1972 (LeBouef *et al.*, 1974). Occasional stragglers appear on the coast of Oregon, and these included in 1969 a male tagged 6·5 years earlier, as a weaned pup on Ano Nuevo. Other Elephant Seals have been recovered from British Columbia, and as far north as Prince of Wales Island, Alaska. No general migration occurs, but in spring there are practically no adult males on the beaches, and it is presumed they have gone far out to sea.

DESCRIPTION

In colour and appearance the Northern Elephant Seal is in general very similar to its relative in the south, although there are differences. The Northern Elephant Seal male reaches only about 4·5m in length and 2–2·5 tonnes in weight – the 15-cm layer of blubber over the body accounting for some 40 per cent of the weight. The females are about the same size as those of the southern species, 2–3m.

The skulls of the Northern Elephant Seal differ from those of the southern species in their lesser sexual dimorphism, the greater amount of suture closure with age, and in such characters as the shape of the maxillo-jugal suture, the shorter length of the palate, the narrower abutment of palatines and pterygoids, and the greater distance of the pterygoid hamulae from the bulla. The width of the skull across the zygomatic arches is less, as is the width of the inter-orbital area. It is considered that with evolution, the skull of *M. leonina* has become foreshortened, and many of the differences in skull morphology are due to this (Briggs & Morejohn, 1976). The lower postcanine teeth show a greater tendency (than in *M. leonina*) to be multicusped and multirooted.

The flexibility of the back of the Southern Elephant Seal, so that it can bend backwards to touch its tail with the back of its head (p. 156), has been mentioned in the

An adult male Northern Elephant Seal on Ano Nuevo Island, California. The very long proboscis of the male can be seen. A juvenile animal occupies the left of the picture. Photo. J. E. King.

chapter on the skeleton; this posture is rarely if ever seen in the northern animal.

VOCALIZATION

The shape, and possibly also the function of the proboscis show some differences. That of the male Northern Elephant Seal is very long, hanging down over the mouth about 30cm when relaxed, and when inflated showing a very deep transverse groove which almost divides it into two. Inflation 'causes it to curve down between the jaws, presses its distal end against the roof of the mouth, and directs the nostrils towards the pharynx' (Bartholomew, 1952). The snorts are thus directed down into the open mouth and pharynx, which act as a resonating chamber. The noise made is rhythmic, resonant and metallic, about three to five pulses a second, and carries for nearly a kilometre. The Southern Elephant Seal does not have such a pendulous proboscis and it overhangs the mouth only about 10cm or so. When erected the transverse grooves are much less deep and the proboscis 'hangs down so that the nostrils open immediately in front of the mouth' (Laws, 1956), but it does not go right down inside the mouth. The noise, a harsh rattling roar is, according to Laws (1956) caused by an expiration by the mouth (not the nostrils), the proboscis acting as a resonator.

The sounds made by male Northern Elephant Seals have been analysed, and there seems to be some difference between the sounds made by animals on different islands. For example, the San Nicolas males emit sound pulses at more than double the rate of males on Ano Nuevo, while the sounds of males at San Miguel and Guadalupe are intermediate in pulse rate between the other two. There are also other properties of the calls that separate the populations, and such dialects have not been recorded in mammals other than man (LeBouef & Peterson, 1969). Analysis of the calls over the years has shown that the pulse rate on Ano Nuevo is rising, probably due to continual immigration by animals from other islands. If this is so, then eventually the pulse rate on Ano Nuevo should be the same as on Guadalupe – the parent island, and dialects will disappear (LeBouef & Petrinovich, 1975).

BREEDING

The general pattern of the breeding behaviour is much the same as in the Southern Elephant Seal. The dominant males arrive at the rookery in early December; they challenge and fight, fasting for the whole of the breeding season. The females, as they arrive in late December, tend to gather in groups, and they too fast during their 34-day stay on land.

The pups are born about a week after the mothers' arrival on shore, most of them during the last half of January. They are about 1·2m long, weigh 30–45kg and have a coat of slightly curly, greyish black hair.

The lactation period is 27 days and at the end of this time the pups have increased their birth weight by at least three times, and sometimes as much as seven times. Towards the end of the lactation period the pups start to moult their black natal coat, and although the time taken to do this is variable, all have finished moulting by the time they are seven weeks old.

At the time of weaning the mother makes her way out of the harem and into the water, leaving her pup behind. The pups then gather in pods. They fast completely during the ensuing 2·5 months, living on their blubber, while they spend most of the time sleeping and playing. Two or three weeks after weaning (about mid-March) they start to enter the water and gradually become proficient at swimming and diving. When they are 3–3·5 months old (April–mid-May) they leave the rookery and by that time are able to stay underwater for 15 minutes.

The pups that are born earliest, obviously are ready to leave first, but their leaving stimulates the later weaners, and all the pups leave during a three-week period. This period of leaving coincides with the main period of coastal upwelling when there is plenty of food available in the sea (Reiter *et al.*, 1978).

After the pups have left the rookery in the spring they disperse northwards and do not appear on land again until about September, remaining resting on land until the beginning of the breeding season in December. They then return to sea while the breeding season is on.

After the cows have left the pups, there is of course no more food available until they learn to catch their own. Some weaned pups make an effort to prolong the good life by becoming milk thieves. They cautiously approach females – who do not always check on the identity of the visitor – and may even displace a young sucking pup in order to steal milk. They do not make any sounds as this would immediately indicate to the cow that it was not her own pup. This milk stealing can be dangerous, as when discovered the cow will bite the visiting pup and even pursue it. On its flight through the harem it could get bitten by several more cows before regaining the safety of the weaner pod, and depending on where it is bitten, some of the bites could be serious.

Some pups are able to suckle successfully from two females. Sometimes the foster mother had lost a pup of the same age as the visitor, and presumably mistook the visitor for her own pup. The foster mother and the real mother fought over the pup for a while, and eventually shared it. Sometimes the pup is adopted and fed by a foster mother after the real mother has left the harem and the pup is supposed to be weaned.

Male pups weigh slightly more than females at birth and they are usually suckled for one day longer.

Obviously the more milk a pup can obtain the bigger it becomes. Males are more adept and persistent at milk stealing, and very large weaned pups are always males. Reiter *et al.* (1978) have correlated these sexual differences with the reproductive strategies of the seal – the earlier the male becomes larger than his peers the sooner he will have the edge over his rivals in aggressive competition, and eventually in acquiring females.

Male pups tend to moult later, and take longer over it than females. The time of eruption of the permanent canines is different too. They erupt before weaning in females, but not until four weeks after weaning in males. Reiter *et al.* (1978) have suggested that as an unmoulted weaner has the same pelage as a suckling pup, this 'disguise' may help the male milk thieves, and too soon an eruption of sharp canines would easily alert the cow from whom they were suckling.

The fighting of the males is to establish the dominance of the individual and his place in the hierarchy rather than to retain a piece of territory or a harem of females. Aggressive encounters may not always lead to actual fighting, sometimes a stare, a threatening movement or a vocal challenge from a dominant bull is enough to deter an opponent. When fighting the animals rear up, chest to chest, rocking back and forth until the opportunity occurs for one of them to deliver a downward blow with open mouth, tearing the opponents skin with the strong canine teeth. Many of the wounds occur on the thickened shield of skin on the chest, but not infrequently the proboscis gets torn. Death from such fighting is rare, and even severe wounds heal quickly.

A dominant male stays close to a group of females, while lower members of the hierarchy are found towards the periphery. A group of about 40 females can be controlled by a single dominant male; with more females, and they may be in groups of 300 or more, he has difficulty in keeping out the lower ranking males (LeBouef, 1971b). Relatively few dominant males on a beach are responsible for most of the mating, but having reached their reproductive peak they can only maintain it for a year or two before they die. Many mature animals die without mating as they never reach the necessary level of dominance.

The females are ready to mate again about 24 days after parturition, just before the end of lactation, and

Northern Elephant Seal, Mirounga angustirostris *on Ano Nuevo Island, California. The very long proboscis of this adult male can be seen, and both it and the skin of the neck bear scars from fighting. Photo. J. E. King.*

most of the females have given birth, mated, and left the rookery by the middle of March. Most females produce their first pup in their third or fourth year. The blastocyst is delayed for four months before it is implanted. Seals have been seen moulting in May, June and July, older animals moulting later than the younger ones. Moulting takes place in the same way as in the Southern Elephant Seal.

MORTALITY, PREDATORS AND PARASITES

Little is known about their natural enemies, but young animals have been seen with the unmistakable prints of sharks' teeth on them, and two killer whales have been observed stalking and fatally wounding an adult male Elephant Seal, although they did not appear to eat the carcass. The sharp edges of the pieces of black lava that are swallowed damage the walls of the stomach, and ulcers result, which may be aggravated by the resident roundworm population. Roundworms in the bronchi may cause death from verminous pneumonia. Other internal parasites are tapeworms and thorny headed worms in the intestine, and liver flukes in the liver. Mites are found in the nares and pharynx. Pups suffer from the usual mortality due to starvation or to being bitten or crushed by adults. Most dead pups are injured in some way, from punctures in the skull the width apart of the adult canines, to ruptured livers. Adult females are responsible for most of the biting, and adult males for most of the crushing. Most of the deaths, however, occur within about two weeks of birth. Yearlings have come ashore with green algae and goose barnacles on them, and sometimes sting ray spines cause abscesses in the mouth.

FEEDING

Stomach contents show that the seals feed on squid and elasmobranch and teleost fish. Ratfish or ghost shark (*Hydrolagus*) and other small sharks and rays have been identified, and also teleost otoliths and vertebral columns. Most of the fish were benthic and slow swimmers, and as ratfish are never taken in water of less than 91m (50 fathoms), the seals presumably feed in deep water.

DEMOGRAPHY

The recovery of the Northern Elephant Seal from near extinction has been mentioned. The animals are now completely protected, and their numbers have increased from about 13 000 animals in 1957 to an estimated 48 000 in 1976 (LeBoeuf, 1977). There is doubt, however, as to whether this remarkable increase in numbers denotes a really healthy population. They have all descended from the very small number of individuals left in about 1890 on Guadalupe, and are without doubt very inbred. This has resulted in a remarkable lack of genetic diversity in their blood proteins, all the seals examined being exactly the same in this respect. The effect on a population of this impoverished genetic variability is not known, but it could well reduce its ability to adapt to changing circumstances (LeBoeuf, 1977).

FOSSILS AND EVOLUTION

Otarioid seals

When considering the story of the evolution and fossil history of the otarioid seals, four or possibly five inter-related groups are involved. The first, and earliest of these is the ancestral group – the Enaliarctidae – from which all other groups could have been derived. There are also the Desmatophocidae, which became extinct in the early part of the late Miocene; the Odobenidae, one branch of which leads to the modern walrus; and finally the Otariidae, leading to the modern fur seals and sea lions. It is possible that the Desmatophocidae may be divisible into two lineages – Desmatophocidae and Allodesmidae (Fig 3.1).

Ancestral group – the enaliarctids

The Enaliarctidae have been recently described (Mitchell & Tedford, 1973) from two incomplete skulls and two natural endocranial casts (representing three individuals) and some isolated teeth. These specimens, given the name *Enaliarctos mealsi*, are from Kern County, California, and are about 22·5m years old – from the beginning of the Miocene. The family was also present in Japan, though the specimens are as yet undescribed (Repenning *et al.*, 1979). No lower jaw or post cranial material was found, but the Californian skull fragments indicate a medium-sized animal, probably about the size of the modern *Callorhinus*. Sufficient characters can be seen in the crania to indicate with reasonable certainty that these animals were derived from amphicynodontine ursids, and were on the way to evolving into otarioids. The presence of carnassial teeth shows that they undoubtedly belong to the order Carnivora. There are resemblances to ursids in the form of the bulla and the degree of its inflation, and in the size and detail of many parts of the middle ear, and in the 'ursid lozenge' of the brain. There are resemblances to otarioids in the short deep snout, the large orbits, the great interorbital constriction, and the mastoid-paroccipital crest. Aquatic adaptations may be seen in the large narial chamber, reduced olfactory bulbs, and evidence of an increased vascular supply to the brain.

These animals were not ursids – they differed in their aquatic specialization from their probable ancestors in the terrestrial ursid subfamily Amphicynodontinae, and because they still had carnassials they were not yet otarioids. Their feet are as yet undescribed but they are definitely pinniped. Thus it is reasonable at the moment to separate these presumably aquatic proto-otarioids as a separate family Enaliarctidae. The Enaliarctidae have, however, been described (Repenning in litt.) as an artificial group that recognizes a transitional stage between otarioids and primitive ursids. It is believed that with further knowledge the need for this group will disappear.

Desmatophocids

At some as yet unknown time in the early Miocene, possibly about 21–20m years ago, a group of animals diverged from the ancestral group and became specialized. This family Desmatophocidae achieved maximum diversity in the early middle Miocene, about 14m years ago, and disappeared in early late Miocene, about 10–9m years ago. Although not many desmatophocids are known, they were quite specialized when compared with contemporary primitive otariids such as *Pithanotaria*, or primitive odobenids such as *Neotherium* and *Imagotaria* that lived at the same time.

Desmatophocids were found in the coastal waters of the North Pacific – from California, Oregon and Japan. There are two genera – *Desmatophoca*, with its single species *D. oregonensis* which is earlier and probably ancestral to *Allodesmus*, which embraces at the moment three species, *A. courseni*, *A. kernensis*, and *A. packardi* – though there are other desmatophocids not yet studied and named.

Relatively large numbers of bones of *Allodesmus* are known from the middle Miocene of California – jaws, skulls in various states of completeness, and incomplete skeletons (Mitchell, 1966, Barnes, 1972). A tentative reconstruction of *Allodesmus* based on the study of an almost complete skeleton, indicates a fairly large animal (*c.* 2·5m in length) of typical sea lion shape, with flippers similar in general shape to those of modern sea lions. Males are thought to have had a proboscis like that of a young male Elephant Seal – indicated by muscle scars and the shape of the skull, and a degree of sexual dimorphism is suggested. The rather large body and small flippers suggest that the animal did not have a thick coat of fur (Mitchell, 1966).

Desmatophocids did not give rise to any of the modern lines of pinnipeds. They were an isolated specialized group, most probably arising from the enaliarctid stock. In many respects they seem to be intermediate between otariids and odobenids, though they have many characters of their own. The distinct knob-like jugular processes of the exoccipital are different from the condition in either odobenids or otariids. They lack supraorbital processes, and have enlarged auditory ossicles like the odobenids; they also have narrow basioccipital bones and a moderately arched palate like the otariids. These, and many other more detailed characters that separate all suprageneric groups of modern and fossil otarioid seals are to be found in Repenning & Tedford (1977) – a paper which is warmly recommended to the serious student of fossil pinnipeds.

Odobenids

The origin of walruses took place in the North Pacific, probably along its eastern coast. The earliest specimen so far known is from the early middle Miocene, but the family Odobenidae was, at least by late Miocene, divisible into two groups. The subfamily Dusignathinae seems to have become extinct before the end of the Pliocene, while the subfamily Odobeninae leads on to the modern walrus. One of the major differences between the two subfamilies is that in the Dusignathinae the upper and lower canines are about equal in size, while in the Odobeninae the upper canines are elongated, and the lower ones reduced.

THE DUSIGNATHINE WALRUSES

At the moment, the earliest recorded odobenid is *Neotherium*, poorly known from very few bones from Kern County, California. It is from the early middle Miocene, about 14m years ago. The characters of its calcaneum indicate that although it is definitely odobenid, it is primitive, and it has been suggested that it may be an enaliarctic ancestor to the odobenids.

Dusignathine walruses include such animals as *Imagotaria*, *Pontolis*, *Dusignathus*, *Pliopedia* and *Valenictus*. Members of this subfamily were abundant in the North Pacific from about 14–11m years ago – they were the dominant seals there for about 5m years, co-existing first with desmatophocids, and later with the early otariids and the early odobenine walruses.

Imagotaria is probably the most generalized of the dusignathines; it has been found in Californian sites of late middle and early late Miocene, about 11–9m years ago. It was the size of the modern walrus, had teeth very like a sea lion (Mitchell, 1968), and has been estimated, from the structure of its ear, to have fed at deeper depths than does the modern walrus (Repenning & Tedford, 1977).

Pontolis was described from a shattered skull of very late Miocene age, but apart from its similarities to

Imagotaria, its affinities are uncertain. *Dusignathus*, from the late Miocene was more advanced; *Pliopedia*, from much the same age as *Dusignathus*, was probably only found in the warm inland sea that occupied the Central Valley of California; and *Valenictus*, extending into the late Pliocene, also lived in this inland sea. Probably all the dusignathine walruses had become extinct by about 3m years ago.

THE ODOBENINE WALRUSES

The earliest, and most primitive known member of the subfamily Odobeninae is *Aivukus*, from the late Miocene (6·5m years ago) of Cedros Island, Baja California, Mexico. Although some of its bones are different from, and more primitive than those of the modern *Odobenus*, others are indistinguishable. In addition it shows a slight increase in size of the upper canines, reduction of the lower canines, and the flattened peg-like cheek teeth are typically walrus-like. *Aivukus* is so far the most primitive, and the only odobenine fossil that has been found in the North Pacific (other than *Odobenus* itself) – all the other odobenine fossils having been found on both sides of the North Atlantic. The southerly distribution of *Aivukus*, and its closeness to the Central American Seaway* makes it possible that it was a species of this walrus that migrated, perhaps some 9m years ago, through this sea passage into the Atlantic.

Further odobenine evolution took place in the Atlantic. *Prorosmarus* is an early Pliocene walrus from Virginia; it is possibly as much as 3m years younger than *Aivukus*, and was becoming increasingly walrus-like in its dentition. *Alachtherium* is from the early and late Pliocene of Antwerp, and the alveoli in the skull now suggest the typical long upper canines of a walrus. The story, however, is not really as simple as this. There is not yet enough adequately dated material from before the Pleistocene, and it is quite possible that *Prorosmarus* and *Alachtherium* may turn out to be forms of the same genus.

In the Pleistocene, all walrus finds are referable to the modern genus *Odobenus*. Walrus remains, probably of Pleistocene age, have been found on both sides of the Atlantic as far south as the Carolinas in the west and Paris in the east. Not until the late Pleistocene, possibly as late as 70 000 years ago, did walruses re-enter the Pacific, this time via the Arctic.

OPPOSITE

Fig. 3.1 Diagram to show the probable fossil ancestry and relationships of the Otarioidea. (Based, with grateful thanks and acknowledgements on Repenning & Tedford, 1977.)

---

* An open sea passage between Central and South America. It closed, at least sufficiently to prevent passage of pinnipeds, in the late Miocene – about 5m years ago.

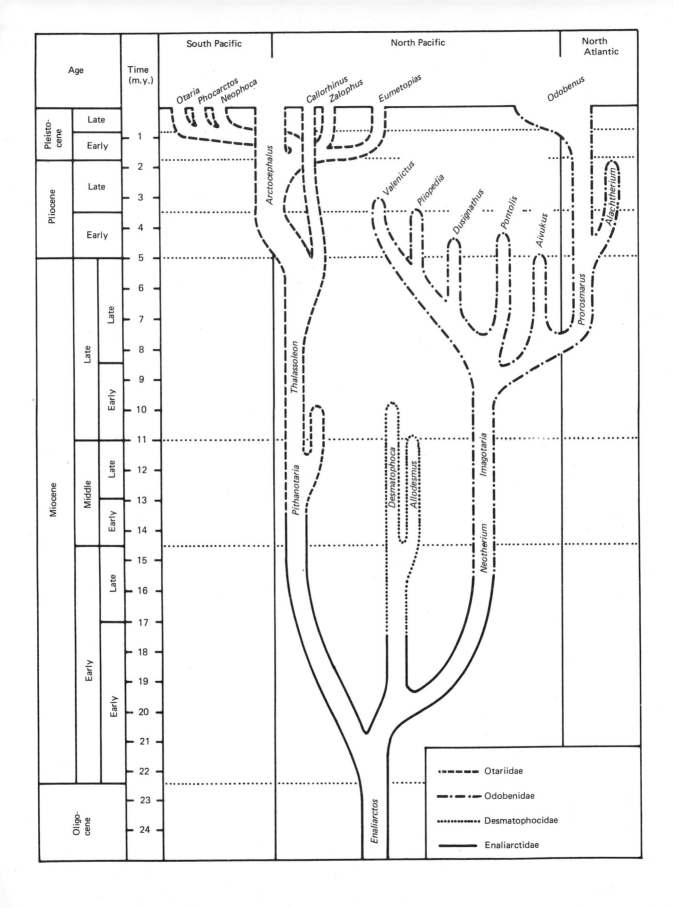

The otariid line

Probably somewhere around 15–13m years ago the ancestral enaliarctids gave rise to otariid seals, distinguished by the presence of supraorbital processes on the skull, and homodont dentition instead of carnassial teeth. *Arctocephalus* is probably the genus most like the original ancestral otariid, and the line leading to it is considered to be the basal lineage of otariids, both *Callorhinus* and the sea lions being divergences. Thus, small body size and the possession of a thick undercoat are considered primitive features when compared with sea lions.

The earliest known otariid is *Pithanotaria,* a small animal, probably only 1·5m in length, from late middle Miocene deposits (*c.* 11m years ago) of California. Although very early in the otariid line, this animal shows very few differences from later otaries. The cheek teeth are all double-rooted – there is no sign of the tendency to develop single roots as in later forms, and there are some primitive features of the limb bones. It is not considered as being on the direct line of succession to the modern fur seals because of an unusual reduction in the number of teeth, having five cheek teeth instead of the usual six.

Thalassoleon is a recently described genus of fossil otariid (Repenning & Tedford, 1977). One species, *T. mexicanus* is from Cedros Island, Baja California, Mexico, and is about 8–6m years old. It was a big animal, about the size of the modern South African Fur Seal, and has a number of characters of its skull and skeleton that are considered primitive. A second species, *T. macnallyae,* was described from material found at Point Reyes, California, and at some other Californian localities, and is probably about 6–4m years old. This is also a large animal, but differs from *T. mexicanus* in characters of its teeth and ectotympanic bone. In these characters it has similarities to the modern *Callorhinus* and it has been suggested that *T. macnallyae* may be nearer the origin of the *Callorhinus* line as it diverges from the main otariid line. However, further specimens may alter this hypothesis.

The divergence of the *Callorhinus* line from the main otariid lineage thus occurred long before the final major split of this line into sea lions and the modern *Arctocephalus.* Studies of the morphology of the baculum and the endemic louse fauna support this (Kim *et al.,* 1975). Little is known of the fossil history of *Callorhinus* after the early Pliocene, though Repenning & Tedford (1977) briefly record remains from San Diego from the middle of the late Pliocene, and from Alaska from the late Pleistocene.

Similarly with *Arctocephalus* – practically nothing is known of the evolution of this line after the appearance of *Thalassoleon mexicanus* in the late Miocene.

Some material from the lower Pliocene of Sacaco, Peru is from an *Arctocephalus* which is not believed to be on the main line of evolution (Muizon, 1978). It is, however, the first to be recorded from South America and possibly indicates that otariids moved into the southern hemisphere about 5m years ago (Repenning & Tedford, 1977). Of all the eight species of *Arctocephalus* that are recognized today, only *A.p. pusillus* and *A. townsendi* have even a meagre history, the former being known from the late Pleistocene, and the latter possibly from the early Pleistocene, but these are virtually modern animals.

Sea lions are distinguished from fur seals by the absence of abundant underfur – a character which is not, of course, determinable in a fossil, by an increase in body size, and by an increase in the rate of development of single rooted cheek teeth. Little is known of their fossil history, but they are believed to be a relatively recent group, having diverged from the main *Arctocephalus* line probably later than 3m years ago. A large jaw from Japan possibly from the late Pliocene, and identified as *Eumetopias* sp. is one of the very few early sea lion fossils available. Otherwise, such Pleistocene fossils as are available are almost all recognizable as modern genera within the limits of modern distribution.

Both fur seals and sea lions dispersed from their place of origin in the North Pacific. Fur seals are more widely dispersed in the southern hemisphere than are the sea lions, but as there are no fur seals or sea lions in the North Atlantic, both groups must have dispersed after the closure of the Central American Seaway in the late Miocene (6–5m years ago).

In the three families of otarioid seals that developed from the Enaliarctidae, there has been a tendency to develop single rooted cheek teeth from the multiple rooted teeth of earlier forms. This development took place at different times in the different groups, but was approximately coincident with their greatest taxonomic diversity. In the Desmatophocidae this took place about 14m years ago, and in the Odobenidae about 12m years ago. The Otariidae were extremely conservative in this respect until about 5 or 6m years ago, and the taxonomic diversification that appeared to go with this stage of evolution seems to be still in progress. All the early otariids had multiple rooted cheek teeth and even today the teeth of fur seals are not completely single rooted. The sea lions have a greater tendency towards a single rooted condition.

Phocoid seals

Our knowledge of fossil phocids, until very recently, was scanty in the extreme. In the last few years most known phocid fossils have been re-studied, and their affinities re-allocated, and many new fossils have come to light. A reasonable story of phocid evolution may now be made, but it has been emphasized that the finding of a few well-preserved bones in certain key localities could still alter our whole understanding of phocid geography (Ray, 1976).

The origin of phocids took place somewhere around the margins of the North Atlantic, probably about the beginning of the Miocene, about 20m years ago. The earliest known phocids are from the early middle Miocene (c. 14m years ago), and are thus some 8m years younger than the earliest otarioid seals. But even then the phocids were divisible into phocine seals and monachine seals, both these groups being equally old, and presumably there must therefore have been an even earlier common form as yet unknown.

The earliest phocids are *Leptophoca lenis* (phocine) and *Monotherium wymani* (monachine) from the Calvert Formation of Maryland and Virginia (c. 14m years ago). It is known that at about this time phocines (*Leptophoca, Prophoca*) were present on both sides of the North Atlantic, and a monachine (*Monotherium*) was on the western side, so it is estimated that perhaps both groups extended throughout this range. Our knowledge of all these early fossils, however, is based on very few bones.

Phocine seals

In the middle Miocene some of the North Atlantic phocines entered the Paratethyan Sea* presumably via an ancient Rhine River, as phocines are not known from the Tethys Sea*. There the seals radiated and evolved in isolation throughout the late Miocene and Pliocene. It is possible that the modern genus *Pusa* originated here as some of the Paratethyan fossil seals are similar to living *Pusa* (Grigorescu, 1976). The ancestor of the future *P. caspica* then presumably remained in that part of Paratethys which is present today as the Caspian Sea. Some 3m years ago *Pusa* could have escaped from Paratethys to the Arctic

Tethys – an equatorial sea which, some 20m years ago connected the Atlantic Ocean with the Indian Ocean. The main body of this sea occupied that area which is now called the Mediterranean, and a southerly arm connected with the Indian Ocean.

Paratethys – a northern arm of Tethys stretching across the area which now houses the Black, Caspian and Aral Seas. Between 20 and 15 million years ago uplifting of mountains isolated the Paratethys which became a brackish inland sea. About 6–5m years ago tectonic movements led to a great drying up of the Mediterranean and both it and Paratethys were reduced to a series of lakes so the Black, Caspian and Aral Seas became separate bodies of water (Hsü, 1978).

Ocean during a time when the Arctic Ocean stretched southwards close to Paratethys in the area west of the Ural Mountains. Perhaps some 300000 years ago the future *P. sibirica* entered the Baikal Sea from the Arctic, via large lakes at the southern margin of the Siberian ice sheet. Much of this though is conjecture, based on what is known of the geologic history of the area, as there are virtually no fossils to suggest the exact history of the distribution of *Pusa* (Repenning *et al.*, 1979).

Analysis of skull characters by various authors over the years has indicated various relationships between the three species of the genus *Pusa*. Recent work (Timoshenko, 1975) indicates a close relationship between the Caspian and Ringed Seal, with the Baikal Seal being different from both. It is apparent that there is still much to be discovered about the origin and relationships of these animals.

Phocine seals were animals of the middle to high north latitudes, and possibly this was why they never entered the Pacific through the Central American Seaway when it was open. In the Pliocene the climate of the North Atlantic began to deteriorate. Phocines apparently reacted to these adverse conditions by adaptation, and modern forms of phocines began to appear. The rather late radiation of these animals may possibly explain the great variety of phocines, and the consequent taxonomic difficulties (Burns & Fay, 1970, Ray, 1976).

In the Pleistocene all the seals extended their ranges, moving northward and southward in the North Atlantic in response to alternate periods of warm and cold climate. The exact history of the various genera of phocines is not known, but it seems that some groups at least are still actively evolving. *Phoca*, for instance, appears to be still differentiating and while some suggest that this genus should include *Pusa, Histriophoca* and *Pagophilus* (Burns & Fay, 1970), others feel that these three have evolved sufficiently to deserve generic distinction. The local populations of *Pusa* and *Pagophilus* may be the signs of the species differentiating.

Monachine seals

Monachine seals are known to be present on both sides of the Atlantic – from North Carolina, Virginia and Maryland, and from the Antwerp Basin in Europe – by the middle Miocene, but relatively few fossils are yet known. Although in the eastern North Atlantic, there is no certain evidence that they invaded Paratethys since the modern genus *Monachus* did not reach the Black Sea until much later – in the Pleistocene.

At the end of the Miocene, and in the Pliocene, monachines were the dominant seals of the North

Atlantic, several genera and species being known from the Antwerp Basin and from North Carolina. From both areas the smaller *Callophoca obscura* and the larger *C. ambigua* are known, though whether these are two species or one sexually dimorphic species is not known. It is possible that *Callophoca* is the ancestor of *Manachus* or *Mirounga*, or both (Ray, 1976).

The monachines present during the early Pliocene on the American Atlantic coast and Caribbean area were *Callophoca* and *Monotherium*. A form rather like *Callophoca* or *Monotherium* must have been the common ancestor of *Monachus tropicalis* and *M. schauinslandi*, some of which must have passed through the Central American Seaway, eventually to become *M. schauinslandi*, while others remained in the Caribbean as *M. tropicalis*. *M. schauinslandi*, however, has several skeletal features that are more primitive than those of *Monotherium*, the oldest known fossil monachine which is approximately 14·5m years old. It has been suggested that *M. schauinslandi* separated from its ancestral population more than 15m years ago – a timing that coincides well with the warm temperatures and favourable currents in the North Pacific (Repenning *et al.*, 1979, Repenning & Ray, 1977). The ancestor of *Mirounga* was probably of the same stock as *Monachus*, and also passed through from the Caribbean to the Pacific, eventually becoming extinct in the Caribbean. Some of the *Mirounga* stock must have moved south along the Pacific South American coast to give rise to *M. leonina*.

Prionodelphis was described from a mixture of cetacean and seal teeth from Argentina, and the name *P. rovereti* has now been restricted to the cetacean (Muizon & Hendey, 1980). The seal teeth are monachine, but not yet named. Late Pliocene monachine remains from South Africa, previously called *Prionodelphis capensis* are now known as *Homiphoca capensis*. The origin of *Homiphoca* is not known, but it could have come from *Monotherium*. Abundant remains of *Homiphoca* have now been found in South Africa (Muizon & Hendey, 1980) and examination of the skulls and skeletons suggests that the genus is morphologically intermediate between the Monk Seals and the Antarctic Seals (Monachini and Lobodontini, excluding *Mirounga*). The structure of the auditory region suggests that *Homiphoca* is more closely related to the Lobodontini, and possibly to *Lobodon*, though direct ancestry is not implied and it may have been a parallel line with similar adaptations.

The Monachinae originated in the North Atlantic, and it is possible that the ancestors of *Homiphoca* moved down the African coast, across the Atlantic and down the east coast of South America and back across the Atlantic again to reach South Africa, the oceanic trips following the major currents (Hendey, 1972, Ray, 1976, Hendey & Repenning, 1972, Muizon & Hendey, 1980).

On the eastern side of the Atlantic the Pliocene descendants of Monotherium were *Pristiphoca* and *Pliophoca*. These eventually gave rise to *Monachus monachus* which then occupied the Mediterranean after it was open to the Atlantic again.

It must be emphasized again that the story of phocid evolution is still in its infancy. New discoveries are being made, and much work is being done that could well alter the story as set down here.*

---

* As an example of this a recently published account describes an interesting phocid population from a Lower Pliocene site at South Sacaco, some 550 km south of Lima, Peru (Muizon, 1981). Two new seals are described *Piscophoca pacifica* and *Acrophoca longirostris*, both belonging to the Monachinae and Lobodontini.

Piscophoca is thought to be a descendant of *Monotherium* and shows similarities to *Monachus*. *Acrophoca* was the most common phocid at Sacaco; it had a particularly long neck, and the skull shows a peculiarly long snout. The anatomy of its limbs seems to indicate a more littoral and terrestrial life than both *Pisciphoca* and the living members of the Lobodontini. It is not the ancestor of *Hydrurga*, but is possibly related to that line.

Homiphoca from South Africa, Monachine remains from Argentina, and the phocids of Sacaco indicate the movement south of members of the Monachinae from their origin in the North Atlantic.

RELATIONSHIPS

At the beginning of this chapter it was noted that the earliest otarioids, belonging to the family Enaliarctidae, were first known from the beginning of the Miocene, and the earliest phocids were known from the middle Miocene. Thus there are no pinnipeds known before the Miocene, and at the moment, no fossils which can act as good connecting links to show from where these animals came. But study of the available fossil material gives many clues about the relationships of the two major groups of pinnipeds to other carnivores.

There is no doubt that seals have come from carnivore stock as they have one of the diagnostic characters of the Carnivora, i.e. two of the carpal bones (scaphoid and lunar) are fused together. But from what particular group of carnivores have they come – and have both otarioids and phocids come from the same group?

The order Carnivora is divided into two major groups – the superfamilies Feloidea and Canoidea. (This is the conventional order Carnivora, not including the pinnipeds.) The Feloidea may be divided into three families – Viverridae, Hyaenidae and Felidae, and all are characterized, amongst other things, by a double-chambered bulla that is divided internally by a septum. All members of the Canoidea have many cranial and dental characters in common, including a single-chambered bulla. The four families that make up the Canoidea can be divided into two unequal groups, and again it is characters of the bulla that separate them. The Canidae (dogs, foxes, etc.) have, as a character unique to them, a low septum inside the bulla, formed from the inbent edge of the entotympanic. The carotid canal is short, and hidden in a ventral view of the skull. Members of the Canidae may be known as the cynoid carnivores.

The other three families in the Canoidea – the Ursidae, Procyonidae and Mustelidae are known as the arctoid carnivores. The single-chambered bulla is without any internal septum, and always has a very prominent carotid canal. In the Ursidae the entotympanic part of the bulla is one third or less of the entire bullar structure. In the Mustelidae, while there is considerable variation, the entotympanic is expanded relative to the other elements, and usually forms more than a third of the bulla. Members of the Feloidea, and also cynoid and arctoid carnivores, are known by the structure of their bullae as far back as the beginning of the Oligocene.

The single-chambered bulla of pinnipeds indicates that their affinities lie with the Canoidea rather than with the Feloidea (Tedford, 1976). A similar relationship is shown by the structure of the baculum which is more canoid-like than feloid-like (McLaren, 1960), and this also agrees with the immunological findings (Sarich, 1969).

Within the Canoidea, the prominent carotid canal found in the arctoid carnivores is similar to that found in pinnipeds.

If the ear region of otarioid seals is 'stripped' of its aquatic specializations, its primitive anatomy may be compared with that of other arctoids, and it has been shown that when this has been done, there are undoubted ursid resemblances. *Enaliarctos* lacks many of the aquatic specializations of later otarioids, and ear and tooth characters of this fossil show that it shares most characters with the Oligocene ursid subfamily Amphicynodontinae. These animals were present in Eurasia in Oligocene times, and there are some repre-

sentatives from North America. There were 10m years between these early ursids and the first recorded fossil of the enaliarctids, and it has been suggested that the evolution of the otarioids was relatively quick, and probably within the last 5m years of this period (Tedford, 1976).

Less is known about the relationships of the phocids, as fossils are less plentiful and less well known. Again it is necessary to look under the overlying aquatic specializations of the ear to search for relationships, and what one must deduce is whether the primitive ear region is more like that of the early members of the Mustelidae or Procyonidae, rather than the Ursidae. The absence of an alisphenoid canal, and the separation of paroccipital and mastoid processes in phocids also suggest similar affinities.

In fact, the hypothetical early phocid is not unlike *Potamotherium*, a possible early Miocene mustelid carnivore which occurs both in Europe and North America. *Potamotherium* has a mixture of primitive, mustelid, pinniped, and indeed phocid characters in its skeleton (Savage, 1957, Tedford, 1976), though no specifically phocid characters of its ear. Reconstruction of the musculature of *Potamotherium* suggests that it swam rather like a phocid. *Potamotherium* itself, and also the late Miocene *Semantor* from the Caspian Basin are too late to have been direct phocid ancestors. It is at the moment believed that 'it might be possible to regard *Potamotherium* as occupying an intermediate structural position between terrestrial mustelids and pelagic phocids in much the same way as does *Enaliarctos* with respect to the ursids and the otarioids' (Tedford, 1976).

To recapitulate briefly the history of otarioids and phocids – arctoid carnivores were present in the Oligocene; a subfamily of Oligocene ursids (Amphicynodontinae) probably gave rise to the Enaliarctidae, which in turn gave rise to the otarioids. The Oligocene bears were present in North America and Asia and the major evolution of the otarioids took place in the North Pacific.

Phocid ancestry is not so well known, but they probably came from a stock of primitive mustelids closely allied to *Potamotherium*. The major evolution of the phocids took place in the North Atlantic.

The pinnipeds have thus arisen from two different groups of arctoid carnivores and have evolved in two different oceans. One cannot therefore group the otarioids and phocids *together* to form the order Pinnipedia.

It is thus better to follow the older authors, and also the most recent authors in leaving the pinnipeds in the order Carnivora.

It now becomes obvious that if the pinnipeds are

included in the order Carnivora, then the various rankings within this order will have to be re-drafted. The usual first division of the order Carnivora into the two superfamilies Feloidea and Canoidea makes the use of the hitherto conventional pinniped terms superfamily Otarioidea and Phocoidea invalid.

Mitchell & Tedford (1973) suggested the relationships might be shown like this:

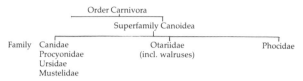

Tedford (1976) reconsidered his opinion, and an abbreviated version of his plan now looks like this:

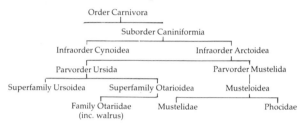

Whatever higher classification is eventually decided upon it has yet to become established by the test of time. Until this has happened, many pinniped zoologists, while appreciating the situation, will have to use *some* form of classification, and it is very likely that many of the 'old' terms will remain in use for some time to come. Repenning (1977) for instance, dealing only with otarioids, starts his classification with superfamily Otarioidea in the time honoured way:

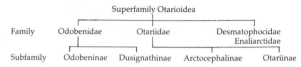

Unless it is necessary to go into the realms of higher classification, it is probably easier at the moment for the 'ordinary' zoologist to start the classification of living pinnipeds at the family level – Otariidae, Odobenidae and Phocidae, but if fossils are being considered as well then the convenience of a classification like that used by Repenning (1977) given above is obvious.

CHROMOSOMES

When studying fossils one is looking at the actual remains of the animals that existed in say, the Miocene. When studying chromosomes one can obviously only look at those of modern animals and one cannot know for certain the relationship between the known modern chromosomes and the hypothetical arrangement of those of the fossil animals.

All otariids so far surveyed have 36 chromosomes, and the walrus has 32. All the monachine phocids have 34, and nearly all the phocines have 32, derived from the 34 type by fusion of two chromosome pairs. The two phocines that are different are *Erignathus* and *Cystophora*, both of which have 34. It seems that the relationships of these last two seals are yet to be settled.

Arnason (1974) suggests that *Erignathus*, with its 34 chromosomes which even show characteristic differences from other 34 chromosome karyotypes, should not even be included in the subfamily Phocinae, showing as it does a separate line of karyotype development. Fay *et al.* (1967) suggest that *Erignathus* has a closer relationship with the Monachinae than is generally recognized.

In spite of the 34 chromosomes of *Cystophora* showing resemblances to those of *Mirounga*, Arnason agrees that *Cystophora* still has a close relationship with the Phocinae. The anatomical differences by which *Cystophora* differs from the *Monachinae* do not tally with the chromosomal similarities, but Arnason notes 'the concordance between the 34 chromosome karyotype of the hooded seal, and the 32 chromosome phocid karyotypes indicates that the karyotype from which the 32 karyotype originated, is still preserved in the hooded seal. The view of a closer systematic relationship between *Cystophora* and the Phocinae is therefore supported.'

There is considerable stability of the karyotypes of both highly aquatic groups – the pinnipeds and the cetaceans. Both phocid and otariid karyotypes have pronounced similarities, but this could be interpreted as indicating the primitive arctoid state. The morphology of some of the pinniped chromosomes corresponds closely to those in procyonid animals, and definitely excludes the lutrines from any place in pinniped ancestry (Arnason, 1974).

SECTION II

Pinniped biology

Chapter 4 Shape, flippers and locomotion

Streamlining

In common with most other mammals that spend the greater part of their life in the water, seals are modified to meet the different needs of a life at sea, and their external shape is changed from that of a typical land mammal so that swimming can be swift and effective. The body is streamlined and spindle shaped with no sharp protuberances to break the even contour (Fig 4.1). The external ear is a very small structure in otariids, and is not evident at all in phocids and odobenids. The teats are retracted, the penis is withdrawn into a pouch and also lies below the body surface, and in phocids and odobenids the testes are inguinal, invisible externally. The phocids and otariids have a small but obvious tail (c. 8cm long, Fig 4.2), and although not used in swimming, it fits neatly into the gap between the hind flippers and helps towards a smooth outline. In the walrus a broad web of skin stretches between the two ankle bones and envelops the tail so that it is not distinct as an appendage.

Flippers

The size and shape of the limbs are perhaps the most obvious ways in which seals differ from a typical land mammal. Leg and arm bones are relatively short and are within the body outline. The axilla in otariids and walrus falls at about the middle of the forearm, and in phocids at the wrist. In all three families the hind flipper is free only from the ankle.

Flippers and grooming

As there is no clavicle the forelimbs have great freedom of movement and this is particularly obvious in those phocids with large claws on the manus, where these claws can be used to scratch almost all parts of the body. An Elephant Seal for example (Fig 4.3), while keeping the palmar surface of the manus against the body, can swivel the flipper forwards to scratch the chin, and then round backwards to scratch the top of the head. Phocid hind flippers are not so mobile, but in Northern phocids the pes can be flexed so that the claws of one pes can scratch the sole of the other. The great width of the outer digits of the hind flipper in some Southern phocids (in *Mirounga, Ommatophoca,* and *Hydrurga* at least) would prevent this flexion. The width of the digits is increased by fibrous tissue, and Elephant Seals for example, have been seen to rub their hind flippers together, but not to flex them.

Fig. 4.1 *The streamlined shape of* Phoca vitulina richardsi *is characteristic of all seals. Photo. J. E. King.*

Fig. 4.2 *The hind end of a female* Phocarctos *showing the small tail. Also shown is the otariid capability of bending the hind flippers forwards at the ankle. Photo. P. G. Poppleton.*

Fig. 4.3 *A young Southern Elephant Seal demonstrating the mobility of the foreflipper while scratching the neck. Photo. J. E. King.*

Otariids can use the middle three claws on the pes to scratch delicately at almost any part of the body, but the shape of foreflippers and the small size of their claws mean that scratching with flexion cannot be achieved, though the whole flipper can be used to rub parts of the body.

Flipper shape

The manus and pes of seals, i.e. the flippers, are larger than the corresponding parts of the body in a terrestrial mammal. The whole flipper, as one unit, is encased in integument, so that the individual digits are not free, and the surface area of the flipper is thus increased. This is a fatty and fibrous tissue pad between the digits of the foreflipper.

Phocid flippers

All phocid flippers have hair on both surfaces. There is a difference in shape between the foreflipper of a typical Northern phocid and that of a typical Southern phocid. A Northern phocid foreflipper is short and broad, each digit armed with a strong claw which protrudes over the free edge of the flipper. The first digit is the largest, but there is only a moderate decrease in length between the first and fifth digits. Within the Southern phocids there is some gradation – *Monachus, Mirounga, Leptonychotes* and *Lobodon* having fairly large claws, but *Hydrurga* (Fig 4.4) and *Ommatophoca* have foreflippers that differ most from the pattern of those of the Northern phocids. The flipper is much more elongated, with a greater difference in length between first and fifth digits. The claws are slim and small – *c.* 8–15mm long in a Leopard Seal, and *c.* 3mm long in a Ross Seal. These lengths should be compared with those for a Baikal Seal where the claws are some 30mm long and correspondingly stout. It is possible that there may be less use of the foreflipper claws for grooming in these Southern seals. In both Leopard and Ross Seals the claw of the first digit is set back from the free edge of the foreflipper, and in the Ross Seal certainly, the extension of soft tissue is supported by *c.* 20mm of lanceolate-shaped cartilage (King, 1969b).

Phocid hind flippers have the first and fifth digits elongated, so that the posterior margin of the flipper is concave, the concavity being more pronounced in Southern phocids (Fig 4.5). The individual digits are more obvious than on the foreflipper, and between them is fine expansible skin, covered with small hairs. All the Northern phocids have large claws on each digit of the hind flipper, but these claws are reduced in the Southern phocids, culminating in the very small nodules that are all that remain in *Ommatophoca*. The lateral expansion by fibrous tissue of the outer digits of at least some Southern phocid hind flippers has already been mentioned.

Otariid and odobenid flippers

Otariid foreflippers are elongated structures with the claws reduced to small nodules. The dorsal surface is

Fig. 4.4 *The right foreflipper of a Leopard Seal – a Southern phocid. The flipper is elongated, with a considerable difference in the length between the first and fifth digits. The small claws are also shown. Photo. J. E. King.*

Fig. 4.6 *The right foreflipper of a* Phocarctos *pup showing the typical otariid shape. The flipper is elongated, (and will be more so in the adult), the first digit is longer than the others, and the claws are very small and remote from the free edge. Photo. J. E. King.*

Fig. 4.8 *Dorsal view of the right foreflipper of a Walrus as the flipper is folded over the belly of the animal. The flipper is broad, but of general otariid form, with small claws. Photo. J. E. King.*

Fig. 4.5 The expanded hind flippers of a Southern Elephant Seal. The dorsal surface of the right flipper is visible, lying directly on top of the left flipper. The concave posterior margin of the flipper can be seen, in addition to the positions of the digits and the expansion of the lateral digits. Photo. J. E. King.

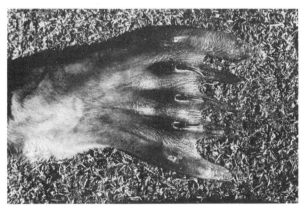

Fig. 4.7 The right hind flipper of a Phocarctos *pup showing the typical otariid shape. Long, grooming claws are present on the three middle digits, while those on the outer digits are very small. The grooming claws are particularly long and sharp here as the pup is newborn. The increased length of all the digits is supported by cartilage internally. Photo. J. E. King.*

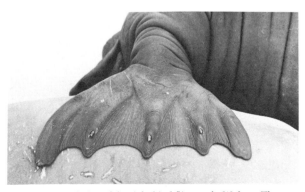

Fig. 4.9 Dorsal view of the right hind flipper of a Walrus. The larger grooming claws can be seen. Photo. J. E. King.

covered with short hair, but the palm is of naked black wrinkled skin. The first digit, forming the leading edge, is conspicuously longer and stronger than the other digits. An extension of the flipper border beyond the normal tips of the digits is supported by cartilaginous rods. The small claws are thus remote from the flipper edge and cannot be used for grooming (Fig 4.6).

The hind flippers of otariids have hairless black soles, and the length is increased by predigital cartilages as in the foreflipper. The distal edge of the flipper has an indented outline because of the projection of these cartilaginous extensions. These extensions can be folded back and the long claws of the three middle digits used for grooming. The two outer claws are very small (Fig 4.7).

Walrus foreflippers are rather short and square, with all five digits much the same length. The five claws are small nodules and the cartilaginous extensions are short (Fig 4.8). On the hind flippers the three middle claws are longer and more functional than the two outer ones, as in otariids, and again the cartilaginous extensions are short. The flippers are the same greyish-brown colour as the rest of the animal (Fig 4.9).

Locomotion
(See also the section on the vertebral column.)

Swimming

OTARIIDS

While in water the head and long, strong sinuous neck of an otariid are moved about in search of food, and the hind end of the body acts as a rudder. In water the foreflippers are the normal means of propulsion and are used as oars with a forceful stroke towards the body, so that the animal 'flies' along under water rather like a penguin. The hind limbs are not used in swimming. During rapid swimming and also during play, otariids often 'porpoise' i.e. they leap clear of the water for a short distance.

WALRUSES

In water most of the power for propulsion comes from the alternate strokes of the hind flippers. The foreflippers are also used to a certain extent, working both together, and they are also used as hydroplanes and for manoeuvring (Ray, 1963, Gordon, 1981).

PHOCIDS

In phocids the propelling mechanism is at the hind end of the body and, because for mechanical reasons a long neck is impracticable, the head merges into the trunk without a clearly defined neck region. Swimming is by alternate medially directed strokes of the hind flippers, the flipper itself being expanded during the propulsive stroke and contracted during recovery. The movement is assisted by lateral swinging of the hind end of the body, and there is consequently great development of the spinal musculature. When swimming fast the foreflippers are usually held close to the body, but are used when changing direction, and are sometimes used in a paddling movement when the animal is going very slowly. Male Harbour Seals have been seen to 'porpoise' at times of excitement during the mating season (Hewer, 1974). As far as observations have allowed, no differences have been seen between the swimming methods of Northern and Southern phocids except that *M. schauinslandi* has been seen to use its foreflippers rather a lot, moving them rapidly up and down in a vertical plane.

Terrestrial locomotion

Otariids have considerable agility on land, as any circus-goer will know (see section on *Zalophus*). The hind flippers bend forwards at the ankle and the fore flippers extend laterally with a right angle bend between the two rows of carpal bones, and the animal supports the weight of the body on the palms and soles. When walking on land the whole of the belly is held clear of the ground and from the short-legged silhouette it is easy to see why the old sealers used to refer to them as 'sea-bears'. The foreflippers are moved alternately when the animal is going slowly, the whole of the manus being placed flat on the ground, and the heavy head and neck region are swung over the weight-bearing flippers so that the other one may be raised and moved forward. It has been suggested that if the neck were only half its length the otaries would find it impossible to move on land. The hind flippers are also moved alternately, the pes raised so that only the heel rests on the ground. As the animal moves more quickly, a gallop results with both foreflippers being moved forwards together, followed by both hind flippers. Walruses also move on land in this manner.

WALRUS
On land the movements of the flippers are much as in *Zalophus* but most of the body weight is taken by the ventral surface and it is a strong body lunge that raises the body from the ground and moves it forward (Gordon, 1981).

PHOCIDS
On land a phocid progresses in a relatively clumsy fashion, but even so it manages to move considerable distances over both flat and rocky ground. The hind flippers are held up clear of the ground and are not used, so the chief method of progression is an effective 'humping', the weight being taken alternately on the sternum, while the hind end is brought forwards, and then on the pelvis while the front end of the body is extended (Tarasoff *et al.*, 1972).

The extent to which the foreflippers are used in terrestrial locomotion varies a good deal. Northern phocids such as the Grey Seal may make considerable use of them, particularly when moving over rocky ground as the terminal phalanges are capable of a powerful haulage grip and may assist in pulling the body forwards (Backhouse, 1961). Harbour Seals put both foreflippers down on the sand simultaneously and then hump the body forwards. The foreflippers leave obvious tracks in the sand and the width between the two flipper marks can give some indication of the size of the animal (Vaughan, 1978).

Among the Southern phocids, the Weddell Seal for example, hitches along without any assistance from its flippers. *Mirounga* puts its foreflippers out at right angles to its body and uses them for balance and possibly a certain amount of traction with the whole of the leading edge against the ground. The Crabeater Seal when moving on ice can attain a high speed of up to *c.* 25km per hour using a sinuous movement – with alternate backward strokes of its foreflippers against the substrate, and a vigorous flailing movement of its hind end (O'Gorman, 1963). When not on ice it uses the usual phocid method. The Leopard Seal will also use a sinuous movement when agitated, but its large foreflippers seem to be more adapted for manoeuvring than for a driving force (Bryden & Felts, 1974). Ross Seals do not use a sinuous movement, but make much use of their foreflippers against the substrate, rather like a turtle. Ribbon Seals can move over ice as fast as a man can sprint, and they also extend their foreflippers laterally and make sinuous movements of the hind end. The neck is also extended and moved from side to side (Burns, 1981).

PUPS
While still very young, the movements of pups are less co-ordinated than those of adults. For the first few days, Grey Seal pups for example, may use the foreflippers alternately, and then later, together, to drag the body along.

SPEED
There is not much information on swimming speeds under natural conditions. Otariids have been estimated to achieve *c.* 28km per hour for short periods (Ray, 1963), phocids may reach *c.* 19km per hour, and the faster movement of the Crabeater on ice has just been mentioned. Walruses are much slower and only rarely cruise above about 6km per hour, but from the sedentary nature of their food one would not expect them to move very fast. If chased, however, they are reported as being able to reach as much as 30km per hour for a very short burst.

Skin

The integument of seals consists of an epidermis and a dermis, these two layers forming the skin, and below this is a hypodermis, the latter forming the blubber.

In all phocids the epidermis is dark coloured, while it is light in colour in otaries and walruses, except, of course, on the flippers (Scheffer, 1964). The outer layer of the epidermis (stratum corneum) is composed, possibly in all phocids (Montagna & Harrison, 1957), (Ling & Button, 1975), and certainly in *M. leonina* where a great deal of work has been done (Ling, 1968), of flattened, solid, keratinized cells. These are lubricated by lipids from the sebaceous glands so that a waterproof, pliable horny layer is formed over the animal, this layer being shed only at the annual moult.

This may be contrasted with such a layer in man for example, where enzymatic digestion of the non-keratinous parts of the outer cells of the stratum corneum results in these cells becoming mere flakes continuously shed. Such a condition in a seal could lead to the skin becoming waterlogged.

The thickness of the seal epidermis may have some inverse relation to the number of underfur hairs or may be an adaptation to prolonged immersion. The epidermis is thicker in sea lions and phocids (c. 100–200μm, through varying between 150–500μm in the Southern Elephant Seal), and thinner (c. 50μm) in fur seals. The thickest epidermis has been reported from the walrus where it may be as much as 3·3mm (in Ling, 1968).

The dermis is formed of dense connective tissue, well supplied with blood vessels including arteriovenous anastomoses. It houses the hair follicles and is closely connected with the underlying blubber. The dermis, as in other mammals, forms the bulk of the 'hide' of the animal, which when removed can be tanned to form leather. The dermis is 3–4mm thick in an adult *Callorhinus*, and is particularly thick in the walrus where it may be over 2·5cm, and may reach 6cm on the neck.

The hypodermis or blubber is closely connected to the overlying dermis, but only loosely connected by connective tissue to the underlying muscular layer. Thus when a seal is 'skinned' the epidermis, dermis and blubber are removed as one layer from the rest of the body. The fat cells that make up the blubber are supported in loose connective tissue and many blood vessels pass through it to reach the dermis. The thick-ness of the blubber depends on the age, size and general condition of the animal, but tends to be thickest in the chest region. In a walrus the blubber in this region may be 4–10cm thick, and in adult male Southern Elephant Seals it may also be 4–10cm thick. The ventral surface of a Northern Fur Seal has barely 3mm of blubber at birth, but this increases to a thickness of 6cm on the belly of an adult bull (Scheffer, 1962).

It is interesting that the histological structure of the skin of a mummified Weddell Seal estimated to be about 1400 years old has been satisfactorily determined (Orr, 1971).

Sebaceous and sweat glands

Epidermal structures that are located in the dermis are the hair follicles, sebaceous glands and sweat glands. The hair follicles and their hairs will be mentioned later.

Sebaceous glands are associated with each pilary canal. In *C. ursinus* each guard hair is flanked by a pair of sebaceous glands (Scheffer, 1962), in *M. leonina* there is a large bilobed gland for every guard hair, and in *Phoca* the glands are in clusters (Ling, 1974). The secretions of these glands are probably primarily to keep the epidermal cells pliable and in good condition rather than to waterproof the hair.

On the human body liquid sweat for cooling purposes is produced by eccrine sweat glands found all over the body. Apocrine sweat glands are also found, but in restricted areas. They are usually developed in close association with hairs, their ducts normally open into the hair follicle, and they have a thicker secretion than do the eccrine glands.

Only the apocrine sweat glands, one to each guard hair follicle, have been identified in seals. They certainly occur on the body where they produce a rather viscid secretion which is not thought to have much value in thermoregulation. The secretion may possibly be responsible for the curious musky smell possessed by many seals. There is great increase in the activity of the sweat glands at the time of moulting, and also in the breeding season, when they may act as sexual scent glands. Sweat glands have not been found on Harbour Seal flippers (Tarasoff & Fisher, 1970) but have been found on *Callorhinus* flippers (Scheffer, 1962), and on the flippers of some *Arctocephalus* (Ling, 1965) and *Zalophus* (Whittow *et al.*, 1975). Sweat glands are small in phocids and larger in otariids and walruses. Those of the walrus are said to atrophy during autumn (Ling, 1965). Some of the largest sweat glands are between the mystacial vibrissae of the walrus where they possibly produce material for olfactory recognition between mother and young (Ling, 1974). As only the apocrine sweat glands are found in seals it seems likely that they have more of an olfactory than a thermoregulating function, the latter function being adequately served by other means.

Fur

The pelage of seals consists of outer protective guard hairs and soft underfur hairs. No hair erecting muscles have been found in seals. The hairs grow in groups or units, a longer stiffer flattened guard hair growing anterior to a group of shorter finer wavy underfur hairs, and lying on top of them when the pelage is lying naturally (Fig 5.1). All the hairs of one unit grow from separate follicles, that of the guard hair being deeper in the skin than the others, but all hairs of each unit emerge to the surface through the one pilary canal whose walls grip the hairs and prevent water entering the canal. The guard hairs have a pith or medulla in otariids, but not in phocids and walrus, and the underfur hairs are always unmedullated (Scheffer, 1964).

The hair units grow on the skin in various patterns. They are spaced regularly, in an arrangement that has been likened to trees in an orchard in all otariids and in the Monachinae; and in small groups in the walrus and Phocinae, though the pattern of the groups varies. The number of underfur hairs per unit varies from none at all in walrus, elephant and monk seals, about one to three hairs in sea lions and the rest of the Monachinae, about five in the Phocinae with the Bearded Seal having rather more (c. 7+), and higher numbers of 17 in the Northern Fur Seal (Scheffer, 1964). In this latter animal it has been shown that some of the underfur hairs remain trapped in the pilary canal from one year to the next, and are not shed from the body during the moult, giving the impression of a much greater proportion of underfur. In this way up to about 68 underfur hairs per hair unit may be found in an adult male, and even in a commercially prepared skin most of these remain firm (Scheffer & Johnson, 1963). Possibly this may also be the situation in other fur seals.

The hairs units are more closely packed on a young animal and become more widely spaced as the seal grows – 1296 units cm^{-2} were counted on a three-year-old male *Callorhinus* and 790cm^{-2} on a 15-year-old female animal. It has been estimated that an average adult female *Callorhinus* with a total haired area of 0·62m^2, and 52 fibres per hair unit, would have approximately 303 million hairs on her body (Scheffer, 1962). This is about 40–60000 hairs cm^{-2}, and it is interesting to compare this figure with that of about 125000 hairs cm^{-2} in the sea otter *Enhydra*, which has the densest fur covering of any mammal. It should, however, be remembered that the sea otter has virtually no blubber and relies on its fur coat for insulation (Tarasoff, 1974).

Whiskers

The mystacial whiskers are useful tactile organs in seals, but these are not the only whiskers present, and the number and arrangement of all whiskers varies considerably. Mystacial whiskers along the sides of the snout are the most obvious and abundant, superciliary whiskers over the eyes are present, and rhinal and submental whiskers are present in some seals. No tactile body whiskers have been reported.

The mystacial whiskers are arranged in horizontal rows along the sides of the snout. There are about 15–20 on each side in Leopard and Ross Seals, about 20–30 in otariids, about 38 in the Southern Elephant Seal, about 40–50 in *Phoca* and *Pusa*, and about 300 in the walrus.

The whiskers are smooth in outline in otariids, walrus, Bearded and Monk Seals, and 'beaded' with a wavy outline which may be more or less obvious in all other seals. In size they range from the short stout quills of the walrus moustache which are 8–9 cm in length and 2–3 mm in diameter to the very long whiskers of fur seals. A Northern Fur Seal bull had a whisker 33cm long (Scheffer, 1962), and the record seems to be held by the South Georgia fur seal bull

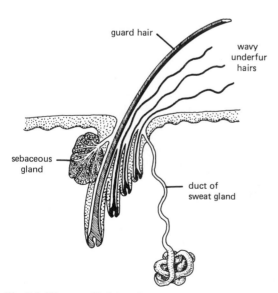

Fig. 5.1 Diagram of hair bundle.

whose longest whisker measured 48cm (Bonner, 1968). Usually only a few of the posterior whiskers reach any great length.

Superciliary (= supraorbital) whiskers are usually better developed in phocids, though as far as is known, otariids and Antarctic phocids are again alike in having smaller numbers. None to two or three whiskers over each eye have been recorded in otariids and Leopard Seals, whereas the Southern Elephant Seal has seven on each side in this position (Ling, 1966).

Rhinal whiskers, one to two on each side, just posterior to each nostril are only known so far from Elephant (Ling, 1966), Leopard and Ross Seals (King, 1969b), and submental whiskers from Bearded (Kosygin, 1968) and Leopard Seals. Whiskers other than mystacial are usually short and may be worn down so they are hardly visible.

The mystacial whiskers are tactile structures, well supplied with muscles and nerves. Each follicle is surrounded by a tough connective tissue capsule, and between it and the shaft of the whisker most of the tissue resembles erectile tissue – venous sinuses set in a network of connective tissue. A ring sinus surrounds the shaft in the middle of this cavernous tissue and as in many other mammals a ringwulst protrudes into the ring sinus. This ringwulst is probably composed of fine collagen fibres with their long axes running between the shaft and the sinus. The function of the ringwulst is not certainly known – it is small in the Southern Elephant Seal (Ling, 1966) and larger in *Zalophus* (Stephens *et al.*, 1973). In the latter animal that part of the ringwulst that faces the nose is narrowest, and this tapered structure suggests specific function.

Nerves enter the base of the follicle and in *Zalophus* in particular, different specialized nerve endings have been found. An onion-like lamellated corpuscle occurs at intervals, and these protrude into the spaces of the venous sinuses. These corpuscles bear a resemblance to pacinian corpuscles which respond to vibrations. There are also two types of specialized mechanoreceptor endings arranged in a ring about the shaft at the level of the ring sinus.

It has been postulated that engorgement of the erectile tissue would increase the sensitivity of the touch receptors to compression. Vibrations received by the vibrissa would cause it to move, exciting the receptors and causing messages to be transmitted. Although the exact function of this system has yet to be determined, it is obvious that this complex arrangement of nerves, sinuses and receptors must have a definite part to play in the life of the animal. Reception and interpretation of waterborne vibrations could play an important part in aquatic life (Stephens *et al.*, 1973).

Fig. 5.2 *A 20mm-long fragment of the moulted skin of the Southern Elephant Seal. The tips of the hairs are above the epidermis with the roots protruding below.*

Moult

The first coat of hair that a seal has is different from the adult coat. This first (or natal) coat may be moulted in the uterus before the pup is born, or at varying periods up to a few months old according to the seal, and these details are mentioned under the individual seal. After this there is an annual moult of the hair. In the Southern Elephant Seal a fusion between the hair and the outermost layer of the skin (stratum corneum) has been observed (Ling & Thomas, 1967), with the result that when the hairs are shed the stratum corneum is shed as well, and patches of shed skin with hairs attached are found where the animal has been lying

Fig. 5.3 *A young male Southern Elephant Seal moulting.*

(Fig 5.2). In the Southern Elephant Seal at least (Fig 5.3), the moult seems to cause irritation which is alleviated by wallowing in mud. This type of moult is particularly obvious in Elephant Seals and patches of shed skin about 20cm^2 may be collected, but also occurs in the Hawaiian Monk Seal, possibly in the walrus, and has been reported in the captive Harp Seal (Ronald *et al.*, 1970).

The moult of fur seals and probably sea lions as well, is a much slower, unobtrusive affair that may take several months. Most, but not all the guard hairs are moulted every year, starting about a week after the underfur, so at all times the animal retains a warm and waterproof coat. The retention of some of the underfur hairs in the pilary canal has already been mentioned. The order in which the areas of the body are moulted is variable, but usually the fore and hind flippers moult first, followed by the face and neck, dorsal surface, then ventral surface.

Temperature regulation

Seals live in many climates – across the Arctic and round the perimeter of the Antarctic, and on the shores of Australia, the Galapagos and the Hawaiian islands, and, of course, in many places in between. A newly-born Weddell Seal may leave its mother's warm internal environment and, wet and blubberless, face an Antarctic blizzard at −35°C – a change for it of some 70°C in a few minutes. On the other hand the sandy beaches of Laysan or the shores of the Galapagos Islands may be hot and brilliantly sunny. Not every seal, of course, faces these two extremes, but seals do have to keep warm in adverse conditions, and also have to be able to lose excess heat.

The average internal body temperature of seals is between about 36·5–37·5°C (97·7–99·5°F) but it can fluctuate a certain amount. That of pups is slightly higher, and that of sleeping adults slightly lower.

Phocids

Phocid seals, as in sea lions and walruses, have a sparse underfur that is relatively little use in insulation, especially in water when the skin and hair get wet. In such circumstances, in ice water for example, the blood vessels in skin and blubber contract, and just enough blood is allowed through to prevent the skin from freezing and to maintain the skin temperature at about 1°C. The insulating power of the blubber is comparable with that of asbestos fibre, and a temperature gradient is maintained between the central body core at the normal body temperature, and the outside of the animal which need only be a degree or so above freezing. Most of the animal is thus warm, and the gradient is only in the outer 4–5cm. Those phocids that normally live in the colder areas, such as Harp Seals, have a lower critical temperature than those seals living in the more temperate areas, and are thus adjusted in this respect to their environment (Irving & Hart, 1957, Hart & Irving, 1959). Weddell Seals, for example, may have a lower critical temperature of −40°C while in air, and although they can always escape to the water, they are obviously well able to survive these extreme low temperatures while on land (Ray & Smith, 1968).

The cells of seal skin (phocid, otariid and odobenid) have been shown to have a greater power of survival at low temperatures than non-pinniped cells (Feltz & Fay, 1966) and this is correlated with the low temperatures the skin has to survive while in polar conditions. However, full use is made of periods on shore and all seals spend much time basking in the sun, when the blood vessels in the blubber expand and the skin is warmed from within and without and is able to undergo repair or moulting activities.

As with all seals, the hind flippers of phocids are useful for heat regulation. The deep vessels are reduced and there is an increase in superficial vessels on the plantar surface (Tarasoff & Fisher, 1970). This surface in phocid hind flippers is somewhat concave and if heat is to be conserved the blood is restricted and the two flippers are held with their plantar surfaces together, and possibly also held out of water if the animal is in the sea. When it is necessary to lose heat the blood supply is increased and the flippers spread widely.

Though the hind flippers are useful in heat regulation, the whole sparsely haired body of a phocid can be used for this purpose. Simple arterio-venous anastomoses are present in the superficial layers of the dermis (i.e. external to the blubber) and they are in much the same density on the body as on the flippers, and in greater numbers than in normal terrestrial animals (Molyneux & Bryden, 1978, Bryden, 1978). These anastomoses allow more blood to flow at a faster rate than would be possible through capillaries, so greater heat loss is possible, and also more blood can thus be exposed to the warming effect of the sun.

The testes of phocids, while outside the abdominal cavity, lie between the blubber and the abdominal muscles, and might be expected to be at the deep body temperature. The temperature at the centre of the testes of Southern Elephant Seals has been taken and found to be 6–7°C lower than the body temperature. It is possible that the temperature difference is maintained by vascular control (Bryden, 1967).

The Northern Elephant Seal and the Hawaiian Monk Seal are probably the phocids that live in the hottest places all year round. The Northern Elephant Seal cannot cope physiologically with too high temperatures on land and will ultimately escape to the sea. Before this point is reached it will lie in the surf, flip wet sand over its body, seek shade or remain very inactive. Monk Seals while on land are as inactive as possible and do not spend energy in flipping sand or waving flippers or even in much social behaviour. They spend the night high up the beach on dry sand, and move towards the sea so that they spend the hot day on wet sand or in the surf. They also spend much time asleep, when their metabolic rate and temperature will be lower (Whittow, 1978).

Sea lions

Sea lions while in water are obviously adequately insulated against the cold by their blubber, and regulation of the blood flow in their flippers prevents too much heat loss this way. On land, where they may spend a considerable time while resting, and during the breeding period, they try to remain in as pleasant an environment – meteorologically – as possible. Lying with their flippers tucked under the body will conserve heat, and a considerable degree of thigmotaxis is shown. All sea lions are known to gather in groups, or sometimes in piles, as in *Phocarctos*, and such close bodily contact would prevent much heat loss. Walruses and Elephant Seals are the other sparsely haired pinnipeds that huddle together, but fur seals do not normally do this (Gentry, 1973).

In wet and windy weather *Neophoca* tends to congregate in the centre of its island habitats, and will lie on vegetation if it is present, but do not apparently take much shelter behind rocks (Marlow, 1975, Stirling, 1972a). Dominant *Neophoca* have been seen to evict a subordinate animal and to lie down in the same place. This happens mainly in cold wet weather and could provide the dominant animal with a dry sleeping spot (Marlow, 1975).

As the temperature increases, various strategies are used to remain comfortable. Remaining in the splash zone of the waves, so the body is always damp, and the winds cause evaporation and cooling, retreating to the shade, lying on the wet sand, flipping damp sand over the body; and keeping on the windward side of an island. Rock pools are used to immerse the hindquarters, especially after strenuous activity such as fighting or copulating. The exact methods of keeping cool do, of course, depend to a certain extent on the nature of the terrain.

Hot sun may increase the temperature of the substrate up to 50°C, and sea lions that live in hotter areas may be subject to air temperatures of *c.* 30°C. Such temperatures may be found on the Galapagos Islands and Mexican coast where the Californian Sea Lion lives, and this animal has been watched carefully to see how it is able to live where it does. The results of observations and experiments have indicated that *Zalophus* is relatively poorly equipped physiologically to be on land in hot places, and these sea lions seem unable to withstand the hot conditions for more than a few hours.

A whole beach, crowded with these sea lions, may become deserted in a few minutes when the cloud cover thins, the hot sun comes out, and the sea lions make for the sea. Soon after they come out of the water, while their coats are still wet, they appear to be comfortable. When their coats get dry they become restless, exposing their flippers, holding them up in the air and fanning with them. They will then seek shade, and if possible lie down on wet sand, pressing their large naked flippers on to the cool substrate and even urinating on the sand to dampen it (Whittow, Matsuura, Lin, 1972).

Experiments on these sea lions have shown that their body temperature rises with the increase in ambient temperature, and that in air temperatures over 30°C they are unable to maintain thermal equilibrium. At high temperatures they breathe through their mouths and salivate copiously, but do not pant, as there is no increase in respiration rate. Small sweat droplets appear on the naked flippers. It has been calculated that on a hot day over half the heat produced is lost in non-evaporative ways – by conduction and convection from the body to the air, which would be increased by sea breezes, and by conduction of heat from the flippers to damp sand. There is relatively little effective evaporative loss from respiration or sweating, and it has been suggested that perhaps conservation of water can be more important than temperature regulation (Whittow, 1974, Matsuura & Whittow, 1974).

Arterio-venous anastomoses in *Zalophus* are present, but are in the deeper regions of the dermis, not as in phocids (Bryden & Molyneux, 1978). There are a greater proportion of them in the flipper skin than in the body skin, and blood passing through these channels would help considerably in the loss of body heat to the environment. Sea lions are more heat tolerant when they are asleep, and their heat production is then about 24 per cent less. Conversely, of course, their heat production increases with activity (Whittow, Ohata, Matsuura, 1971).

It is likely that the conclusions from all these experiments on *Zalophus* apply at least to most other sea lions, and possibly to other pinnipeds as far as the climatic conditions are comparable.

Fur seals

The Northern Fur Seal breeds on the Pribilof Islands, where in summer the air sometimes warms up to 15°C – which is more than 20° colder than the body temperature of the seal. In winter the seals move out into the Pacific, meeting waters just above freezing, but only rarely as warm as 15°C which occurs at the level of Lower California. Fur seals seem to be most frequent, and are thus most comfortable in waters about 11°C.

The thick layer of underfur effectively insulates the fur seal from its environment. Even in the coldest water it is likely to meet, the air trapped in the underfur keeps the skin dry and also warm. The temperature gradient between the warm body and the outside environment is in the fur – a gradient of some 25°C through some 20mm of fur (Irving *et al.*, 1962).

Strangely, the blubber seems to provide little insulation, as the skin may only be 5°C cooler than the body core, whereas in a similar situation the skin of a phocid would be almost at the temperature of the cold environment and well insulated from the warm body core.

The large flippers have no insulation by either fur or blubber, and the tips of the flippers are barely warmer than the surrounding water. The hind flippers are well supplied with blood vessels, with a concentration of them in the superficial layers of the plantar surface. In the tarso-metatarsal region branches from the main veins (saphenous) divide up into venules which are closely associated with the arteries in this region (Tarasoff & Fisher, 1970). Arterio-venous anastomoses are particularly numerous in the hind flippers where they occur in the superficial layers of the dermis (Bryden & Molyneux, 1978).

In cold water the vessels to the flippers would be constricted, and the small amount of blood returning to the body would be warmed up in the tarsal region by the proximity of the veins to the larger arteries. Flippers are also held up in the air, where the heat conduction is less than in the water.

Fur seals, however, have more difficulty in dissipating heat than in conserving it. Excessive activity on land, as would occur on the Pribilofs during the breeding season, causes the body temperature to rise. Herding of the animals for commercial use can cause their temperature to rise from the normal 37·7°C to 41 or even 43°C, when they are likely to die of heat exhaustion. Because of the fur, little heat is lost from the body, but the flippers heat up quickly. The large superficial blood supply and the use of the arterio-venous anastomoses would allow large amounts of blood to pass near the surface of the flippers to lose heat. Fur seals spend much time fanning with their hind flippers and this too would assist in heat loss.

Although even on the Pribilof Islands where 12°C is considered as a warm day and the fur seals have difficulty losing heat when on land, they are still able to breed off Southern California, where presumably they keep as close to the water as possible.

Other fur seals (*Arctocephalus* spp.) are present on oceanic islands of the southern hemisphere, and on some mainland coasts. The conditions they meet are summer temperatures within a degree or so of freezing at South Georgia, to sea temperatures up to 20°C and air temperatures as high as 43°C in the South African summer. On the whole though, sea temperatures up to 15°C are preferred by the seals (Bonner, 1968, Rand, 1967). In high temperatures on land, activity is curtailed, shade and windy places are sought, flippers are spread out and waved, seals breathe through their mouths and salivate, they keep their hind flippers wet, and eventually escape to the sea. Flippers may be pressed together and held out of the water in cold weather, and will be tucked under the body when on land. Fur seals, however, do not huddle together or lie in heaps as do the sparsely haired sea lions. This would probably compress the insulating layer of air in their underfur too much.

It is an interesting speculation that the extensions of tissue that increase the length of an otariid hind flipper well beyond the nails, may have as their major function the increase of the temperature regulating surface of the flipper (Bonner, 1968). Certainly the hind flippers have no great locomotory function in water, and the very flexible extensions would hardly increase this much.

Otaries and walruses have descended testes which lie in a naked scrotum. In the Northern Fur Seal this scrotum has been seen to become particularly pendulous while the animal is fanning, and presumably hot. Scrotal temperatures of this seal have been taken and were approximately 6°C below the deep body temperature (Bartholomew & Wilke, 1956).

Walrus

Living in the Arctic, walruses can tolerate fairly severe conditions, and they may be found in places with air temperatures between −20 and +15°C. In fact, the animals are considered to be thermoneutral between these temperatures, without change of metabolic rate and without being inconvenienced by the temperature. They are, however, capable of withstanding colder temperatures and have been seen sleeping on the ice in a high wind at temperatures of −35°C.

Walruses are very gregarious, and are rarely found singly. They huddle together in large groups, particularly when on land, and many hundred may congregate together. When huddling, about 20 per cent of the animal's body surface is in contact with a neighbour, and this would reduce heat loss. They do, however, huddle in all climatic conditions and show signs of heat stress in warm weather.

A cold walrus on land will curl into a foetal position with the flippers tightly pressed against the body, the skin becomes pale as the blood is withdrawn to deeper layers, and the animal is insulated by its blubber. If the weather becomes too cold, or conditions too uncomfortable, they retreat to the sea.

As the temperature increases the animals gradually unfold, exposing their bellies and spreading their flippers. They also fan, usually with their foreflippers. They still, however, remain in closely packed groups. The skin becomes pink and suffused with blood, and both it and the flippers are warm to the touch, the flippers usually being warmer than the body, and in fact they are only cooler than 25°C when wet or touching ice or snow. Between air temperatures of 0–15°C the skin on the body of a dry walrus is usually higher than 20°C, but does not normally rise above 32°C, without the animal showing signs of obvious discomfort and escaping to the water.

In water the core of the walrus is at the normal body temperature of about 36·6°C; there is a temperature gradient through the blubber and skin, and the skin and flippers are about 1–3°C warmer than the water.

Although obviously comfortable in water, walruses spend a lot of time on land and expose themselves to sunshine which will allow the skin to grow and heal. Above air temperatures of about 15°C walruses prefer to stay in water and this obviously curtails their distribution as they would have difficulty in satisfactorily continuing the terrestrial part of their life (Fay & Ray, 1968, Ray & Fay, 1968).

Pups

Seal pups have a less efficient thermoregulatory mechanism than adults. Elephant Seal pups are particularly restless in the sun, and *Zalophus* pups seem to be less sensitive to both heat and cold than adults.

Some phocids, for instance pups of Harp Seals and Weddell Seals, in the polar regions of the world, may be born into air temperatures of −30°C, and a drop of 70° at the moment of birth must be quite a shock. Many phocid pups are born with a dense fluffy coat which, as long as it keeps dry, will insulate the body sufficiently until the pup has laid down a layer of blubber. At the moment of birth, of course, the pup is wet, and the fluffy coat sticks closely to the skin and has no value in insulation. There is a certain amount of vasoconstriction of the vessels to the flippers even at this stage, and while the pup is still wet it will shiver violently. This short-lived period of shivering uses up all the glycogen that is present in the muscles of the newborn seal, and this store will be replenished on suckling (Blix *et al.*, 1979).

Brown, or thermogenic adipose tissue has been found under the skin of the back of newly-born Harp Seals, and also internally on the abdominal walls and around the venous plexuses of kidneys, pericardium, and in the neck. This tissue produces heat when the pup is under thermal stress, and the heat production is by oxidation of lipid stores through the action of the large loosely coupled mitochondria. As suckling takes place and blubber is laid down the thermogenic tissue on the back loses its heat-producing function and changes into ordinary blubber after about three days. The internal tissue round the plexuses remains at least until the pups enter the water at about four weeks (Blix *et al.*, 1979, Grav *et al.*, 1974), and probably throughout the life of the animal (Blix *et al.*, 1975). As well as the

brown adipose tissue, it is possible that the skeletal muscles also produce heat through the action of loosely coupled mitochondria, without shivering. Such mitochondria have been found in skeletal muscle of cold-stressed *Callorhinus* pups (Grav & Blix, 1979). Dark fibres with mitochondria and lipid droplets suggesting non-shivering thermogenesis, have been found in skeletal muscle of adult Harp Seals, and also in the muscle of the diaphragm and caval sphincter (George & Ronald, 1975).

Without blubber to insulate it, the skin of a pup is at the same temperature as the body core. It has been shown that during the first 40 days of the life of a Harp Seal pup, as the blubber is laid down, it gradually takes over the insulation of the body. As this happens the skin temperature decreases, and the temperature gradient between the body core and the skin increases. The high metabolic rate of the pup also decreases as the animal becomes more tolerant of low temperatures, but they do not normally go to sea until the fluffy coat has been shed and the thermoregulatory blood vessels in the blubber are functional (Davydov & Makarova, 1965).

A similar situation exists in Northern Elephant Seal pups, and possibly in other seal pups too. The suckling pup has a high oxygen consumption and high metabolism to counteract the greater heat loss due to the absence of blubber. The high metabolic rate is kept up by very frequent suckling until the blubber is laid down (Heath *et al.*, 1977), and the high fat content of the milk is obviously correlated with the need to lay down blubber quickly.

Fur seal pups with their underfur layer thinner than in the adult, may get their bodies wet in sea or rain, when they become bedraggled and shiver. Young pups of both fur seals and sea lions are much more terrestrial than older pups, but tolerate a certain amount of cold, shake themselves like dogs to remove the water from their coats, and huddle together to keep warm. They may also take shelter from the natural features of their environment. *Phocarctos* pups for instance, on the sandy beach of Enderby Island take shelter in rabbit burrows. The diameter of the burrow, however, does not give the pup room to turn round and it may get pushed deep down a burrow by other pups crowding after it. Many pups die from being trapped in this way.

Walrus calves are almost invisible in bad weather when they shelter under the mother's chest, between her foreflippers, obtaining warmth from her body. Without the shelter of their mothers, calves try to seek other shelter, or assume a foetal position and shiver. Mothers brooding young calves are very reluctant to enter the water should they be disturbed by hunters (Fay & Ray, 1968).

Chapter 6 Skull and skeleton

Skull

The pinniped skull varies widely in shape and proportions from genus to genus, but is, in general, characterized by a large rounded cranium which is abruptly delimited from the elongated, approximately parallel-sided, constricted interorbital region; (Fig 6.1) by the large orbits, the relatively short snout, and the unossified space in the orbit where the palatine does not meet the frontal. The detailed characters of the ear which characterize pinnipeds are noted in the chapter on senses.

The differences in proportion between the skull of a domesticated dog (Foxhound) and the skulls of a Harbour Seal and a Californian Sea Lion can be illustrated by the table below of measurements and the proportions of these measurements to the total skull length (condylobasal length).

Though from only three individual skulls, the proportions show the larger and wider cranium, and the shorter snout of the pinnipeds when compared with the dog.

The major skull characters that are used in the classification of the various groups of seals are listed below (but see also the chapter on classification).

Otarioidea
1. Tympanic bulla small and flattened Fig 6.2 (but see chapter on classification).
2. Bulla formed almost entirely by the ectotympanic, the entotympanics forming the bone surrounding the carotid canal. Fig 6.3.
3. Alisphenoid canal present.
4. Frontal bones project anteriorly between the nasal bones. Fig 6.4.
5. Great development of the mastoid process. Fig 6.5.
6. Jugal–squamosal joint overlapping. Fig 6.6.

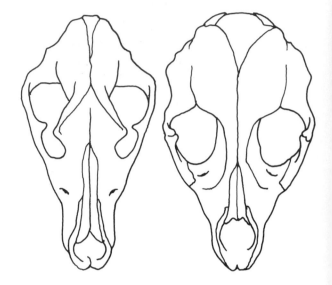

Fig. 6.1 Dorsal views of the skulls of a foxhound (left), and Harbour Seal.

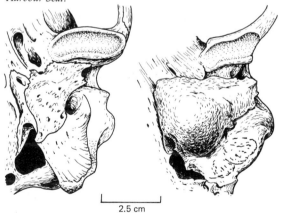

2.5 cm

Fig. 6.2 Ventral views of left tympanic bulla of Zalophus *(left) and* Phoca vitulina *to show the small flattened bulla of the otariid and the inflated bulla of the phocid.*

	Condylobasal length		Cranium length		Cranium width		Snout length	
	mm	%	mm	%	mm	%	mm	%
Foxhound D.102	221	100	82	37·1	65	29·4	101	45·7
Phoca vitulina 1891.12.18.7	231	100	99	42·9	96	41·6	70	30·3
Zalophus californianus ♂ JEK 81	292	100	128	43·8	110	37·6	96	32·8

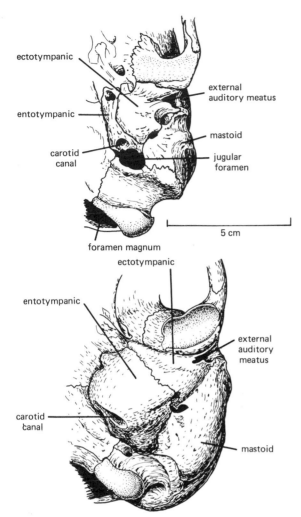

Fig. 6.3 *Ventral views of the left ear region of young* Eumetopias *(top) to show the entotympanic forming the carotid canal and young* Mirounga leonina *(above) to show the composition of the bulla.*

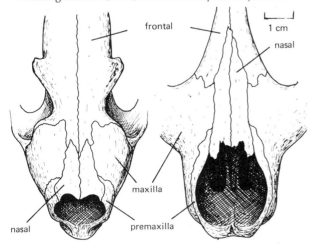

Fig. 6.4 *The arrangement of the nasals and the neighbouring bones in Otariidae and Odobenidae (left), and Phocidae.*

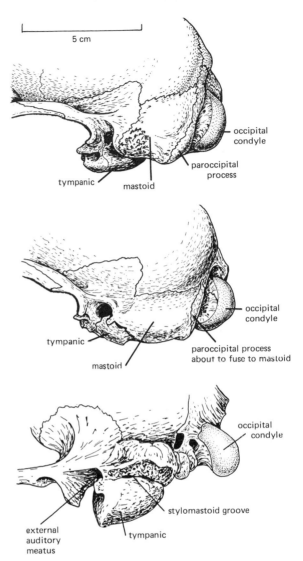

Fig. 6.5 *Lateral view of the left mastoid area of a* Eumetopias *pup (top), and a juvenile* Zalophus *(middle), to show how the mastoid area increases in size with age, fuses with the paroccipital process, and tends to obscure the bulla in lateral view. In an adult* Phoca *(above), the mastoid is well dorsal to, and not obscuring the inflated tympanic.*

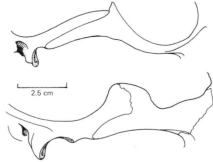

Fig. 6.6 *Right jugal-squamosal joints of* Callorhinus *(top) to show overlapping and* Phoca *to show interlocking.*

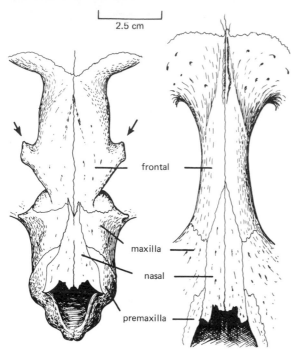

Fig. 6.7 Dorsal views of interorbital regions of A.forsteri (left) to show postorbital processes (arrowed) and Phoca showing lack of postorbital processes.

Phocoidea

1. Tympanic bulla inflated. Fig 6.2.
2. Ectotympanic forms small (⅓) part of bulla, while entotympanic forms ventral floor and medial half of bulla. Fig 6.3.
3. No alisphenoid canal.
4. Nasal bones extend posteriorly between the frontal bones. Fig 6.4.
5. Mastoid process less developed. Fig 6.5.
6. Jugal–squamosal joint interlocking. Fig 6.6.
7. No supraorbital processes.

Otariidae

7. Supraorbital processes present. Fig 6.7.

Odobenidae

7. No supraorbital processes.

(Other skull characters are listed in the chapter on classification.)

Most of the above characters when considered in conjunction with the diagrams are self explanatory. In phocids the ectotympanic forms the ventral rim of the external auditory meatus and about a third of the adjacent region of the bulla. In very young skulls the line of fusion can be seen between this and the entotympanic which forms the rest of the bulla. In otarioids the bone surrounding the carotid canal is formed by the entotympanic, while the rest of the

bulla is formed by the ectotympanic. Fig 6.3 (Repenning, 1972, Hunt, 1974).

In phocids the paroccipital process is small and is not fused to the mastoid part of the periotic. The mastoid lies almost horizontally dorsal to the inflated tympanic, and separated from it by the stylomastoid groove. In lateral view the tympanic is clearly visible. Fig 6.5.

In adult otariids the paroccipital process becomes fused with the mastoid area, and this rather large 'curtain' of bone between the external auditory meatus and the paroccipital process expands ventrally, virtually obscuring the tympanic in a lateral view of the skull. Walruses are similar to otariids in this respect except that there is no fusion between mastoid and paroccipital process.

A traction epiphysis on the roughened part of the mastoid has been described in otariids, but has not been seen in walrus or phocids. This is the position of attachment and of the greatest pull of the sternomastoid and other muscles (Cave & King, 1964).

A lacrimal bone is present in otariids and walrus. It is visible as a separate bone in young animals, but fuses and becomes indistinguishable in adults (King, 1971). It has not been seen in phocids, and no lacrimal duct has been recorded in any pinniped.

The shape of the nasal bones of the walrus conforms in general with the otariid pattern, though the posterior ends tend to be very square in outline. In most seals the usual two nasal bones are present, fusing neither with each other nor with the surrounding bones. The four Antarctic phocids, however, show a pronounced tendency for the two nasal bones to be fused to each other in the mid-line. This fusion shows all degrees of completeness, but no relation to age.

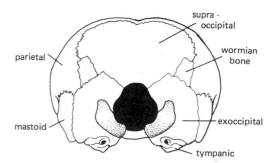

Fig. 6.8 Posterior view of the skull of a Eumetopias pup to show the Wormian bones.

Wormian bones, i.e. small irregular ossifications, are frequently found in seal skulls, though with fusion they become less obvious in adult skulls. The most usual places for them to occur are mid-dorsally, at the junction of parietal and frontal bones, where one or more Wormian bones may occur, and in the occipital region, bounded by the exoccipitals, supraoccipitals, parietals and mastoid part of the periotic. Fig 6.8.

A study of a series of skulls of *M. angustirostris* aged from near-term foetuses to pups of about a year old showed that these extra bones in the occipital region ossify progressively with age. In the foetal skulls there were a number of small ossifications in the area and these small bones gradually fuse together until by about six months there is a single rectangular bone on each side. With increasing age this bone is incorporated into the occipital complex which has fused into a single unit by the time the pup is about a year old. Histological preparations showed that these extraoccipital bones were of endochondral origin (not membrane bone). They probably serve to allow rapid growth of the brain case in the early months of development (Morejohn & Briggs, 1973).

There are no cranial air sinuses in pinnipeds.

A comparison of phocid skulls (King, 1972a) suggests that many of the modifications may be connected with feeding. A scoop-like anterior end to the lower jaw, formed by a long symphysis or the ventral edges of the jaw curving inwards, or both characters together, may have some connection with an arrangement of the muscles that facilitate sucking. Such jaws are found in particular in walrus, Bearded, Crabeater, Ringed and Harp Seals. The two former animals probably suck in mollusc feet and both have in addition rather concave palates. Both sides of the walrus lower jaw fuse solidly together. Small crustaceans form an important part of the diet of the three latter seals, and of these both Crabeater and Harp Seals have been seen to suck in food.

Long facial regions housing large teeth that are important in feeding are found in *Monachus* with its crushing-type teeth, and in *Lobodon* and *Hydrurga* with their remarkable subdivided teeth.

For seeing small prey in dim water, good visual acuity is needed, and seals have large eyes. Fig 6.9. The size of the eyes if reflected in the size of the orbits which take up a much greater part of the skull in seals than they do in a dog for instance (see Fig 6.1). In the smaller phocids in particular the eyes take up a disproportionate amount of room in the head, and the interorbital region is very compressed in order to accommodate them. In the skull of the small Ringed Seal for example, only 3mm separates the two orbits, and through the delicate, almost transparent bone of this region, can be seen the outlines of the reduced ethmoturbinals. The enormous orbits of the Ross Seal may be responsible for the very low setting of the zygomatic arches (Piérard & Bisaillon, 1978).

The Ross Seal is also remarkable in having an open anterior fontanelle in a high proportion of adult skulls,

Fig. 6.9 The 15cm-long head of a Baikal Seal with the skin removed, showing the very large eyes and reduced interorbital part of the skull.

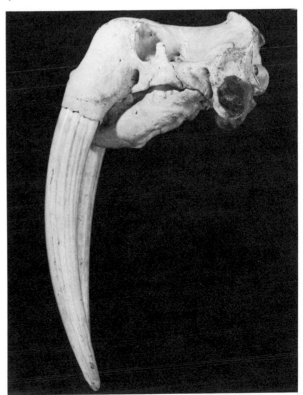

Fig. 6.10 Lateral view of a walrus skull. The solid construction of the skull can be seen, with the truncated snout bearing the elongated canines (59cm from gum line to tip in this specimen).

though it is not known why this should be so (King, 1969b).

Two phocids have an enlarged and mobile proboscis, and in both Elephant and Hooded Seals this is reflected in the shape of their skulls. Both have an increase in the length of the narial basin, and a high fronto-nasal area to give space for attachment of the muscles that move the proboscis. Similar skull characters are shown in other mammals that have a mobile proboscis – such as sirenians, saiga, elk and tapir (King, 1972a).

Otariid skulls in general, have less inter-generic variation in shape than do phocid skulls, and perhaps the two most characteristic individual skulls are those of *Otaria*, and the adult male *Zalophus*. *Otaria* skulls have a particularly long concave palate, the posterior edge of which is almost level with the glenoid fossa. Skulls of adult male *Zalophus* have a very high sagittal crest which may rise some 4cm above the general level of the cranium. This crest starts growing in the animals fifth year and reaches to full size when the seal is about ten years old (Orr *et al.*, 1970).

The skull of a walrus is unmistakable (Fig 6.10), the entire maxillae anterior to the orbits being modified to house the roots of the enormous canines, so that the nasal opening is reduced in size, and the anterior end of the skull has a truncated appearance. The entire skull is heavily ossified, and many of the sutures are soon obliterated.

Hyoid

The pinniped hyoid consists of the same elements normally to be found in the Carnivora. A transverse basihyal is connected with the posteriorly directed thyrohyal, and with the anteriorly directed curved anterior cornu. This cornu is composed of ceratohyal, epihyal and stylohyal, the latter being connected by a ligament to the tympanohyal pit in the tympanic. These three bones are the usual complement in each side of the anterior cornu, but four centres of ossification have been seen in a young Harbour Seal, *P.v. vitulina*, and well-developed tympanohyals have been reported in addition in *P. largha*. In this respect the hyoid of *P. largha* differs from that of the other North Pacific Harbour Seals, where *P.v. stejnegeri* and *P.v. richardsi* have progressively abbreviated and more cartilaginous stylohyals and tympanohyals (Naito, 1974). In Ross and Leopard Seals only two bones – ceratohyal and epihyal – are present in each anterior cornu, and in both these animals the proximal unossified end of the epihyal lies freely inside a fibrous tube. This tube is attached at its proximal end to the tympanohyal pit in the bulla, as is the usual tympanohyal ligament, and at its distal end to the distal ossified part of the epihyal (King, 1969b).

Muscles

A detailed description of pinniped musculature would be out of place here. Complete, or reasonably complete accounts have been given by Brazier Howell (1929) for *Zalophus* and *P. hispida*; Bryden (1971a) for *M. leonina*; Miller (1888) for *P. vitulina* and other phocids, and *A. gazella*; Mori (1958) for *Zalophus*; Murie (1872a) for *Odobenus*; Murie (1872b and 1874) for *O. byronia*; and Piérard (1971) for *L. weddelli*.

Muscles of the head are given by Huber (1934) for *Zalophus* and *P. hispida*, and King (1969b) for *O. rossi*; and muscles of the Harp Seal ear by Ramprashad *et al.* (1971).

The muscles, and also the general structure and function of the otariid foreflipper, are described by English (1976b, 1977).

Bryden (1973) describes the relative growth rate of the muscles of *M. leonina*, and in Bryden & Felts (1974) the muscle groups of various regions of the body of the Antarctic phocids are compared by weight, and some assessment of comparative function made.

Skeleton

Vertebral column

The vertebral column, while basically of the normal mammalian pattern, and with the usual reduction of articulating processes common to aquatic mammals, shows differences between otariid and phocid that reflect their different methods of movement. Fig 6.11. That of the walrus is, in most respects, intermediate though rather more like the phocids. The pinniped vertebral formula is C7 Th15 L5 S3 Cd10–12, though Th14 L6 is said to be more usual in walruses. There is the usual variation, however, and in most pinnipeds Th14 15 or 16 may occur occasionally (Fay, 1967).

The chief differences in the cervical vertebrae are the stronger transverse processes and neural spines in the otariids, associated with the muscles for the more diverse movements of the head and neck. This is also shown in the way the insertions of the muscles on the occipital region of the skull extend from the vertex to the mastoid process, giving great range of movement. The swimming action of an otariid has been likened (Brazier Howell, 1930) to a rowboat with the rower (i.e. the foreflippers) in the centre, and for mechanical reasons a reasonable mass of body both before and behind the foreflippers is necessary. Fig 6.12. Thus the neck remains elongated, and its essential swinging action during terrestrial locomotion has already been mentioned.

In phocids the emphasis is on sagittal and lateral movements in particular; the transverse processes are less stout, the neural spines very small, and the muscles for moving the head are principally confined

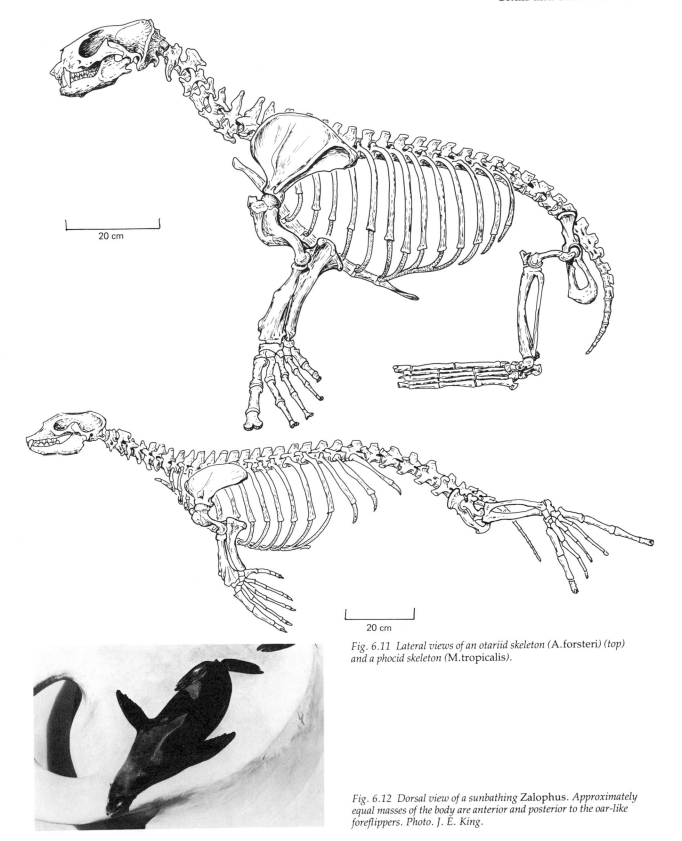

Fig. 6.11 *Lateral views of an otariid skeleton (A.forsteri) (top) and a phocid skeleton (M.tropicalis).*

Fig. 6.12 *Dorsal view of a sunbathing* Zalophus. *Approximately equal masses of the body are anterior and posterior to the oar-like foreflippers. Photo. J. E. King.*

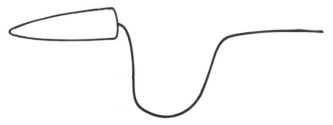

Fig. 6.13 The neck curvature of Hydrurga, *drawn approximately to scale, from measurements on a dead animal.*

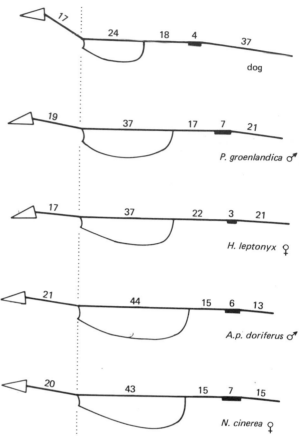

Fig. 6.14 Diagrammatic representation of the body proportions in dog and pinnipeds. The length of each segment is shown as a percentage of the length of the entire vertebral column. The increased thoracic length, and decreased caudal length in seals are particularly noticeable.

to two areas – the supraoccipital and the mastoid. In swimming, phocids propulsion is from the rear and the body shape must be fusiform, so a long neck is mechanically impractical. The length of the cervical vertebrae of a phocid is, however, very little less than that of an otariid. The neck length may be masked by the heaviness of the muscles in this area, or the neck may also be actually retracted. This has been seen in a dead Leopard Seal at least, where there was a deep U-bend of the whole neck. Fig 6.13. The length of the neck of pinnipeds in proportion to the length of the entire vertebral column shows virtually no difference from the proportion of this part in a 'typical' dog.

The importance to the otariid of the musculature of the anterior part of the body is also shown in the larger neural spines of the thoracic vertebrae, while those of a phocid are much lower. The thorax of pinnipeds is much longer in proportion to the entire vertebral length than is that of a dog, and this is correlated with the larger lungs. Fig 6.14.

The otariid lumbar vertebrae show small transverse processes and more closely set zygapophyses that would allow the convexity of the back in the terrestrial position of the animal, but little concavity. Those of a phocid show larger transverse processes for the strong swimming muscles of the posterior end of the body, and wider, more loosely fitting zygapophyses that allow the back considerable range of movement both laterally and vertically. Southern Elephant Seals in particular show this flexibility of the back, and can bend backwards so that the top of the head can touch the tail. Fig 6.15.

The fused sacral vertebrae articulate with the ilium in the normal way, and neither they nor the caudal vertebrae present any peculiarities. As the tail is not used in swimming, it is short, and the caudal vertebrae are small, cylindrical, and without strong processes. In common with other aquatic mammals that are supported in the water for the entire length of the body, the neural spines show little, if any, change of direction, or anticline.

The os penis and os clitoridis are mentioned in the section on reproductive organs.

Fig. 6.15 Southern Elephant Seal (in captivity) demonstrating the flexibility of the back. Photo. J. E. King.

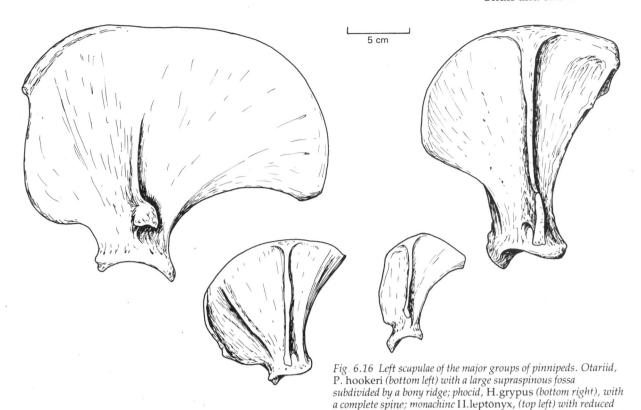

Fig. 6.16 *Left scapulae of the major groups of pinnipeds. Otariid, P. hookeri (bottom left) with a large supraspinous fossa subdivided by a bony ridge; phocid, H.grypus (bottom right), with a complete spine; monachine H.leptonyx, (top left) with reduced spine, and odobenid, the walrus (top right), with a scapula like that of an otariid but without the extra ridge.*

Sternum and ribs

The sternum consists of eight or nine pieces altogether, the number varying according to the animal. In otariids the manubrium is elongated anterior to the attachment of the first pair of ribs and can be seen as a lump on the chest when the animal sits upright. In phocids and walrus the length of the manubrium is increased by cartilage. The attachment of the first pair of ribs is thus near the anterior end of the manubrium in phocids and walrus, but more than halfway back along its length in otariids.

The number of ribs corresponds with the number of thoracic vertebrae and thus has the same variability.

Pectoral girdle and anterior limb

There is no clavicle in pinnipeds. The otariid scapula has a well-developed spine situated at the beginning of the posterior third of the blade, so that the supraspinous fossa is twice the size of the infra-spinous fossa. This greater space is for the origin of the supraspinatus muscle, used in movements of the humerus, and the fossa is further subdivided by a bony ridge, giving greater area of attachment for the muscle. The walrus scapula is like that of an otariid in the size and position of the spine, but there is no extra ridge subdividing the supra-spinous fossa.

The spine of the phocid scapula is smaller and is usually more medianly placed on the blade, and there is a tendency for the inferior angle of the scapula to be extended caudally. In the Phocinae the spine is complete, but it is reduced in the Monachinae so that it may be hardly more than a low ridge with a knob-like acromion process. Fig 6.16 (King, 1966).

The humerus is short and stout, though that of the walrus is characteristically longer, and as long as the ulna. In otariids the lateral tuberosity rises higher than the head and is extended into the strong deltoid crest which runs down about three-quarters of the length of the humerus before curving down to meet the shaft. Fig 6.17. The medial tuberosity is much lower than the head. In phocids it is the medial tuberosity which rises higher than both head and lateral tuberosity, while the latter does not rise higher than the head. The lateral tuberosity is prolonged into the deltoid crest, the shape of which is different in phocine and monachine seals. In the former the crest is strong, with an overturned lateral edge, and its distal end stops abruptly, forming a concavity as it curves to meet the shaft. In the Monachinae the deltoid crest is long, does not end abruptly and curves gradually down to meet the shaft. A supracondylar foramen is usually found in phocines, but not usually in monachines or otariids.

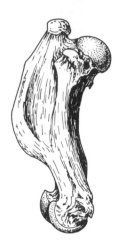

5 cm

As with the humerus, the radius and ulna are short when compared with a land mammal of similar size. They are laterally flattened, the ulna with a posterior expansion of the olecranon process, and the radius with an anterior expansion of its distal end. Fig 6.18.

As in the fissiped carnivores there are six carpal bones – scapholunar, cuneiform, unciform, magnum, trapezoid, trapezium and the associated sesamoid bone, the pisiform Fig 6.19. The scapholunar is the largest and articulates with the lower end of the radius. In the Phocidae the elements of the ulnar side of the carpus are crowded because of the normal position in which the manus is held, but the shape and articulations of the carpal bones still allow all the diverse movements of the manus. In the Ross Seal the carpals are more firmly interlocked so that the whole flipper can be used as a more rigid unit. The carpal elements of the Otariidae are not so crowded. On land the otariid manus is extended more or less at right angles to the rest of the limb, and the 'hinge' comes in the middle of the wrist, between the scapholunar proximally, and the trapezoid and trapezium of the distal row of carpals.

There is no increase in the number of digits or phalangeal bones in the formation of the pinniped flipper. The foreflipper of a phocine seal is typically short and square and the metacarpals of the first two digits are of approximately the same length, the others decreasing slightly. The articulations between metacarpals and phalangeal bones, and between the phalangeal bones themselves indicate that they are reasonably strong, and their shape can most probably be correlated with the strong grip of the terminal phalanges that is possible. The monachine phocids show a tendency towards an elongated flipper, and this is reflected in the first metacarpal being noticeably longer than the second, in the strength of the bones of the first digit, and also in the reduction in the size of

Fig. 6.17 Medial view right humerus showing shape of deltoid crest and height of medial tuberosity (x) in relation to the head of the humerus. A.p.doriferus (left), P.groenlandica (centre) and H.leptonyx.

Fig. 6.18 Drawings of the forelimb skeleton of a greyhound (left) and Monk Seal (M.tropicalis).

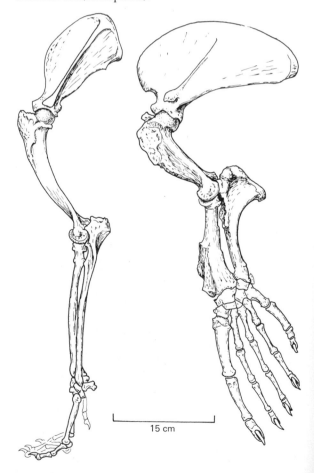

15 cm

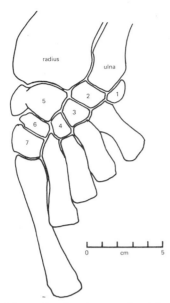

Fig. 6.19 *The left carpal region of a young female Leopard Seal (slightly diagrammatic – taken from a partially dissected specimen).*
1. pisiform, 2. cuneiform, 3. unciform, 4. magnum, 5. scalpholunar, 6. trapezioid, 7. trapezium.

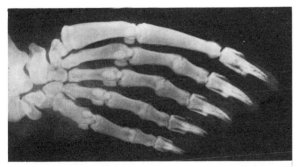

Fig. 6.20 *X-ray of the left foreflipper of a Grey Seal. The typical phocine structure is evident in the rather square shape, the first digit only marginally stronger than the second, and the terminal phalanx shaped to bear the strong claws.*

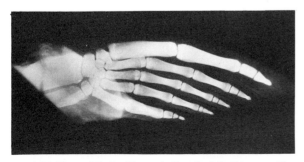

Fig. 6.21 *X-ray of the foreflipper of a Ross Seal. In this monachine flipper the first digit is very strong and the fifth digit is reduced. The flattened terminal phalangeal bones reflect the presence of very small claws. The particularly large proximal epiphyses of the Ross Seal terminal phalangeal bones can be seen.*

the fifth digit. Fig 6.21. The latter effect is achieved in the monachines (with the exception of *Mirounga* and *Monachus monachus*) by the great reduction in size of the second phalangeal bone of this digit. The phalangeal bones are in general much flatter than those in phocine seals, and the articular surfaces themselves are flatter. The end result is that the most highly modified monachine flipper is very like that of an otariid. In otariids the flipper is elongated, the first metacarpal is long, the whole first digit is strong, and the second phalangeal bone of the fifth digit is very reduced. Fig 6.22. The walrus foreflipper is shorter and wider than that of the otariid, but its internal structure is believed to be similar.

The terminal phalanx is shaped to bear the claw and its details vary with the shape of the claw. The groove for the claw insertion is strongly marked and the tip of the bone claw-shaped, when the claws are large, as in phocines. When the claw is very reduced as in the Ross Seal or otariids, the terminal phalangeal bone tends to be flat, only slightly marked for claw insertion. When cartilaginous extensions are borne, as in otariids, the bone ends abruptly, with a flat surface to which the cartilage is attached. Figs 6.22 and 6.23.

Proximal epiphyses on the terminal phalanges of phocids are usually of 'normal' dimensions, though particularly large epiphyses are found on the terminal phalanges of the Ross Seal, where the epiphysis of the terminal phalanx of the first digit may be 24mm long on a shaft of 31mm (King, 1969b). Fig 6.21.

Fig. 6.22 *X-ray of the left foreflipper of a Californian Sea Lion. The strength of the first digit, the diminution of the fifth digit, and the flattened ends of the terminal phalangeal bones that bear the cartilaginous extensions, can be seen.*

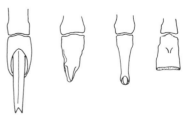

Fig. 6.23 *The terminal phalanx of the middle (3rd) digit of the right foreflipper of, from left to right, Grey, Leopard, Ross and Australian Fur Seal. The size of the bony ungual process is indicative of the size of the claw (not shown) it supports.*

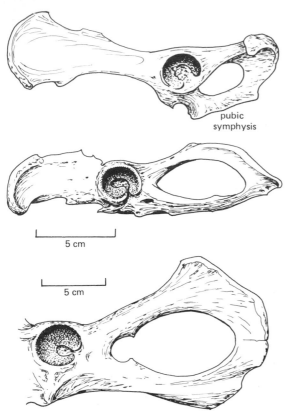

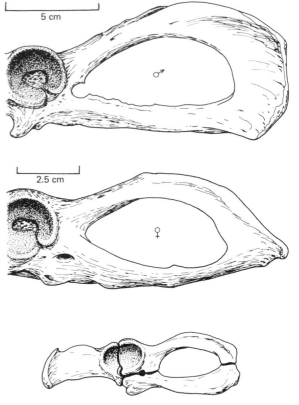

Fig. 6.24 *Left innominate of dog (top left) and Australian Fur Seal (centre left) to show the shortened ilium, increased post acetabular region and dorsally bowed ischium in the seal. Part of the left innominate of* H. leptonyx *(bottom left) to show the ischiatic spine*

of a phocid. Ischium and pubis of male (top right) and female (centre right) fur seals (A.p.doriferus) *to show the difference in shape of the posterior end. Left innominate of a very young* Neophoca *(bottom right) to show the three individual bones.*

Pelvic girdle and posterior limb

Compared with that of terrestrial mammals, the pinniped pelvis has a shortened ilium and an elongated ischium and pubis. There is no fused pubic symphysis. The elongated post-acetabular part of the pelvis produces increased leverage for the muscles used, in the Otariidae for the more detailed movements of the flippers, and in the Phocidae for the stronger and more important movements of adduction and abduction of the flippers in swimming. Fig 6.24.

The pubis is straight and the ischium curves dorsally. A dorsally directed ischiatic spine is present in phocids only, and the muscles attached to this spine help in elevating the hind flippers to produce this very typical phocid posture. The ischiatic spine tends to be higher and more obvious in females. Another sexual difference that is relatively easy to see, particularly in otaries and phocine seals is the shape of the posterior end of the innominate bone. This is much more rectangular in outline in males and more rounded in females.

The shortening of the ilium means that the articulation with the sacrum occurs only just anterior to the acetabulum. The angle the ilium makes with the vertebral column is small, and very much the same in otariid, odobenid, monachine and dog (and also in *Erignathus* – King, 1966), and in these animals the lateral face of the ilium presents a relatively smooth surface. Fig 6.25. This contrasts strongly with this bone in phocine seals where the ilium is bent laterally at a much greater angle, and its lateral surface bears a deep hollow. The great eversion of the phocine ilium means that the medial surface of the bone now faces almost anteriorly. This presumably gives a much greater area of attachment in these seals to the strong ilio-costalis lumborum muscle which is, to a large extent, responsible for much of the lateral body movements used in swimming.

The femur is short and broad, and flattened anteroposteriorly. Fig 6.26. The position of the fovea is barely visible on the head, and the round ligament – between

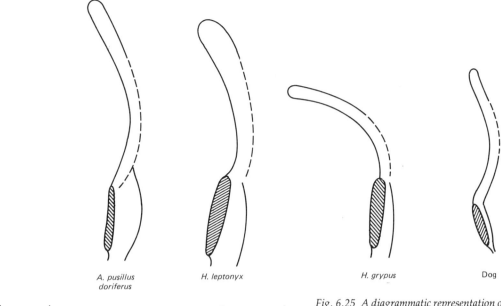

A. pusillus doriferus   *H. leptonyx*   *H. grypus*   Dog

Fig. 6.25 *A diagrammatic representation of a dorsal view of the ilium of, from left to right, an otariid, monachine, phocine, and dog. The great eversion of the phocine ilium is shown. The short pinniped ilium with the sacro-iliac joint close to the acetabulum (shaded) is compared with the arrangement in a dog.*

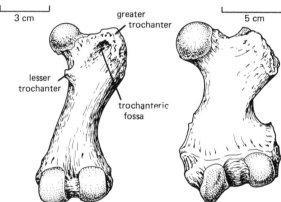

3 cm   5 cm

greater trochanter
lesser trochanter
trochanteric fossa

Fig. 6.26 *Posterior view of the right femur of* A.p.doriferus *(left) and* H. leptonyx.

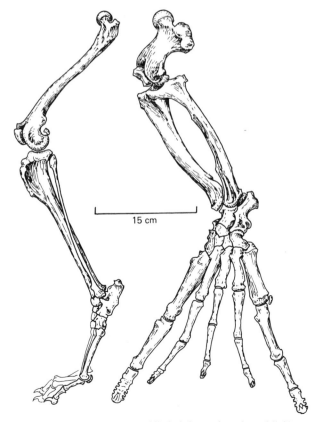

15 cm

Fig. 6.27 *Drawings of the hind limb skeleton of greyhound (left) and Monk Seal (*M. tropicalis*).*

head of femur and acetabulum – does not occur in pinnipeds. The greater trochanter, lateral to the head is broad, but well developed, but the lesser trochanter is present only as a small knob distal to the head in otariids, and is absent in phocids. The trochanteric fossa is small, but present in phocines and otariids, absent in monachines. The head of the femur is distinctly higher than the greater trochanter in the walrus.

The patella of the Phocidae is flatter, while that of the eared seals and walruses is conical.

The tibia and fibula are long, fused together at their proximal ends in phocid (except *M. schauinslandi*), and otariid, and bound by ligament but not fused at their distal ends. Fig 6.27. In the walrus the proximal end of the fibula is not normally fused to the tibia, even in old animals.

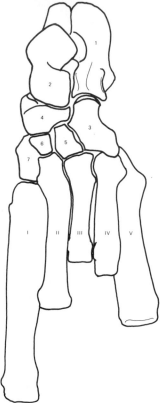

Fig. 6.28 *Diagram of the left tarsal region of an adult female Leopard Seal.*
1. calcaneum, 2. astragalus, 3. cuboid, 4. navicular, 5. external cuneiform, 6. middle cuneiform, 7. internal cuneiform.

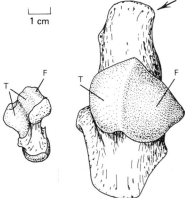

Fig. 6.29 *Anterior view of the left astragalus of an otariid (A.p.doriferus) (left) and a phocid (H. leptonyx) showing the characteristic process, arrowed, of the phocid. The articulation with the tibia is marked T, and with the fibula F.*

There are seven tarsal bones – astragalus, calcaneum, cuboid, navicular, external cuneiform, middle cuneiform and internal cuneiform. Fig 6.28. The phocid astragalus has a strong caudally-directed process similar to the tuberosity of the calcaneum. Fig 6.29. The very strong tendon of the flexor hallucis longus muscle passes over the tip of this process on its way to the digits, and it seems to be the tension on this tendon that prevents the foot turning at right angles to the leg as in an otariid. The astragalus articulates with both tibia and fibula and the differing angles of these articulations in phocid and otariid reflect the different angles of movement of the foot. There are differences between the calcanea of phocids and otariids, and perhaps the most obvious are the presence of a groove for the tendon of Achilles on the tuberosity, and the presence of the medially-directed process (the sustentaculum) in the otariid calcaneum. Both these characters are missing in phocids. The walrus calcaneum is very similar to the otariid bone (Robinette & Stains, 1970).

In all pinnipeds the first metatarsal, and the proximal phalanx of this digit are stronger than those of the other digits, and digits 2, 3 and 4 are the least robust of the pes. In phocids all the bones of the third digit are shorter than the corresponding bones of the other digits, so the distal outline of the flipper is concave. This shortening is particularly pronounced in monachine seals. As in the manus, the bones of the otariid foot tend to be flatter, and the shapes of the articulations are similar. The terminal phalanx is again shaped to bear the claw or cartilaginous extension.

The first metacarpal and first metatarsal have an epiphysis at the proximal end only in the northern phocids, and at both ends of the bones in otariids and southern phocids. The remaining metacarpals and metatarsals have epiphyses at the distal ends only. The phalangeal bones, both fore and hind (except the terminal ones) have epiphyses at the proximal ends only in otariids and northern phocids, but at both ends in southern phocids. The terminal phalangeal bones (fore and hind) have proximal epiphyses in the Phocidae, but no epiphyses in the Otariidae. These statements on epiphyses are, unfortunately, generalizations based on the investigation of a limited number of specimens. It may be that a full comparison might alter the above facts slightly.

Chapter 7 Dentition, age determination and longevity

Dentition

The dental formula* for a typical mammalian dentition may be written $i\frac{3}{3}$, $c\frac{1}{1}$, $pm\frac{4}{4}$, $m\frac{3}{3} = 44$, and pigs are among the few mammals to possess the complete number of teeth. A reduction in the number of teeth is usual amongst most other mammals, from the dog, where only one tooth is lost ($i\frac{3}{3}$, $c\frac{1}{1}$, $pm\frac{4}{4}$, $m\frac{3}{2}$) to the anteaters (*Myrmecophaga*) where there are no teeth at all.

There is reduction of teeth in pinnipeds, the exact number depending on the group concerned. There are never more than three upper and two lower incisors in each half jaw in pinnipeds, and sometimes less than this. There is a total absence of carnassial teeth in living forms, and the nearly homodont condition of the cheek teeth means that it is frequently convenient to refer to them collectively as postcanines or cheek teeth rather than premolars or molars.

Milk teeth and eruption of permanent teeth

The milk dentition is very feebly developed in all seals. The permanent incisors and canines have milk precursors in the normal way, but of the postcanines only the second, third and fourth are preceded by milk teeth. These are in small alveoli in the jaw, external to those of the permanent teeth which, as they erupt, gradually obliterate the alveoli of the milk teeth. Fig 7.1.

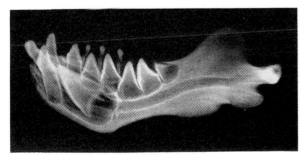

Fig. 7.1 X-ray of the left side of the lower jaw of a young Southern Sea Lion showing the three small milk teeth and the developing permanent teeth.

A large otariid, such as *Otaria*, has deciduous postcanines that are about 7–8mm long, with the crown 2–3mm in diameter, the tooth being rather like a large pin with a laterally flattened head. In a *Eumetopias* pup at birth the deciduous teeth present are the outer incisors in upper and lower jaws ($\frac{i3}{i2}$), the canines, and

upper and lower deciduous postcanines 2, 3, and 4. By the time the pups are about four months old the only deciduous teeth still remaining are the canines, and upper pc1, lower pc4, and of these the canines, may still remain, adhering to gum tissue, until the animal is about nine months old (Spalding, 1966). Some of the permanent teeth may have just pushed through the gum at birth, and they continue to erupt. The first upper and lower incisors are the first permanent teeth to erupt, the second postcanines are rather slower than the other cheek teeth, and the third upper incisors, and the canines are the last to reach full size. By the time the animal is a year old the third incisors are usually, but not always, fully grown, and the canines are less than half grown. In another year the canines are still little more than half erupted (Orr & Poulter, 1967).

In *Callorhinus* the deciduous teeth are very small and simple in shape. Their calcification starts between the seventh and ninth week of post-implantation uterine life, but these teeth are never functional and the smallest of them may be shed before birth and may occasionally be found in the amniotic fluid (Eastman & Coalson, 1974). Most of the others are shed about the time of birth, although the deciduous canines may persist for a few weeks. Calcification of the permanent teeth may start as early as the ninth week of implanted life, and most of them have erupted before birth, (Scheffer & Kraus, 1964).

In the Phocidae the milk teeth are usually resorbed before birth or shed very shortly after birth. They are smaller than those of the Otariidae, and may sometimes be seen in the prepared skull as small nodules in the dried gum tissue. X-ray investigations of foetuses of Weddell, Crabeater, and Elephant Seals have shown that the milk teeth make their first appearance four to six weeks after implantation of the blastocyst, and reach their maximum development when the foetus is about three months old (Bertram, 1940, Laws,

* *Dental formula*: the teeth in the jaws can conveniently be represented by a *dental formula*, in which the different teeth are indicated by their initials (incisor, canine, premolar, molar by i, c, pm, m, respectively). As the two sides of each jaw normally possess the same number of teeth, the dental formula shows those teeth present on one side only of upper and lower jaws, but the final figure (eg 44) gives the total number of teeth in the mouth.

1953). After this they diminish in size as they are resorbed, but may not disappear entirely. In a collection of cleaned and dried Southern Elephant Seal skulls from animals of known ages, remains of the milk teeth can be seen in skulls up to 38 months old. Most of these are only small nodules, but there are one or two examples of quite well-defined little teeth, one of them being 12mm in length. At birth the gums may be smooth, or some of the teeth may have their tips showing. The entire permanent dentition has erupted by the time the animal is approximately a month old. In *M. angustirostris* milk teeth may sometimes remain in position for several years and a female at least 9·5 years old has been seen with lateral milk incisors still in place (Briggs, 1974).

WALRUS – MILK AND PERMANENT TEETH

The adult dentition of the walrus is peculiar in that there are only four postcanines on each side of each jaw, but examination of the milk teeth has led to the permanent teeth being properly interpreted (Cobb, 1933). There are the usual number of milk teeth – three incisors, a canine, and three postcanines in each side of both upper and lower jaws, the first permanent postcanine not being preceded by a milk tooth. The formula for the milk dentition is therefore $i\frac{3}{3}$, $c\frac{1}{1}$, $pc\frac{3}{3}$ = 28. The milk teeth are shed shortly after birth, but not all of them are succeeded by permanent teeth. The first upper milk incisor has no permanent successor, and neither the second upper milk incisor nor the third upper milk postcanine are invariably replaced. It is most usual to find neither of these two milk teeth replaced by permanent teeth, although occasionally there are skulls in which one or other of them has a small, permanent successor.

Thus, apart from the canine, the four teeth usually found in the upper jaw of an adult walrus are the third incisor and three postcanines, and in the lower jaw there are no permanent incisors, and the four teeth are the molariform canine and three postcanines. A small tooth sometimes erupts behind the third milk post-canine. This belongs to the permanent dentition, and is usually shed quite early, although it may sometimes persist into adult life. Because the teeth that erupt after the milk teeth do not all remain into adult life, two dental formulae have been devised to show the successional, and the functional dentition. These are:

Successional $i\frac{2}{0}$ $c\frac{1}{1}$ $pc\frac{5}{4}$ = 26
Functional $i\frac{1}{0}$ $c\frac{1}{1}$ $pc\frac{3}{3}$ = 18

although this rather complicated arrangement may perhaps be more clearly shown in Fig 7.2. There is not entire agreement on the interpretation of walrus teeth, but the above description seems to fit in with the arrangement in other pinnipeds. Further information on walrus teeth is in the chapter on walruses.

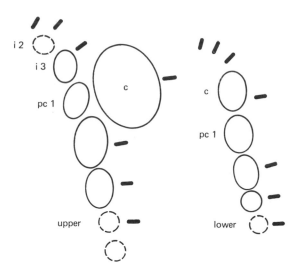

Fig. 7.2 *Diagram to illustrate walrus dentition. i incisor; c canine; pc postcanine. Milk teeth are indicated by short bars, permanent functional teeth by closed circles. Dotted lines indicate teeth that are not invariably present in the adult.*

Permanent teeth

OTARIIDAE

The permanent teeth of the various members of the Otariidae are far less diverse in their shape than are those of the Phocidae. The dental formula is $i\frac{3}{2}$, $c\frac{1}{1}$, $pc\frac{5 \text{ or } 6}{5}$, with some individual variations. Fig 7.3. On each side of the jaw the first two upper incisors have a deep, transverse groove and the third incisor is larger and caniniform. In fur seals the third upper incisor has an oval cross-section and is less conspicuously larger than the other incisors, whereas in sea lions the cross-section tends to be circular and this incisor is much bigger than the others (Repenning *et al.*, 1971). Otariid canine teeth are well developed and for each species, those of males may be distinguished from those of females by their greater diameter at the enamel–root junction. The cheek teeth are simple, approximately cone-shaped structures, slightly flattened parallel to the long axis of the jaw. There is usually a well-developed cingulum on the inner surface, and there may be a small cusp on the anterior edge. Posterior cusps may also be present, and are more frequent on the hinder cheek teeth. The amount of cusping varies – in *Callorhinus* and some species of *Arctocephalus* (e.g. *A. gazella, A. forsteri, A. galapagoensis*) the posterior cusping is minimal, whereas in *A. pusillus doriferus* for example, there are prominent posterior cusps. The teeth of *Arctocephalus* are considered and figured in Repenning *et al.* (1971). A considerable diastema between pc 4 and 5 is characteristic of *Eumetopias*, though it develops with age and is not present in young pups.

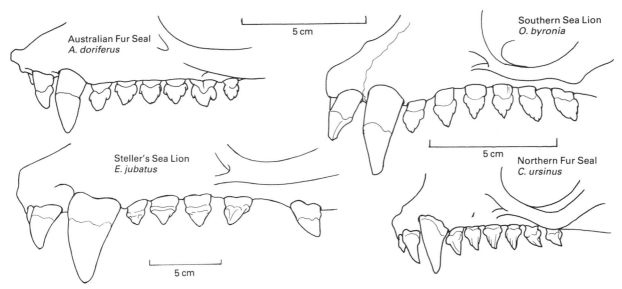

Fig. 7.3 Drawings of typical otariid teeth.

In otariids a double-rooted condition of the cheek teeth is primitive, apart from the first premolar (pc1) which is single rooted. During the evolution of otariids for the last 15 million years, single-rooted teeth have been gradually evolving, but even now the teeth of sea lions are not completely single-rooted, the upper molars (pc5 and 6) being the last to change. There is a greater tendency for the first upper molar (pc5) to be single rooted in sea lions and *A. pusillus* than in other fur seals where a variable number of cheek teeth are double rooted (Repenning *et al.*, 1971, Repenning & Tedford, 1977).

The number of cheek teeth is usually $\frac{6}{5}$ in *Otaria*, *Phocarctos*, *Callorhinus* and *Arctocephalus*; and is usually $\frac{5}{5}$ in *Eumetopias* and *Neophoca*. The number is very variable in *Zalophus*; Californian animals more frequently having $\frac{5}{5}$, Galapagos animals $\frac{6}{5}$, while all variations are found in Japanese animals.

The lower teeth are very similar in shape to the upper, although there is no caniniform incisor, and the incisors do not have a transverse groove.

PHOCIDAE

The two subfamilies of the Phocidae cannot be distinguished by their dental formulae. Phocinae have incisors $\frac{3}{2}$, except in *Cystophora* which has $\frac{2}{1}$, and the Monachinae have incisors $\frac{2}{2}$ except in *Mirounga* which has $\frac{2}{1}$. In all the phocids the rest of the dental formula is $c\frac{1}{1} pc\frac{5}{5}$. In no phocid is there a transverse groove in the incisors. The outer upper incisors are usually larger than the inner ones, the difference being most marked in *Leptonychotes* in which the outer pair is more than twice as large as the inner pair, project markedly forwards, and are used in ice sawing.

The shape of the cheek teeth, and the great variation in the degree of cusping can be seen in Fig 7.4. Cheek teeth of the Bearded Seal are loosely rooted, wear flat very quickly and soon drop out of the skull, and those of the Ross Seal are also weak, frequently loose in the living animal, and lost in the skull, and are also subject to considerable variation in number. All three species of Monk Seal have broad, heavy rugose postcanines, and the complex sieving teeth of the Crabeater have already been mentioned. Fig 7.5. The roots of the canines of the adult male Southern Elephant Seal are circular in cross-section, and can be distinguished from those of the female which show four well-marked longitudinal grooves. The canines of the Northern Elephant Seal also show sexual dimorphism. Those of the adult male may reach a length of about 113mm, and the pulp cavity remains open for at least 12 years. The root of the male canine is stout and parallel-sided with two longitudinal grooves along the lateral surface. The canines of the adult female are smaller, reaching about 91mm in length. The root tapers to a point after seven years, and the opening of the pulp cavity closes when the animal is about 12 years old. There are two longitudinal grooves along the lateral surface of the root (Briggs & Morejohn, 1975).

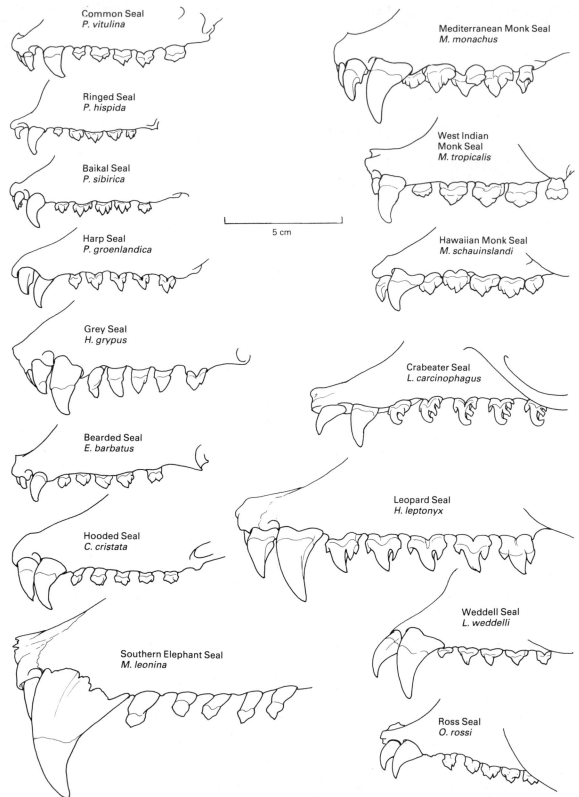

Common Seal
P. vitulina

Ringed Seal
P. hispida

Baikal Seal
P. sibirica

Harp Seal
P. groenlandica

Grey Seal
H. grypus

Bearded Seal
E. barbatus

Hooded Seal
C. cristata

Southern Elephant Seal
M. leonina

Mediterranean Monk Seal
M. monachus

West Indian
Monk Seal
M. tropicalis

Hawaiian Monk Seal
M. schauinslandi

Crabeater Seal
L. carcinophagus

Leopard Seal
H. leptonyx

Weddell Seal
L. weddelli

Ross Seal
O. rossi

5 cm

Fig. 7.4 Drawings of phocid teeth.

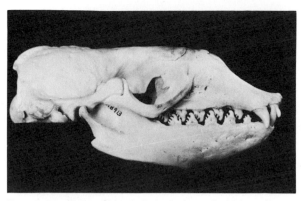

Fig. 7.5 Lateral view of the skull of a Crabeater Seal (Australian Museum M8913) showing the elegantly cusped cheek teeth. Photo. Australian Museum, Sydney.

Age determination

The age of most seals can be deduced, reasonably accurately, from the growth layers in the roots of the canines. Although, after the canine has erupted, there is no further growth of the enamel crown, the root continues to grow by the deposition of dentine in concentric cones on its inner surface. In most seals the pulp cavity eventually closes up, growth ceases, and no more dentine is deposited. This occurs at various ages however, in *Halichoerus* for instance, the pulp cavity closes at five or six years old (Hewer, 1964), but dental layers were still visible in a Weddell Seal canine 18 years old (Stirling, 1969). Growth of the canines continues, probably throughout life in the male Elephant Seal and walrus. If the outer surface of the root is not covered by cement, the ridges caused by the lower edges of the dentine cones can be counted, and the number has been shown by tagging experiments to be the age in years of the animal. In transverse or longitudinal section each one of these annual ridges is usually seen to be made up of alternating rings of incompletely calcified or marbled dentine which show as light rings, and fully calcified dense dentine which show as dark rings, the darkness depending on the density.

The sequence of rings in the canines of a Crabeater Seal gives an idea of the arrangement that seems to be normal to most seals. Under the enamel is a layer of dense dentine laid down when a foetus. A conspicuous neonatal line indicates the discontinuity at birth. A second layer of dense dentine represents the period of suckling, followed by a dark line indicating weaning. A further layer of dense dentine with fine lines showing minor variations in the rate of deposition is followed by a layer of vacuolated dentine showing the end of a year's growth. Once the normal pattern of adult life has been established, this is reflected in the teeth, the animal laying down poorly calcified dentine in September–October at the spring pupping season when the normal feeding regime is interrupted – a wider band in the female than in the male – and dense dentine at other times of the year when feeding is normal. Only one vacuolated layer is laid down annually (Laws, 1958, 1962). In *Callorhinus* variations in the density of the dentine deposited during suckling can be correlated with the suckling cycles of the pup – periods of suckling when the female is ashore, alternating with foodless days when the female has gone to sea (Scheffer & Peterson, 1967).

Similarly in the Ringed Seal, less dense dentine indicates the disturbance of pupping and the period of fasting during the moult in spring and early summer, while the dense dentine of the rest of the year is laid down while the animal is feeding normally.

The canines of the Southern Elephant Seal have been studied more than those of many other seals, and the rings in this seal are particularly clear. As in the Crabeater, dense dentine is laid down in foetal life and during lactation, and continues until the pup first goes to sea, darker neonatal and weaning lines being visible. In the immature animals of both sexes, light marbled dentine is laid down during the initial nine months at sea, though rings of darker dentine occur if the animal hauls out during this time. This marbled dentine is only produced by sexually inactive animals when they are at sea.

In males, the adult pattern of two dark (dense) layers separated by one narrow and one broad lighter (less dense) layers becomes established during the third and fourth years. The dense bands represent the moulting period and the breeding period – or period of testicular activity in those males not yet holding a harem.

In females the start of pregnancy is marked by dense dentine filling in the space between the two denser layers of moult and pupping. An adult breeding female will therefore have a dark band at moulting, followed by a slightly less dark band of pregnancy, followed by a dark band of pupping, followed by a light band of the period at sea just after copulation. The whole cycle then starts again for the next year.

Southern Elephant Seals can thus produce dense dentine in a variety of situations – feeding or fasting, moulting or breeding. Various hormones are produced during these activities – thyroxine when moulting, testosterone and progesterone during breeding for example, and possibly adrenalin during periods of stress such as birth or fasting. Any or all of these, or still other substances may affect the rate of calcification of the dentine, and experimental work is needed to solve this problem (Carrick & Ingham, 1962, Laws 1962). There thus seems to be a difference, in that

poorly calcified dentine is laid down in the pupping season in the Crabeater and Ringed Seals, whereas dense dentine is laid down in the breeding season and also at the moult in the Southern Elephant Seal. In the Weddell Seal too, dense dentine seems to be laid down at the moult.

Deposition of cement on the outside of the root has, of course, no relation to the closure of the pulp cavity. It is thus able to be laid down for much of the life of the animal, although its deposition is believed to cease in very old animals. A young Harbour Seal, tagged when new born, and recovered three years later close to its birthplace in Alaska, showed three narrow layers of cement on a lower canine (Divinyi, 1971). Cement rings on Bearded Seal canines have been seen in animals up to 31 years old (Benjaminsen, 1973).

Claws of some seals have been shown to have annual ridges, or alternating dark and light coloured bands, one dark and one light band being the annual growth layer. The use of these as age indicators is, of course, limited as the tip of the claw will get worn away. However, in Bearded Seals there is good correlation between the annual layers of cement and claws up to about the age of eight years. After this the claw tip is too worn.

A laminated appearance, in section, of some other parts of the skeleton, such as the tympanic bulla and the lower jaw, has also been noted. Although there is some indication that these are annual layers, little work has been done on this yet.

Longevity

The maximum age to which seals may live is usually recorded from the length of time they spend in captivity. Annual rings in dentine and cement now allow the age of wild seals to be known. The following tables of the ages of seals as known from both sources give some indication of their length of life. The tables are not intended to be exhaustive, and the ages given are not necessarily the greatest it is possible for the animal to attain.

Longevity in captivity

Halichoerus			
26 y 3m ♀	1912–1939	Victoria Park, Cardiff	Matheson, 1950
33 y	1924–1957	Edinburgh	Crandall, 1964
41 y	1899–1940	Skansen	Matheson, 1950
Phoca			
19 y 6m ♂		Seattle Zoo	Mansfield & Fisher, 1960
26 y		Skansen	Crandall, 1964
Phoca hispida			
nearly 15 y	1914–1929	Skansen	Crandall, 1964
Mirounga leonina			
15 y ♀	1925–1940	Hagenbecks	Crandall, 1964
Callorhinus			
9 y		Washington	Bonham, 1943
Eumetopias			
17 y 3m	1900–1918	Washington	Crandall, 1964
Otaria			
17 y 6m	1879–1896	London Zoo	Flower, 1931
Zalophus			
19 y 3m ♂	1938–1957	Washington	Crandall, 1964
20 y 2m ♀	1941–1961	Washington	Crandall, 1964
30 y ♀	1926–1956	Bremerhaven	Ehlers, 1957
Arctocephalus pusillus pusillus			
20 y	1884–1904	London Zoo	Flower, 1931
Arctocephalus pusillus doriferus			
16 y +	(still alive 1930)	Taronga Park, Sydney	Flower, 1931
Odobenus			
11 y 10m	1937–1949	Copenhagen	Reventlow, 1951

	Longevity in the wild	
Halichoerus		
26 y ♂	from cement rings	Platt, Prime, Witthames, 1975
38 y ♀		
46 y at least ♀	cement rings single aged individual	Bonner, 1971
Phoca		
35 y at least		Burns & Fay, 1972
Phoca hispida		
43 y at least ♂		McLaren, 1962
Phoca groenlandica		
28 y	from dentine rings	Fisher, 1952
Live to 35 y		Sergeant, 1965
Cystophora		
Live to 25 y		Sergeant, 1965
Erignathus		
25 y ♂	from cement rings	Benjaminsen, 1973
31 y ♀		
Monachus schauinslandi		
20 y ♂	from cement rings	Kenyon & Fiscus, 1963
Mirounga leonina		
20 y ♂		Laws, 1953
14 y + ♂	branded 1945, seen 1959; still not a harem master	Csordas, 1964
12 y + ♀	branded 1945, pupped 1957	
Hydrurga		
16 y + ♂	from dentine rings	Laws, 1957
13 y + ♀		
23 y + ♂	from dentine rings	Øritsland, 1970
26 y + ♀		
24 y ♂	from dentine rings	Tikhomirov, 1975
20 y ♀		
Lobodon		
25 y ♂	from dentine rings	Øritsland, 1970
29 y + ♀		
28 y ♂	from dentine rings	Tikhomirov, 1975
23 y ♀		
Leptonychotes		
18 y ♂	from dentine rings	Stirling, 1971
Ommatophoca		
12 y +	from dentine rings	Øritsland, 1970
21 y ♂	from dentine rings	Tikhomirov, 1975
19 y ♀		
Callorhinus		
21 y ♀		Kenyon & Scheffer, 1954
Zalophus		
15 y +	skull and dental characters	Orr, Schonewald, Kenyon, 1970
Neophoca		
12 y + ♂ and ♀	from cement rings	Stirling, 1972a

Chapter 8 *Viscera*

Alimentary canal

The alimentary canal is uncomplicated in its general arrangement. Salivary glands are present, but relatively small. They produce lubricating mucus to help in swallowing, but as the food is swallowed whole, the saliva has no digestive enzymes (Fava-de-Moraes *et al.*, 1966). The salivary glands of the Otariidae and Odobenidae are slightly better developed than those of the Phocidae. The submandibular gland is generally the largest, situated just posterior to the ramus of the mandible. The sublingual gland, which may be present or absent according to species, extends from the level of the submandibular gland to the posterior margin of the mandibular symphysis. The parotid gland is very small, or may be absent (Eastman & Coalson, 1974). The tongue is mentioned in the chapter on senses.

The oesophagus frequently has deep longitudinal folds, is dilatable to accommodate bulky food, and is well supplied with mucous glands whose secretions would help the easy passage of such food (Eastman & Coalson, 1974). The oesophagus appears to be particularly muscular and dilatable in the Ross Seal (Bryden & Erikson, 1976). The stomach is reasonably capacious, but is a simple curved enlargement of the tube, with the pyloric region sharply bent back on the body of the stomach. The duodenum is only indistinctly recognizable from the rest of the small intestine, and the latter is narrow in diameter and very long (see later, and Table). At its junction with the large intestine is a small caecum, sometimes hardly distinct, sometimes about 2–3cm in length. The large intestine shows the normal increase in diameter, is relatively short, and passes into the rectum. No fat is found in the mesenteries, it is all restricted to the blubber. The normal mammalian pancreas and spleen are present (Eastman & Coalson, 1974).

Digestion

Digestion of food is rapid, the whole process is said to take about 18 hours, and most digestion takes place at sea, so that animals on land usually have empty stomachs.

Elimination of faeces and urine usually takes place whenever necessary. On land, occasionally some movement away from a closely packed group is made for such activity, as in *Neophoca* and *Phocarctos* (Marlow, 1975), and some Steller's Sea Lions and New Zealand Fur Seals have been seen moving to the edge of an overhanging rock and defaecating over the edge (Stirling & Gentry, 1972). As members of the same colony were underneath, no motives of hygiene can be suggested. It may be significant that all the animals mentioned above are otaries. On the other hand, *Zalophus* adopts no special posture when defaecating (Peterson & Bartholomew, 1967), neither do phocids.

As seals eat virtually no carbohydrates or sugars it is not surprising that there is an absence of enzymes for breaking down such materials. Otariids and walrus have no lactose in the milk, so there is virtually no lactose activity in the intestine. Phocids have a small amount of lactose in the milk, and there is a comparable amount of lactose in the intestine (Kretchmer & Sunshine, 1967). The effect this virtual absence of lactose can have on the digestive troubles of captive orphan seals is mentioned in the section on reproduction. The blood glucose level of pinnipeds is normal, the glucose presumably being obtained from protein. Enzymes for converting glucose to galactose are found in the erythrocytes in the normal mammalian way (Kerry & Messner, 1968).

The presence of the enzyme thiaminase in many of the fish that seals eat, both in captivity and in the wild, can give rise to some problems. The enzyme inactivates the vitamin B_1 which occurs in other fish, and if the seal feeds for too long on such a diet it can develop fatal thiamine deficiency (beri beri in man), with symptoms such as cardiac lesions and the loss of the myelin sheaths of the nerves. Seals in the wild probably do not feed on one species of fish to the exclusion of all others for very long at a time, but the situation is different for animals in captivity. Given with the normal diet, vitamin B_1 supplements will be destroyed by the thiaminase, but the supplements have been found to be effective if given some two hours before the normal feeding time (Rigdon & Drager, 1955).

Length of gut

The pinniped small intestine is remarkable for its great length. The gut of a dead seal is of course difficult to measure with any great accuracy, but even allowing for this, the length is remarkable for a carnivore, and there is also enormous variation even within the same species. The small intestine of herbivorous animals is known to be long, and may be some 20–28 times the length of the body, while in a typical carnivore it is only about five or six times the body length.

The small intestine of pinnipeds may be anywhere between about four and 40 times the body length (Table). Perhaps the greatest number of intestine lengths have been measured on *M. leonina*, and in 89 seals it was found that the small intestine varied in length between 20 and 25 times the body length, with one very long specimen 42 times as long. Even two animals with the same body length could have small intestines of very different lengths (Laws, 1953).

It has been suggested that possibly the great length may give a greater area for food absorption related to the high metabolic rate, or perhaps compensate a little for the large areas of gut that are incapacitated by parasites (Eastman & Coalson, 1974). The diet seems to have little bearing on the matter as, for example, *M. leonina* and *O. rossi* are both predominantly squid eaters and yet they have possibly the longest and shortest intestines respectively.

Liver

The liver is deeply divided into five to eight long, more or less pointed lobes which surround the hepatic sinus. A gall bladder is present. The bile duct and the pancreatic duct run parallel for the last part of their length, entering the intestine at the beginning of the duodenum. They continue for a short distance in the wall of the duodenum and open into a lumen between the inner and outer walls of the duodenum. Sections through this intramural chamber show that it resembles the intestine in structure rather than the bile duct, and is thus most probably a diverticulum of the intestine rather than a dilated end of the bile duct. This intestinal sac opens by a papilla into the lumen of the duodenum. It has been described in *P. vitulina*, *O. byronia* and *O. rosmarus*, and thus may well be present in all pinnipeds (Burne, 1909).

Hypervitaminosis

Eskimo and Arctic travellers and their dogs have always relied to a large extent on seal and polar bear meat for their food. They are wary, though, of eating the livers of the bears, and those of Bearded Seals, large Ringed Seals and walruses, particularly the 'rogue' carnivorous walruses (Fay, 1960b), though livers of smaller animals are frequently eaten without ill effect (Rodahl, 1949). The very high amount of vitamin A in these livers makes them toxic, and it has been estimated that about 40g of polar bear liver, or 80g of Bearded Seal liver may be enough to cause quite severe illness. Drowsiness, headache, vomiting and peeling of the skin are some of the symptoms of such hypervitaminosis.

Shipwrecked sailors have become ill after eating livers of *Neophoca* (Cleland & Southcott, 1969a), and analysis of Australian Fur Seal (*A.p. doriferus*) liver indicates that there is more vitamin A in the livers of the older animals, and a single meal of about 500g could be toxic (Southcott et al., 1974). Some of the Antarctic phocids seem to have less vitamin A in their livers, and about 2250g (5lb) of Weddell Seal liver would have to be consumed before the eater showed ill effects (Southcott et al., 1971).

Husky dogs 'on duty' in the Antarctic, when fed on seal meat, tend to store large amounts of vitamin A in their livers, so that even a small amount (100g) of such a liver could be toxic. The illness of Sir Douglas Mawson, leader of the 1911–14 Australasian Antarctic Expedition, and the death of his companion Xavier Mertz have been attributed to eating husky dog livers, as both men had symptoms of hypervitaminosis (Cleland & Southcott, 1969b).

Illness from eating seal liver has, however, been known for some long time. Hamilton (1934) notes comments from the accounts of Loaysa and Alcazaber, written in 1526 and 1535 respectively, and referring to the Southern Sea Lion *Otaria* 'In Loaysa's voyage and again in that of Alcazabar it is stated that the liver is more or less poisonous. "Most of us who ate it suffered from the head to the feet", and "the livers of these seals is so poisonous that they give fevers and headache to everyone who eat them, and presently all the hair on their bodies falls off and some die".' Hamilton himself, however, said that he had eaten a great deal of sea lion liver without ill effect.

Kidney

A truly reniculate kidney is possessed only by cetaceans and pinnipeds. In these animals each kidney is made up of small units – the reniculi. Each reniculus is like a small simple kidney with its own cortex, medulla and calyx, and the duct from it joins with others eventually to form the ureter. Single kidneys of *M. leonina* and *P. groenlandica* have been found to have about 300 and 136 reniculi respectively, not all of which could be seen on the surface. In the *M. leonina* 230 reniculi could be seen on the surface, and the central ones seen by injection and X-ray (Arvy & Hidden, 1973) (*P. groenlandica* – Dragert et al., 1975). The surface reniculi are separated from each other by little more than shallow furrows in the otariids while the covering fibrous coat of the kidney is anchored to them by strands. In phocids the fibrous coat dips down at intervals, separating groups of reniculi, the greatest degree of separation occurring in the Ross Seal (Anthony & Liouville, 1920).

The possession of a reniculate kidney is most probably correlated with the large size of the animals and the ability to concentrate urine (Vardy & Bryden, 1981).

In phocine seals it seems that the ureter leaves the kidney from a funnel-shaped depression in the middle of the medial border (Arvy & Hidden, 1973). Kidneys of Weddell, Crabeater and Southern Elephant Seals have a slit near the medial border of the ventral surface (Vardy & Bryden, 1981). This slit extends into the thickness of the kidney between the reniculi and provides the place of entrance and exit for branches of the renal artery, ureter and branches of the venous plexus respectively. It is possible that the Leopard Seal also has a similar slit. In the Ross and Hawaiian Monk Seal the ureter appears to emerge from the ventral face of the posterior end of the kidney. Possibly closer investigation would reveal a slit here too, in which case it may be a monachine character.

Within the kidney one branch of the ureter drains the reniculi of the anterior end, and a second branch those of the posterior end. These two branches join in a T-junction to form the main ureter. This seems to be the arrangement in most phocids, and in one otariid at least (*A.p. pusillus*, Bester, 1975), while in *Zalophus* and *A. gazella* the ureter is reported to be formed by three and six main branches respectively (Arvy & Hidden, 1973).

In the Weddell Seal, and possibly also in the Harp Seal, the muscles of the calyx wall extend between the cortex and the medulla of the reniculus. This peripyramidal muscle forms a fine open mesh enclosing the medulla. Such a muscle is thought to exist to varying degrees in other mammalian kidneys and is possibly best represented by the Sporta perimedullaris musculosa of the Cetacea (Vardy & Bryden, 1981). The blood vessels of the kidney are mentioned in the section on the vascular system.

Water

It has been shown experimentally that seals can obtain all the water they need from the food they eat. While swallowing they take in virtually no sea water, and this little has no effect on them (Depocas *et al.*, 1971). If, however, sea water is administered experimentally the stomach becomes upset and the excess salts have to be eliminated by using body water (Albrecht, 1950; Ridgway, 1972). Just weaned Northern Elephant Seals, fasting for up to 12 weeks before they catch their own food, have been shown to obtain all the water they need from breakdown of the blubber, and in fact were shown to have an exceptionally low rate of water turnover (Ortiz *et al.*, 1978).

In spite of this, seals *have* been seen drinking sea water. Apart from a captive Leopard Seal (Brown, 1952) and an exhausted walrus that had been trapped in a crevice (Loughrey, 1959), the seals that have been noted as drinking sea water have all been otariids – *Callorhinus, Eumetopias, Zalophus, Arctocephalus forsteri, A. australis* and *A.p. pusillus* (Gentry, 1981). Nearly all the animals that have been seen drinking have been adult males that are fasting during the breeding season. These fasting seals are depending for their water on that produced from fat breakdown. If the weather is hot they are possibly not producing enough water this way to provide the amount needed for sweating and urinating for thermoregulation.

So it is postulated that the animals consume small amounts of sea water at intervals. This would not be enough to cause digestive troubles, but would be sufficient for nitrogen excretion. It may be significant that it is the seals in warmer climates that have been seen drinking. *A. forsteri* in New Zealand does not drink, but those in the Australian colony on the South Neptune Islands do; similarly *Callorhinus* on the Pribilofs and *Eumetopias* in Alaska do not drink, while those on San Miguel and Ano Nuevo respectively, do.

Stones

Stones have been recorded from the stomachs of many seals, and have been found in almost all sizes and quantities. Many hundred small pebbles may be found in a single stomach, or the number may be much smaller, but the individual stones much bigger, up to the size perhaps of a tennis ball. Eleven kilograms of stones have been found in the stomach of a single *Otaria*. Not surprisingly, there is some evidence that the larger stones are found in the larger animals.

Why seals swallow stones still remains a mystery, although plenty of theories have been put forward. A certain small proportion of the stones may be swallowed accidentally, but most authors are agreed that most of them are ingested deliberately. Very young pups, still feeding on milk, have stones in their stomachs, but it is very likely that these are swallowed more or less accidentally as a result of the pups' habit of playing with, and picking up, almost anything.

Most seals are infested with large numbers of parasitic worms, and it has been suggested that the stones serve to grind up some of these worms and allay the possible irritation that they cause, though there is no evidence of this function. It has been suggested that the stones may help break up the food which is swallowed whole, but again there is no evidence. The stones may give extra weight and thus assist the stomach muscles in ejecting masses of light fish bones – but not all seals vomit all the fish bones. Sea lions seem to have a tendency to vomit, and it is a common thing in a colony of *Phocarctos*, for instance, frequently to see animals heaving repeatedly, with the end production of a small amount of liquid and hard skeletal remnants of fish and cephalopods, and a few stones.

Comparative dimensions of gut in seals

	Sex & age	nose to tail length m	stomach to anus m	small intestine m	caecum cm	large intestine m	Ratio small intestine to nose to tail length	Reference
O. rosmarus	♀ juv	1·22	22·65	22·55	3·8	0·305	17·1	Owen, 1853
O. rosmarus	♂ ad	3·66	–	42·39	–	–	11·6	Brooks, 1954
Z. californianus	♂ ad	2·07	34·61	32·61	1·2	2·00	15·8	Forbes, 1882
E. jubatus	♀ ad	2·13	80·51	77·71	7·6	2·80	36·5	Engle, 1926
O. byronia	juv	1·70	19·82	18·30	1·2	1·52	10·8	Murie, 1874
P. hookeri	♀ pup	0·77	12·08	11·34	2·5	0·74	14·7	Ph5*
P. hookeri	♀ ad	2·00	36·30	34·00	5·0	2·30	17·0	Ph7*
P. hookeri	♀ ad	1·78	21·50	19·75	5·0	1·75	11·0	Ph11*
A. forsteri	♂ ad	1·95	19·50	–	–	–	–	Miller, 1975b
A. forsteri	♂	1·40	14·59	–	–	–	–	Miller, 1975b
A. forsteri	♀ old	1·36	20·05	–	–	–	–	Miller, 1975b
A. forsteri	♀ yg	1·24	15·90	–	–	–	–	Miller, 1975b
P. vitulina	new born	0·91	–	10·40	–	–	11·4	Ball, 1930
E. barbatus		'large'	25·00	21·64	–	3·04	–	Burns & Frost, 1979
M. monachus	♀ yg	1·54	9·14	8·74	–	0·40	5·7	Schnapp et al., 1962
L. carcinophagus	♀ 5y	2·27	20·65	20·00	–	0·65	8·8 ⎫	Bryden & Erickson, 1976
L. carcinophagus	♀ 4y	2·23	30·00	29·00	–	1·00	13·0 ⎬	
L. carcinophagus	♂ 9y	2·22	28·15	27·00	–	1·15	12·1 ⎭	
H. leptonyx	♂ ad	2·60	23·40	22·20	3·5	1·20	8·5	JEK41*
H. leptonyx	♀ old	3·06	23·80	21·66	c.2·0	2·14	7·1	JEK83*
O. rossi	♀ ad	2·30	12·27	11·27	–	1·00	4·9	King, 1969b
O. rossi	♂ 4y	2·00	5·05	4·50	–	0·55	2·3	Bryden & Erickson, 1976
M. leonina	♂ ad 11y	4·80	–	122·00	–	–	25·4	Laws, 1953
M. leonina	♂ ad 11y	4·80	–	201·91	–	–	42·0	Laws, 1953
M. leonina	♀ ad 11y	3·23	–	86·92	–	–	26·9	Laws, 1953
M. leonina	♀ ad 15y	2·89	–	73·20	–	–	25·3	Laws, 1953

* These measurements were made by the present author.

Scavenging birds soon remove the organic material, but the stones remain, sometimes to confuse the geologist. Basaltic pebbles from the Auckland Islands have been ingested by *Phocarctos* and then regurgitated on the Snares, where the local stone is granite (Fleming, 1951).

A ballasting function used to be a popular theory amongst the sealers, who thought that seals have a 'ballast bag' full of stones in addition to the stomach. Two theories occur here. The first is that the stones are taken in when the animals are thin, and the second is that when the animal is at its fattest it will be more buoyant, and in both instances the stones are thought to be needed for adequate balancing and swimming. It is obvious that big male sea lions in particular have very light hindquarters when compared with the very heavy anterior end of the body, although it is difficult to see how stones in the stomach would do anything to redistribute the weight.

Many seals may spend two months or more without food during the breeding season, and sometimes during the moult as well. It has been suggested that the presence of stones at this time will provide bulk upon which the stomach muscles may contract and help relieve the theoretical hunger pangs. A certain amount of evidence in favour of this theory has been put forward from the examination of Elephant Seal (*M. leonina*) stomachs in the summer, a time when the animals spend a long period on land without feeding (Laws, 1956). However, the number of stomachs examined in relation to the time of year is still relatively small. This annual fasting is normal to the life of the animal, and although it would be very difficult to prove, there is no evidence that it causes the seal any discomfort.

It is interesting to note that stones are found also in crocodile stomachs (Cott, 1961). Work on the Nile crocodile (*Crocodylus niloticus*) has shown that all animals acquire stones by the time they are mature, and the stones usually form about one per cent of the body weight.

It is assumed that the stones must be taken in deliberately as even crocodiles in swampy areas manage to collect stones, and sometimes they must have travelled a few kilometres to find them. There is no reason why the stones should serve a triturating function, and as the crocodile feeds throughout the year they need not provide bulk for the stomach. A suggestion is that the stones have a stabilizing effect on the animal, rather like the cargo in the hold of a ship. A young crocodile, without its ballast, appears to have difficulty in lying level at the surface. The extra weight has also apparently a value when struggling prey is being held under water, or when the animal lies on the bottom in a strong current.

Chapter 9 *Respiratory system*

Nostrils

In most pinnipeds the nostrils are situated in the normal position on the end of the snout, although the details of their position vary. In the Otariidae, and most of the Phocinae the nostrils are more or less vertical, in the Odobenidae and Monachinae they are more horizontal and towards the dorsal surface of the snout, while in the Hooded Seal and Elephant Seals their position is complicated in the adult males by the nasal appendage, with the result that the nostrils face downwards. While the animal is under water the nostrils are, of course, closed, and this is believed to be the relaxed position, maintained by the tone of the muscles and the pressure of the moustachial pad. Contraction of the nasolabialis and maxillonasolabialis muscles results in movements of the pad and the opening of the nostrils.

Externally, in the Otariidae, the naked dark brown or black tip of the nose is rather dog-like in shape, as is the curved, comma-like shape of the nostrils. The Phocidae tend to have straighter nostrils, with varying degrees of approximation of their lower ends.

The maxilloturbinal bones, in the anterior part of the nasal cavity, are finely divided and convoluted, well filling the cavity. In the Ross Seal, for example, they are so densely packed that when in life they are covered with epithelium, there would seem to be little room for the passage of air.

The breathing rhythm of seals on land is fairly irregular, and, of course, breathing stops while the animal is diving – a period that can last about an hour. There is some evidence that pups have a higher breathing rate than adults. Weddell Seal pups, for instance, take 16 breaths per minute, while adults take eight breaths; pups of *Neophoca* and *Phocarctos* take 13 breaths a minute, while adults take three to five. Northern Elephant Seals may take three breaths a minute, but do not do so regularly; they take several rapid breaths, then hold their breath for 1–20 minutes. In the sea lions *Zalophus*, *Phocarctos* and *Neophoca* at least, the respiratory cycle is a short exhalation, then a short inhalation, followed by a longer period of breath holding (apnea). In *Zalophus* the periods of exhalation and inhalation each last about a second, and the period of apnea is about 12 seconds.

The pharyngeal air sacs of the walrus are mentioned in the chapter on walrus.

Larynx

Vocal cords are present, and also the normal number

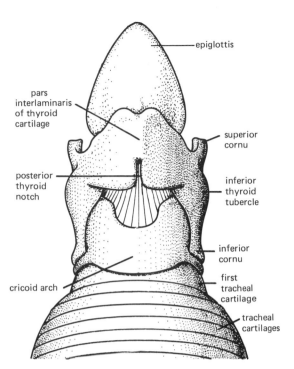

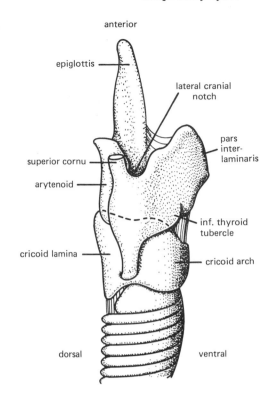

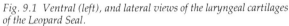

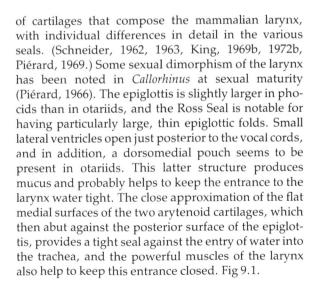

Fig. 9.1 Ventral (left), and lateral views of the laryngeal cartilages of the Leopard Seal.

of cartilages that compose the mammalian larynx, with individual differences in detail in the various seals. (Schneider, 1962, 1963, King, 1969b, 1972b, Piérard, 1969.) Some sexual dimorphism of the larynx has been noted in *Callorhinus* at sexual maturity (Piérard, 1966). The epiglottis is slightly larger in phocids than in otariids, and the Ross Seal is notable for having particularly large, thin epiglottic folds. Small lateral ventricles open just posterior to the vocal cords, and in addition, a dorsomedial pouch seems to be present in otariids. This latter structure produces mucus and probably helps to keep the entrance to the larynx water tight. The close approximation of the flat medial surfaces of the two arytenoid cartilages, which then abut against the posterior surface of the epiglottis, provides a tight seal against the entry of water into the trachea, and the powerful muscles of the larynx also help to keep this entrance closed. Fig 9.1.

Lungs

In phocids and walrus the trachea divides into the two primary bronchi immediately outside the substance of the lung. In otariids the bifurcation is more anteriorly, at approximately the level of the first rib, and the two elongated bronchi run parallel until they diverge to enter the lungs dorsal to the heart. Fig 9.2.

The trachea, in most pinnipeds, is supported by cartilaginous rings that either form complete circles, or are incomplete and overlap dorsally. The degree of overlap may vary considerably between one end of the trachea and the other. The cartilaginous support is reduced to ventral bars only in Ross, Leopard and Weddell Seals (King, 1969b, Tarasoff & Kooyman, 1973b, Sokolov *et al.*, 1968), and this condition also occurs at the posterior end of the trachea of the Ribbon Seal, and in most of the trachea, particularly the central region in the Bearded Seal (Burns & Frost, 1979).

A curious air sac has been described in the Ribbon Seal (Abe *et al.*, 1977). A longitudinal slit, some 30mm long occurs between the ends of the cartilage rings on the dorsal surface of the trachea just before it bifurcates to form the bronchi. A membranous sac, sometimes as long as 43cm, extends both anteriorly and posteriorly on the right side of the body. The anterior part passes through the sternomastoid muscle, and the posterior part lies external to the ribs and dorsal to the pectoralis muscle. It has been suggested that this sac could easily be missed if the animal were only skinned. Male Ribbon Seals possess this sac, and it

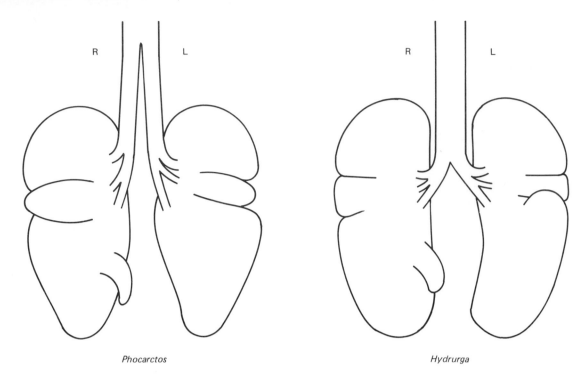

R   L   R   L

Phocarctos   *Hydrurga*

Fig. 9.2 *Diagrams of ventral views of lungs of an otariid (left) and a phocid showing the different points of bifurcation of the trachea, and the degree of lobulation.*

increases in size with age. The tracheal slit is present in females, but the sac is small or absent. Little is known about this sac. It may provide buoyancy for this very pelagic animal, or it may have some function in sound production.

The lungs of marine mammals, pinnipeds included, are slightly larger than those of terrestrial mammals, but the whole animal tends to be more buoyant because up to 30 per cent of the body weight is blubber (Kooyman, 1973). This state of affairs may be compared with that in the sea otter (*Enhydra*) where the lungs are particularly large in relation to the body size, forming about 4 per cent of the body weight (c. 1·5 per cent in Harp Seal). As the sea otter has virtually no blubber the buoyancy must be obtained from the large lungs (Tarasoff & Kooyman, 1973a).

The lungs of most pinnipeds are about equal in size to each other, and are lobulated approximately as in normal terrestrial mammals – both lungs have three main lobes, and the right lung has in addition a small intermediate lobe. There does, however, seem to be a tendency to reduce the lobulation, and lungs of *Odobenus*, *Phoca fasciata*, *Phoca groenlandica* and *Phoca*

largha have been recorded as having practically no lobulation (Tarasoff & Kooyman, 1973a, Sokolov *et al.*, 1968, Fay, 1981).

The bronchi subdivide within the lung to form bronchioles and eventually end in alveoli, but the details vary in the three families. In all pinnipeds the terminal airways are supported and reinforced with either cartilage or muscle, such reinforcement being found in cetaceans, sirenians and the sea otter *Enhydra*, but in no other semi-aquatic or terrestrial mammal (Denison & Kooyman, 1973). Fig 9.3.

In phocids the distal two or three generations of small airways have their walls reinforced with a layer of oblique muscle which runs right up to the alveoli. The cartilaginous rings are also present but become irregular in shape. They are present to within 2mm of the alveoli. The alveolated endings of the terminal airways are in groups of about ten, surrounded by a very thin tissue layer, and separated by a gap from the neighbouring lobule. Occasional alveoli emerge from the muscle layers of the terminal airways.

In otaries there is less muscle, but the cartilaginous rings extend right up to the alveoli. The terminal

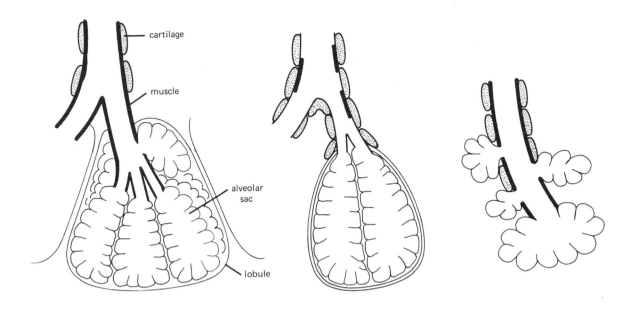

Fig. 9.3 Diagram to show the structure of the alveolar sacs and the distribution of cartilage and muscle in the terminal airways of (from left to right) phocid, otariid and odobenid. (After Denison & Kooyman, 1973).

airways end in paired alveolated sacs, each pair being enclosed in a definite thick connective tissue sheath. No alveoli occur on the airways.

In walruses the arrangement is somewhat intermediate as some alveolar sacs branch from both the muscle and the cartilage supported parts of the terminal airways, but the part of the tube that is supported by muscle only is very short and under 1mm in length.

In both phocids and otariids the development of the reinforcement of the terminal airways takes place late in foetal life and may not be complete before the pup is several months old.

As the terminal muscular reinforcement of the phocid lung has been shown to be sufficient to prevent nitrogen absorption on diving, the presence of the thicker terminal cartilaginous strengthening of the otariid lung is puzzling. Its function, however, may be to keep the terminal airways wide open so that the animal may take in as much air as possible in a short time. Otariids tend to surface briefly and frequently while they are swimming, and need to complete their respiration quickly. Phocids on the other hand, dive

deeply for long periods, but also stay at the surface longer in between dives.

Diaphragm

The diaphragm, while attached in the normal mammalian fashion, that is from the dorsal face of the xiphisternum, curving upwards and backwards to the level of, approximately, the second lumbar vertebra, is rather more oblique than in, for example, a terrestrial carnivore. This is achieved by the sternal elements being slightly shorter, particularly in the Phocidae; and the normal number of fifteen ribs, compared with thirteen in say, a dog, means that the thorax is slightly longer. The diaphragm of the Harp Seal has been described as 'conspicuously fleshy with the central tendon extremely reduced'. A dorsal notch accommodates the oesophagus and a foramen ventral to this allows the passage of the dorsal aorta. The vena cava is guarded by an annular caval sphincter, (George & Ronald, 1975).

Chapter 10 **Vascular system**

Heart

The normal mammalian heart anatomy is present, though pinniped hearts tend to be broader and flatter than those of terrestrial mammals. A heart of this form would accommodate well to the diminution of the chest cavity due to pressure while diving. Occasionally the tips of the ventricles are separated by a deep groove so the tip of the heart appears to be bifurcated. The right ventricle tends to be thin-walled, long and narrow, particularly in deep diving seals, and this shape is thought to be able to respond better to the greater resistance of the lung tissues while diving (Drabek, 1975).

The ascending aorta increases greatly in width by some 30–40 per cent or even more, between its beginning and the level of the brachiocephalic artery, to form an aortic bulb. After all the great vessels have left the aorta, at about the level of the ductus arteriosus, there is a sudden diminution in size of the aorta by about 50 per cent, and it then continues posteriorly as a relatively slender dorsal aorta. This enlarged aortic bulb has been seen in phocids, otariids and walrus (Drabek, 1975, 1977, King, 1977). Limited measurements indicate that Weddell Seals probably possess the greatest aortic enlargements, the diameter being c. 70 per cent greater than the aorta at its base. There is some correlation between the size of the bulb and the diving habits of the seal, the more shallow water Leopard Seal for instance having a smaller bulb than the deep diving Weddell. The significance of the aortic bulb has hardly been studied yet, but it has been suggested that this highly elastic enlargement has some function in helping to maintain normal arterial pressure while diving.

Foramen ovale and ductus arteriosus

Investigation of the age of closure of the foramen ovale in the atrial septum has shown that it closes when the seal is about five or six weeks old, rather later than in terrestrial mammals. The ductus arteriosus, between the pulmonary artery and the aortic arch normally closes by the time the pup is about six weeks old. Both apertures in the Harbour Seal may be anatomically, if not physiologically open until the pup is about three months old. Diving by the very young seal with possible consequent pressure in the pulmonary artery has been suggested as a possible explanation for the delay (Slijper, 1961, 1968).

Heart rate

The normal heart rate of seals varies considerably, for example 75–120 beats a minute have been recorded in *P. vitulina*, 60 beats a minute in *M. angustirostris*, 150–210 in *C. ursinus*, and 76–114 in *Z. californianus*. In pups the rate is slightly faster. The great reduction of the heart rate on diving, and the blood volume of seals are mentioned in the chapter on diving.

Arterial system

The arterial system of pinnipeds conforms very closely with the normal mammalian arrangement as seen in a dog. There is variation in the arrangement of the main vessels that leave the aortic arch. Two branches – the brachiocephalic and the left subclavian arteries – arise separately in some seals (as in the dog); in other animals the left carotid artery also arises separately (as in man), and in others both right and left carotids and subclavians arise separately, but there seems to be no taxonomic significance in the arrangement. Fig 10.1.

Venous system

Most of the modifications of the pinniped vascular system are to be found in the venous system, and many of these can be explained in relation to the animals diving habits, although some of them are also found in other groups of mammals. Most work has been done on the venous system of phocid seals and this system will be described first (Harrison & Kooyman, 1968, Ronald *et al.*, 1977). In many respects the venous system of otariid seals and walruses is different from the phocid arrangement, and these differences will be noted later. In all seals, however, the body is well supplied with numerous anastomosing networks of veins (plexuses) in many parts of the body, for example in the cervical region, in the ventral abdominal and lateral body wall, in the pelvic region, and in the musculature of the limbs. In this way nearly all parts of the venous system are in communication with all other parts.

Phocid venous system

The venous system (Fig 10.2) is more complex than in 'normal' mammals, the vessels are large, thin-walled, and usually without valves. The lack of valves has been demonstrated by the swift passage of injected X-ray-opaque material both anteriorly into the cranial

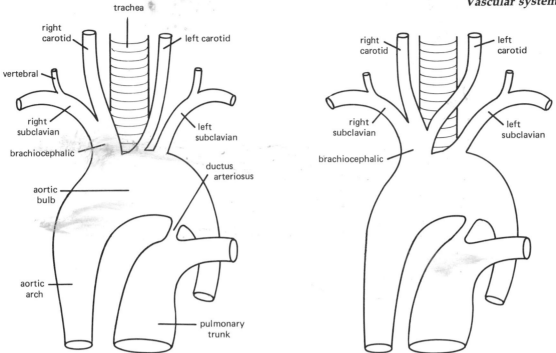

Fig. 10.1 *Diagram to illustrate the variability in the main arteries as shown in* Neophoca *(right) and* Phocarctos.

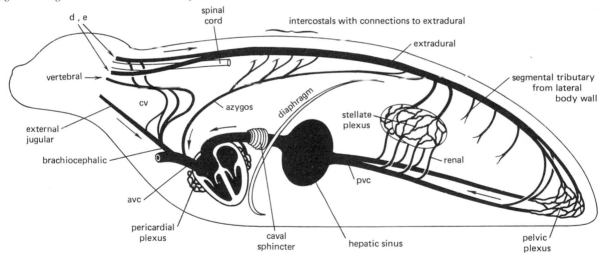

Fig. 10.2 *Very diagrammatic representation of the main venous trunks of a phocid. For clarity the connections of all the many plexuses are not included, and some of the more obvious larger* tributaries – *from foreflippers etc. are omitted. The right azygos only is included avc – anterior vena cava, pvc – posterior vena cava.*

venous sinuses and posteriorly into the abdominal veins.

Both internal and external jugular veins are small, though the external jugular is enlarged by its connections with the cervical vertebral venous system before it joins the brachiocephalic vein. At this point, in the Harp Seal at least, there are thin valves in the external jugular, concave towards the heart, which would prevent venous overflow into the head and neck (Hol *et al.*, 1975). Similar valves are found in the anterior vena cava.

Most of the blood from the cranium leaves via the extradural intravertebral veins which, on leaving the cranial sinuses are duplicated and ventral to the nerve cord, but soon become lateral to it. They are connected by segmental anastomoses, but while still in the cervical region they pass dorsally and join to form a single large crescentic vein. This vein, which is the main pathway for venous blood from the head lies dorsal to the spinal cord, and between the dura mater and the periosteum of the vertebral canal. It is single for most of its length, though a thin septum divides it internally in places. The importance of this vein can be judged from its large size, particularly in the thoracic region, where it can be 3×1cm in a very young *P. vitulina*, and 5·5×4·5cm in an *Ommatophoca*; and also by its communication with almost every other part of the venous system. In the neck and anterior part of the thorax there are connections between the extradural veins and the vertebral veins, and from this area many vessels extend from the vertebral veins, uniting and reuniting before they enter the brachiocephalic veins. This system of vessels forms the 'cervical vertebral venous system' (Ronald *et al.*, 1977). Posteriorly there are connections with the renal and pelvic plexuses, with the dorsal muscles, and also with the veins of the abdominal wall. In addition, the more posterioly situated intercostal veins are connected with plexuses in the abdominal walls and with the renal plexus. In the sacral region the extradural vein is double and extends caudally into the pelvic plexus.

The small veins draining the blood from the individual reniculi of the kidney join to form interlobular veins, which rise to the surface of the kidney between the reniculi and there join the stellate plexus of veins that runs on all surfaces of the kidney in the grooves between adjacent reniculi. From this plexus the blood usually drains by three main vessels into the appropriate limb of the posterior vena cava. As mentioned above, the stellate or renal plexus is also connected with the extradural vein and thus with much of the venous system of the body. It is probable that most, if not all of the venous plexuses of the phocid body are embedded in brown adipose tissue and thus have a probable thermoregulatory function (Blix *et al.*, 1975).

From the posterior end of the body the blood drains by way of a plexus of veins into the limbs of the posterior vena cava which, in the region of the kidneys, is usually, though not invariably, duplicated. The two veins here are very large, over 2·5cm in diameter in a Harbour Seal.

Just caudal to the diaphragm, and covered by the lobes of the liver, the hepatic sinus, formed from the enlarged hepatic veins, receives blood from the posterior vena cava and conveys it to the heart through the short thoracic part of this vein. The hepatic sinus is well formed at birth in phocids and holds a considerable quantity of blood in the adult animal. In a Harbour Seal the hepatic sinus may hold over a litre of blood, and that of the Ross Seal holds three litres.

Immediately anterior to the diaphragm the vena cava has a muscular caval sphincter surrounding it. This is funnel shaped, its anterior end being narrower than its posterior end; the striated muscle fibres of which it is composed being separated from the wall of the vena cava by connective tissue, and from the diaphragm by a narrow tendinous band. The caval sphincter is apparently not exactly the same in all phocids. In the Harp Seal it is thin and annular rather than funnel shaped (George & Ronald, 1975). The caval sphincter is supplied by branches from the right phrenic nerve. Anterior to the sphincter the veins from the pericardial plexus enter the vena cava. This convoluted mass of anastomosing veins and brown adipose tissue is in the form of a ring round the base of the pericardium, and sends out leaf-like projections into the pleural cavities. Branches of the phrenic nerve are found in the plexus, which has venous connections with the intercostal and internal thoracic veins, and also with plexuses caudal to the diaphragm. The vein walls are relatively thick and contain coiled collagenous fibres, elastic fibres, and spirally arranged smooth muscle fibres, all of which would suggest the possibility of considerable dilatation. There may also be a thermoregulatory function, as mentioned in the chapter on diving. Although typically present in phocids, the Hawaiian Monk Seal has been recorded as lacking a pericardial plexus (King & Harrison, 1961).

In the anterior part of the abdomen paired azygos veins are formed from lateral tributaries of the extradural vein. In the thorax, where these tributaries enlarge to form the intercostal veins, the left azygos becomes very reduced and the blood from the left intercostal veins drains through the left azygos and into the larger right azygos and thence into the anterior vena cava.

Otariid venous system

Although much less comparative work has been done on the otariid vascular system, it does appear that in the venous system there are some differences from the arrangement in phocids (King, 1977).

Cranial drainage in otariids seems to follow the normal mammalian pattern, with most blood returned to the heart by the external jugular vein. The presence of an internal jugular vein is at the moment uncertain. Blood also leaves the cranium by the condyloid veins which then become paired extradural veins lying ventro-lateral to the spinal cord – also the normal mammalian arrangement.

There is no obvious stellate plexus on the surface of the kidney in otariids, and it seems likely that there may be a less obvious sub-surface plexus. The main direction of the venous blood is thus outwards towards this plexus, but then via inter-renicular veins to a conventional renal vein that leaves the kidney to join the posterior vena cava (King, 1977, Bester, 1975).

The drainage of the posterior end of the body seems to be much as in phocids, and the posterior vena cava may again be either double or single.

A large hepatic sinus is present in adult otariids, but in contrast to the situation in phocids, newborn pups have very little, if any indication of such an enlargement. At what age it develops is not known. A caval sphincter is present but its muscle fibres are not separated from the diaphragmatic muscle by any tendinous area. No pericardial plexus has been found in otariids. A single right azygos vein is present, as in the majority of mammals.

Blood is carried to the hind flippers of both phocids and otariids by the saphenous artery, and returned to the body by the saphenous veins. The branches of the saphenous vein are more numerous on the plantar surface of the flipper and lie ventral to the branches of the artery, i.e. nearer the surface of the flipper. They can thus be of use in temperature regulation (Tarasoff & Fisher, 1970). The arteriovenous anastomoses of the skin and flippers are also mentioned in the chapter on thermoregulation.

Walruses are said to resemble phocids in having a large hepatic sinus and a well-developed caval sphincter, but to resemble otariids in the single azygos vein, no well-developed pericardial plexus and no prominent stellate plexus (Fay, 1981).

Many of the modifications of the venous system of phocids have been described as aquatic specializations. Some of these modifications are also present to some degree in other aquatic and partially aquatic mammals, and even in sloths. It is suggested that they increase the efficiency of the circulation during difficult circumstances such as diving. Barnett *et al.* (1958) should be consulted for a full discussion on this matter. It is, however, noticeable that only in the Phocidae are the greatest number of venous modifications combined – pericardial plexus, renal stellate plexus, large extradural vein receiving the main cranial drainage, hepatic sinus and caval sphincter.

By contrast the Otariidae, and most probably also the Odobenidae, are less specialized in this respect, and their venous system is much more of the typical mammalian pattern, though they do, when adult, have a large hepatic sinus and a type of caval sphincter.

Compared with otariids, phocids have higher Hb concentration and haematocrit values; they have a larger blood volume, giving them greater O_2 storage capacity, and the muscle myoglobin concentration is twice as high (see chapter on diving). The average diameter of the red cells in *M. leonina* is $10 \cdot 06 \mu m$ (human $7 \cdot 40 \mu m$) (Harrison & Kooyman, 1968, Wells, 1978, Lane *et al.*, 1972, Lenfant *et al.*, 1970b, Lenfant, 1969, Clausen & Ersland, 1969). All this may be correlated with diving adaptations.

Chapter 11   *Reproduction*

Male reproductive organs

In all seals the penis is retractable within a cutaneous pouch, the external opening of which is mid-ventrally, between the anus and umbilicus (Fig 11.1). Seals possess a baculum, or os penis, which is the ossified anterior end of the corpus cavernosum. There are differences in shape of the apex or distal end of the baculum between phocids and walrus on the one hand, and otariids on the other. Within the Otariidae a sequence of genera with a possible indication of phylogenetic significance has been described (Kim *et al.*, 1975, Morejohn, 1975). (See chapter on classification.)

The baculum shape of *Callorhinus*, while similar to *Arctocephalus* and *Zalophus* while it is developing, has an adult shape peculiar to itself.

The largest pinniped baculum belongs to the walrus, where it may be about 50–60cm in length, and

Fig. 11.1 Ventral view of a male Southern Elephant Seal pup. The posterior dark spot on the belly is the penile opening. The anterior spot is recently healed umbilical scar, more obvious in this pup than it would be in an adult. Photo. J. E. King.

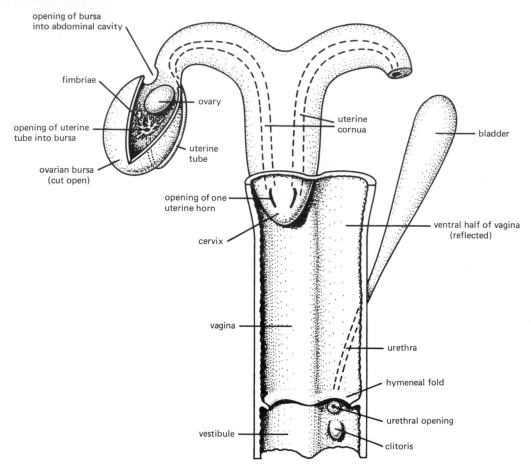

opening of bursa
into abdominal cavity

fimbriae

opening of uterine
tube into bursa

ovarian bursa
(cut open)

ovary

uterine
cornua

bladder

uterine
tube

opening of one
uterine horn

cervix

ventral half of vagina
(reflected)

vagina

urethra

hymeneal fold

urethral opening

vestibule

clitoris

Fig. 11.2 Diagrammatic representation of the female reproductive organs, based on those of an old H. leptonyx. *The uterine cornua were separate in this specimen.*

about 1kg in weight (Scheffer & Kenyon, 1963). In *Callorhinus* the baculum grows at a steady rate throughout life, while in *P. vitulina* an increase in the rate of growth has been shown at puberty. The urethra lies in a groove on the ventral face of the baculum.

In the Phocidae and walrus the testes are inguinal, lying ventral and lateral to the pelvis, and just posterior to the level of the knee (Harrison *et al.*, 1952). They lie outside the abdominal muscles, but are covered by part of the skin muscle, and there is no external indication of their presence. In *H. grypus* the testes have fully descended about five months before the birth of the pup (Hewer & Backhouse, 1968).

In the Otariidae the testes are scrotal and lie in the normal position just anterior to the tail and anus, covered with thin, wrinkled, black, naked skin. In *C. ursinus* the testes descend in the third or fourth year (Scheffer, 1962). Some cryptorchid fur seals (*C. ursinus*) have been reported from the Pribilof Islands. These adult males have only infantile testes, and the animals tend to be more slender and finely built than normal bulls, and lack their aggressiveness (Scheffer, 1951).

Female reproductive organs

The ovary is enclosed in a double fold of peritoneum which forms an ovarian bursa, a sac which is many times the size of the ovary in a mature animal. The ovaries usually function alternately. The uterine (or fallopian) tube opens into the bursa, and then enlarges to become the cornu of the appropriate side (Harrison *et al.*, 1952).

The undissected female reproductive system has the conventional Y form (Fig 11.2) – the vagina and body of the uterus diverging cranially into the two cornua. But on dissection it can be seen that within the body of the uterus the two cornua with their separate cavities are separated by a thick medial partition, and that the common part of the uterus, immediately cranial to the cervix is very small – about 3cm or less. A narrow cervical canal traverses the cervix and opens into the vagina. As an individual variation, sometimes the cornu are completely separate, and each opens separately into the vagina.

The vagina is long and is separated from the vestibule by a fleshy hymeneal fold. The bladder lies ventral to the vagina and the urethra opens by a large urinary papilla just caudal to the hymen. The vestibule may be 10cm long in a large seal and it opens to the exterior just ventral to the anus. Both anus and vestibule open into a common furrow which is surrounded by muscle fibres acting as a sphincter.

An os clitoridis has been recorded from members of all three pinniped families, but its appearance, even within a single species is irregular. Usually it is hardly more than 1cm in length.

Mammary glands

The Otariidae, Odobenidae and the phocid genera *Erignathus* and *Monachus* have four mammary teats, one pair in front of, the other pair behind and closer to, the umbilicus (Fig 11.3). The remaining phocids have two teats, corresponding to the posterior pair of the otariids. When not in use the teats lie retracted beneath the body surface, the opening concealed by the hair, and in the walrus there is a sphincter-like arrangement of the skin around them. Supernumerary teats have been recorded.

In phocids the mammary glands remain distinct, each enclosed in a connective tissue sheath and communicating with the appropriate teat. In otariids the mammary glands coalesce to form a sheet-like layer under the blubber, over most of the ventral surface of the body. Each mammary gland in an actively lactating Elephant Seal (*M. leonina*) has an approximate area of 58×20cm, it is about 7·5cm thick and its volume is about five litres (Laws, 1956).

In the Weddell Seal each mammary gland is *c.* 30×16cm in area, 2cm thick, and has a volume of *c.* 600cm³ (= 0·6 litre) when not lactating. During lactation the width and length of the gland remain much the same, but it increases in thickness to *c.* 6cm, and in volume to *c.* 2820cm³ (= 2·8 litre). Histologically the gland resembles that of other eutherians, though there are differences in the arrangement of the collecting ducts. Small collecting ducts at the periphery of the gland join together to form progressively larger ducts, so that eventually a fan of ten large ducts join to form 3 or 4 bigger ducts, each with a diameter of 11–14mm which then open into the large gland cistern (2·5×3cm) at the base of the nipple. Sinusoidal blood vessels have not been seen in the mammary glands of other mammals, but are present in the lactating gland of the Weddell Seal. These may serve to increase the total volume of blood within the gland and permit a more efficient exchange of nutrients than could be achieved by just a capillary network (Tedman & Bryden, 1981).

Milk

Seal milk has been described as white and creamy, slightly viscous, with an oily and, to humans, rather unpalatable taste. A slight production of milk by newborn animals, both male and female, has been recorded (Bertram, 1940). Seal milk is particularly rich in fat, has a high level of protein and a low level of water (Table p. 186). The lactose level is very low, being just detectable in phocid milk, and virtually absent in milk of otariids and walrus (Pilson, 1965). In these characters it is very like whale milk, and differs markedly from the milk of such mammals as the dog, domestic cow or human.

The high fat and protein content can be correlated with the fast rate of growth of the pup, the frequently short lactation time, and the need to lay down an insulating layer of blubber as soon as possible. It is also correlated with the need for water conservation by the mother, as concentrated milk would facilitate its storage by the mother during the periods when she is away at sea feeding and the pup is therefore not suckling (Peaker & Goode, 1978).

The need for water conservation, particularly in those seals that fast during the lactation period, may also be correlated with the change in the proportions of the milk constituents during lactation. In this way the milk is correct for the physiological requirements of the pup at the various stages of its growth. For this to be demonstrated it is necessary to analyse the milk throughout the period of lactation, and this has been done on relatively few occasions. In *M. angustirostris*, during the first 21 days of lactation the fat content of the milk rises from 15 to 55 per cent, while the water content falls from 75 to 35 per cent and then the composition remains fairly constant until weaning at 28 days (Riedman & Ortiz, 1979). Similar changes have been shown in the milk of *M. leonina* (Bryden, 1968)

Fig. 11.3 *Ventral view of a female Southern Elephant Seal pup. The dark spot in the centre of the belly is the umbilical scar. Lateral to the umbilicus are the less obvious positions of the two teats. The vulva is just anterior to the bases of the hind flippers. Photo. J. E. King.*

and *L. weddelli* (Kooyman & Drabek, 1968), although in these animals the fat content of the milk decreases just before weaning. Water conservation by the pups may be correlated with the production of a large amount of cystine by the pituitary gland (see chapter on endocrine glands).

Daily increase in weight by phocid pups is high – 2·5kg in *P. groenlandica* (Stewart & Lavigne, 1980), 1·3kg in *H. grypus* (Boyd & Campbell, 1971) and 5–6kg in *M. leonina*, with one individual gaining as much as 11kg in a day (Laws, 1953). Southern Elephant Seal pups may double their birthweight in about 11 days. There is a certain correlation between the short lactation period (e.g. *c.* 21 days in *H. grypus*; 23 days in *M. leonina*), the fact that the mother remains with her pup on land during this period, and the rapid growth rate. In the Weddell Seal the pattern of weight gain is different. The pups will increase from about 25kg at birth to 93kg at five weeks, but their increase during the second week of life reaches a peak of some 16–17kg (*c.* 2·4kg per day) and the weekly increase is less in the succeeding four weeks of lactation. The start of the reduced weekly increase coincides with the pup entering the water with its mother and consequently having to use much of its energy in swimming and heat production instead of blubber production. The lactation period in the Weddell Seal (five to six weeks) is longer, possibly to compensate for this, and the pup learns to swim and catch food before weaning so that it is well able to survive when it is finally weaned (Tedman & Bryden, 1979).

Otariid pups have a slower gain in weight which may be correlated with a tendency towards a slightly lower percentage of fat in the milk, this percentage appearing to remain fairly constant throughout the period of lactation, and, in general, the much longer lactation period when compared with phocids (see Table, p. 188). *A. forsteri* pups between 8 and 60 days old have an average daily increase of 69g and their birth weight is double by the time they are 55 days old (Crawley, 1975). Other rates of increase of *A. forsteri* pups give 40g per day (in Payne, 1979a).

Most otariids, as far as is known, have a 12-month lactation period. Only *A. gazella* and *C. ursinus* have shorter lactation periods of four and three months respectively, and both of these animals have similar growth rates which are faster than those of the other fur seals.

Male *Callorhinus* pups put on 93g per day, and male *A. gazella* pups 98g per day (Payne, 1979a).

Rearing pups in captivity

The lack of, or the very small amount of lactose in seal milk is correlated with the near absence of lactose activity in the intestine. Normally the seal gut will barely tolerate lactose, and this would account at least in part for the digestive troubles of orphan and captive seal pups, particularly otariids and walruses who cannot cope with such usual standbys as domestic cow's milk. Many young seals in captivity in earlier times must have lived in spite of their diet rather than because of it.

A very young walrus (two to three months) was fed on tinned milk during its trip between the Bering Sea and captivity in San Diego. This diet was continued in the zoo and after a while the young animal had skin lesions, mucous discharge and difficulty in breathing. After some trial and error, the milk was eliminated from the diet and recovery started immediately (Schroeder, 1933).

A special feeding formula based on fish flour and whale oil, with suitable supplements has been used successfully for raising captive fur seal pups (*Callorhinus*) (Keyes, 1968). Geraci (1975) gives recipes for suitable mixtures for raising sea lion pups (*Eumetopias* and *Zalophus*), and also one that is suitable for Harbour Seals (*P. vitulina*). Young walruses appear to thrive in captivity on a blended mixture of clams, whipping cream and brewers yeast, with vitamin supplements added (Brown & Asper, 1966). This diet would seem to be a great improvement over that offered to some of the earlier animals in captivity who were given oatmeal, milk and water, and strips of boiled pork or whale blubber.

Placenta

The pinniped chorio-allantoic placenta, typical of all Eutheria, has the zonary development of the villi as in the Carnivora. Also as in this latter group the uterine epithelium and underlying connective tissue have broken down so that the epithelium of the chorion comes into contact with the endothelial walls of the maternal capillaries. Thus the placenta of pinnipeds is of the zonary endothelio-chorial type, but with additional features of its own. In the placenta the maternal blood vessels are very dilated, forming sinusoids, and both maternal and foetal components are very thin. Thus the barrier between the maternal and the foetal circulation is very thin and may be as little as 1μm. Much of this barrier may be non-cellular and there is some indication of secretory activity so hormones may be produced (Harrison & Young, 1966, Harrison & Kooyman, 1968).

Foetal gonads

Examination of the gonads of late foetal and newborn phocids has shown that both ovaries and testes, and also the prostate, are almost the same size as those of adult animals. This enlargement is due to the abundant interstitial tissue and increased vascularity, and the organs diminish in size within about a fortnight after birth. The increase in size is thought to be caused by the action either of maternal oestrogens or of hormones from the foetal pituitary, which is also precociously developed. Enlargement of the gonads has been recorded in *H. grypus, P. vitulina, M. leonina* and *L. carcinophagus,* but not apparently in any otariid (Amoroso *et al.,* 1965).

Gestation period

The total gestation period, that is the entire period between fertilization and parturition varies. It may be, for instance, about nine months (*P. vitulina*), 11·5 months (*H. grypus*), nearly 12 months (*Zalophus*) or 15 months (*O. rosmarus*) depending on the time-lapse between pupping and mating. In spite of the length of the *total* gestation period, the actual time during which the embryo is growing (the *active* gestation period) depends on the period of delay in the attachment of the blastocyst. This varies according to the species of seal involved. Periods from six weeks to five months have been quoted for different species, but the overall average is about 3·5 months.

Detailed information on the British population of *Halichoerus* has shown that the blastocyst has developed within about eight to ten days after fertilization. It then remains dormant, lying freely in the anterior part of the uterine horn, for the next 100 days (c. three months). Development then recommences and after three or four days the embryo becomes attached via the placenta to the uterine wall and normal growth continues for about another 240 days. This, including the ten days after fertilization gives 250 days (*c. eight months*) of active gestation, and a total gestation period of c. 350 days (c. 11·5 months) (Hewer & Backhouse, 1968).

This delay in development has been called 'delayed implantation', but 'suspended development' has been suggested as a better term. This phenomenon is not restricted to seals, and occurs in other mammals such as roe deer, badger and armadillo. The endometrium of the uterus, in its preparation for the attachment of the blastocyst, shows the normal mammalian reactions to progesterone and oestrogen in spite of the delay in attachment (Boshier, 1979). The relationship between pregnancy and diving is mentioned in the chapter on diving.

Twins

Pinnipeds normally produce a single young at a birth. There are, however, sufficient records to indicate that although twins are relatively rare, they are by no means unknown (Rae, 1969). Two pups may often be seen to be associated with or even suckling from the same female, but this does not necessarily indicate twinning. Dissection revealing twin blastocysts or twin foetuses is of course incontrovertible as is actually seeing the two pups being born. Such evidence is available for *P. vitulina, P. hispida, Halichoerus, Mirounga, Lobodon, Leptonychotes*; and for *Odobenus, Eumetopias, Otaria, Zalophus, Callorhinus* and *Artocephalus pusillus.*

The twin foetuses have been both male, both female, and one of each sex, and even unborn triplets have been seen in *P. hispida* (Kumlien, 1879). Conjoined 'Siamese' twins have been reported in *Mirounga leonina* (Laws, 1953). This still-born monster was composed of two male pups joined from the umbilicus forwards.

Some of the twin foetuses have been of a size and degree of development that had the mother lived it seems likely that both pups would have been born. Three *Callorhinus* females were noted as they each gave birth to twin pups (Peterson & Reeder, 1966). In all three instances if appeared that the mother found it difficult to achieve adequate communication with both pups and very soon rejected one of them. It seems unlikely that a cow would be able to produce enough of the rich milk to raise two pups to weaning, even if communication did not break down. Even when twin *Zalophus* were produced in captivity and the female given extra vitamins, one of the pups soon died (Uchiyama, 1965).

An extra-uterine foetus, where the placenta had most probably been attached to parts of the stomach and duodenum, has been recorded in *Eumetopias* (Talent & Talent, 1975). This female was carrying, in addition, a normal foetus in the uterus. Both foetuses were male, the uterine one being 74cm in nose to tail length, while the extra-uterine foetus was 90cm in length. Both appeared to be fully developed, though slightly smaller than a normal pup. It was nearing the normal pupping season, but the cause of death of the female was not obvious.

Hybrids

Understandably, all recorded hybrids between different species or genera of seals have occurred in the unnatural conditions of captivity. One can be less certain of what happens in the wild although a young adult male *Halichoerus* was seen attempting to copulate with young female *Phoca vitulina concolor* on Sable Island, Novia Scotia (Wilson, 1975b). On this sandy beach groups of male Grey Seals were hauled out in the middle of a herd of 'adolescent' Harbour Seals, both groups of seals largely ignoring the other. One of the Grey Seals, however, harassed the young Harbour Seal females, pursuing, stroking and mounting them, though as the females struggled there was no successful copulation.

Another attempted copulation was witnessed between a male *Phocarctos* and a dead female *Arctocephalus forsteri*. When disturbed the *Phocarctos* took the Fur Seal by the back of the neck and swam out to sea, holding the Fur Seal under water. It was suggested that the Fur Seal had been killed earlier in the day, possibly by a previous attempt at copulation (Wilson, 1979).

On Marion Island in the South Indian Ocean the Fur Seal *A. tropicalis* has large breeding colonies, but *A. gazella* is now occurring there in increasing numbers. Adults of the two species are not difficult to distinguish by their colour and other characters. *A. gazella* bulls have been seen with *A. tropicalis* cows in their harems and *A. gazella* cows have been seen in the harems of *A. tropicalis* bulls. Bulls with a mixture of the external characters of both species have been seen on several occasions, suggesting some interbreeding, but no specimens have yet been taken, or unequivocal observations made (Condy, 1978).

In captivity a dead pup was born as the result of a mating between a male *Halichoerus* and a female *Phoca hispida*. The pup was said to resemble its mother, but had many skull characters of the father (Lonnberg, 1929). Also in captivity, male *Arctocephalus p. pusillus* and female *Zalophus* have produced pups on several occasions. One pair produced a dead female pup whose skull was said to be more like *Zalophus* while its coat was more like that of the Fur Seal (Schlieman, 1968). Another similar pair produced pups in each of three successive years. The first two pups died within 24 hours, but the third, a male, lived for at least six months and was said to bear a close resemblance to its father (Jennison, 1914). Another female *Zalophus* was in an enclosure with two male *A.p. pusillus* and a male *Otaria*, and produced pups in two successive years. The first was a miscarriage, and the second pup lived only 24 hours. As the *Otaria* was the only one that courted the female, defended the female and pup, and was tolerated by the *Zalophus*, it was presumed to be the father of the pups (Kirchschofer, 1968).

PUPPING TABLE

The main pupping season of most seals is in the spring and summer of the appropriate hemisphere (see Table opposite). Several of the northern phocids start their breeding season in late winter and their breeding activities are associated with ice. *Halichoerus* is unusual amongst the northern phocids in the timing of its breeding season.

The apparently long breeding seasons of *P.v. richardsi* and *P. largha* are related to their extensive distribution through latitudes, but in any particular region the season is of normal duration. Very long breeding seasons in animals such as *Zalophus* and *Arctocephalus* on the Galapagos Islands, *Monachus schauinslandi* and *M. monachus* may be correlated with the equable climate.

Milk analysis

	% fat	% protein	% water	% lactose	Reference
P. groenlandica	42·6	10·5	45·3	0*	Sivertsen, 1941
P. groenlandica	46·9	6·8	45·1	0·77	Cook & Baker, 1969
P. groenlandica	53·2	6·0	40·2	0·32	Cook & Baker, 1969
P.v. vitulina	45·0	9·0	45·8	0·2	Harrison, 1960
H. grypus	52·2	11·2	–	2·6	Amoroso *et al.*, 1951
M. leonina	49·0	8·5	–	–	Bryden, 1968
M. angustirostris	54·4	9·0	32·8	–	LeBouef & Ortiz, 1977
Z. californianus	36·5	13·8	–	0	Pilson & Kelly, 1962
Z. californianus	31·1	13·3	–	–	Kretchmer & Sunshine, 1967
C. ursinus	43·0	–	–	70·1	Ashworth *et al.*, 1966
A.p. pusillus	18·6	10·0	–	–	Rand, 1956a
A.p. doriferus	49·0	12·0	–	trace	Kerry & Messner, 1968
A. gazella	26·4	22·4	51·1	–	Bonner, 1968
A. gazella	30	5–10	60	–	Fay, 1981
Blue whale	38·1	12·7	47·1	0	Sivertsen, 1941
Dog	9·3	9·7	77·0	3·1	Sivertsen, 1941
Bovine	3·4	3·3	88·0	4·4	Pilson & Kelly, 1962
Human	3·5	1·0	88·8	6·5	Marshall, 1922

* author notes lactose possibly missed in analysis.

N.B. It should be remembered that the fat and water content of the milk of some seals have been shown to change in their proportions during the period of lactation. Most of the above analyses were carried out before this fact was known.

Pupping table*

NORTHERN HEMISPHERE

	Spring			Summer			Autumn			Winter		
	Mar.	Apr.	May	June	Jly	Aug.	Sept.	Oct.	Nov.	Dec.	Jan.	Feb.
E. barbatus												
H. grypus	Baltic							British			Canada	Baltic
P.v. vitulina E. Atlantic												
P.v. concolor W. Atlantic												
P.v. stejnegeri												
P.v. richardsi								**				
P. largha			**									
P. hispida												
P. caspica												
P. sibirica												
P. groenlandica												
P. fasciata												
C. cristata												
M. schauinslandi†												
M. monachus†												
M. tropicalis												
M. angustirostris												
E. jubatus												
Z. californianus (Calif.)												
C. ursinus												
A. townsendi												
O. rosmarus												

** depends on latitude † peak pupping periods indicated by thicker lines

SOUTHERN HEMISPHERE

	Spring			Summer			Autumn			Winter		
	Sept.	Oct.	Nov.	Dec.	Jan.	Feb.	Mar.	Apr.	May	June	Jly	Aug.
H. leptonyx					††							
L. weddelli												
L. carcinophagus												
O. rossi												
M. leonina												
O. byronia												
N. cinerea												
P. hookeri												
Z. californianus (Galap.)												
A. galapagoensis												
A. philippii												
A. australis												
A. tropicalis												
A. gazella												
A.p. pusillus												
A.p. doriferus												
A. forsteri												

†† ----- uncertain information * This table is a summary of information and should not be used without reference to the text.

Summary of reproductive facts

	E. jubatus	Z. californianus	O. byronia	N. cinerea
nose to tail length ad.♂	3m	2·4m	2·3m	2–2·5m
nose to tail length ad.♀	2·2m	1·8m	1·8m	1·7–1·8m
nose to tail length pup	1m	75cm	85cm	70cm
weight pup	18–22kg	c. 6kg	–	6–8kg
pups born	mid May–mid Jun.	Jun.	Dec.–Jan.	? Oct.–Jan.
peak of pupping	beg. Jun.	Jun.	Jan.	? Oct.
length lactation	c. 12 mths	c. 12 mths	c. 12 mths	c. 12 mths
oestrous*	10–14 days	c. 14 days	few days	4–9 days
delay in attach.	3–5 mths	–	–	–
total gestation period	11·5 mths	11·5 mths	11·75 mths	11·75 mths
active gestation period	c. 8 mths	–	–	–

* time between parturition and oestrous

	P. hookeri	A. townsendi	A. galapagoensis	A. philippii
nose to tail length ad.♂	2–2·5m	1·8m	1·5m	2m
nose to tail length ad.♀	1·6–2m	–	–	–
nose to tail length pup	75–80cm	–	–	–
weight pup	–	–	–	–
pups born	Dec.–Jan.	Jun.	? Aug.–Dec.	beg. Dec.
peak of pupping	Dec.	Jun.	–	–
length lactation	c. 12 mths	–	–	–
oestrous*	7 days	–	–	–
delay in attach.	–	–	–	–
total gestation period	11·75 mths	–	–	–
active gestation period	–	–	–	–

* time between parturition and oestrous

	A. australis	A. tropicalis	A. gazella	A.p. pusillus
nose to tail length ad.♂	1·9m	1·8m	1·8m	2·3m
nose to tail length ad.♀	1·4m	1·3m	1·3m	1·6m
nose to tail length pup	–	60cm	65cm	65cm
weight pup	3–5kg	4–5kg	6kg	5–7kg
pups born	Nov.–Dec.	Nov.–Feb.	Nov.–??	Nov.–Dec.
peak of pupping	–	–	beg. Dec.	beg. Dec.
length lactation	c. 12 mths	c. 12 mths	4 mths	c. 12 mths
oestrous*	–	8–12 days	8 days	6 days
delay in attach.	c. 4 mths	–	–	4 mths
total gestation period	c. 11·75 mths	c. 11·75 mths	c. 11·75 mths	c. 11·75 mths
active gestation period	c. 7·75 mths	–	–	c. 7·75 mths

* time between parturition and oestrous

	A. forsteri	C. ursinus	O. rosmarus
nose to tail length ad. ♂	2m	2·13m	3·2m
nose to tail length ad. ♀	1·5m	1·5m	2·6m
nose to tail length pup	55cm	65cm	1·2m
weight pup	3·5kg	5kg	60kg
pups born	Nov.–Jan.	Jun.–Jly	Apr.–Jun.
peak of pupping	mid Dec.	Jun.–Jly	May
length lactation	c. 12 mths	3 mths	2 years
oestrous*	8 days	7 days	post partum†
delay in attach.	–	3·5–4 mths	4–5 mths
total gestation period	c. 11·75 mths	c. 11·75 mths	15 mths
active gestation period	–	c. 7·75 mths	c. 10 mths

* time between parturition and oestrous

† but see chapter on walrus

	H. grypus	P.v. vitulina	P. hispida	P. caspica
nose to tail length ad. ♂	2·2m	1·5–1·8m	1·5m	1·5m
nose to tail length ad. ♀	1·8m	1·2–1·5m	1·5m	1·4m
nose to tail length pup	76cm	85cm	65cm	70cm
weight pup	14kg	11–12kg	4·5kg	5kg
pups born	Sept.–Apr.	Jun.–Jly	mid Mar.–mid Apr.	Jan.–Feb.
peak of pupping	Oct., Jan., Mar.	June	beg. Apr.	end Jan.
length lactation	16–21 days	4–6 wks	2 mths	c. 1 mth
oestrous*	16–21 days	2·5–3 mths	peak mid Apr.	1 mth
delay in attach.	3·5 mths	2 mths	81 days	–
total gestation period	c. 11·5 mths	c. 9 mths	11·75 mths	c. 11 mths
active gestation period	c. 8 mths	c. 7·5 mths	9 mths	–

* time between parturition and oestrous

	P. sibirica	P. groenlandica	P. fasciata	C. cristata
nose to tail length ad. ♂	1·3m	1·6m	1·5m	2·6m
nose to tail length ad. ♀	1·3m	1·6m	1·5m	c. 2m
nose to tail length pup	70cm	90cm	90cm	1m
weight pup	3kg	6–10kg	10·5kg	15kg
pups born	Mar.	Feb.–Mar.	Apr.–May	end Mar.
peak of pupping	mid Mar.	Mar.	beg. Apr.	–
length lactation	2·5 mths	8–12 days	3–4 wks	12 days
oestrous*	c. 2·5 mths	8–12 days	3–4 wks	12 days
delay in attach.	–	4·5 mths	–	4 mths
total gestation period	c. 9 mths	c. 11·5 mths	11 mths	11·5 mths
active gestation period	–	c. 7 mths	–	7·5 mths

* time between parturition and oestrous

	E. barbatus	*M. monachus*	*M. tropicalis*	*M. schauinslandi*
nose to tail length ad. ♂	2·5m	2·8m	2·4m	2·1m
nose to tail length ad. ♀	2·6m	2·8m	2·4m	2·3m
nose to tail length pup	1·5m	80cm	–	1m
weight pup	29kg	20kg	–	16kg
pups born	mid Oct.–mid Nov.	May–Nov.	beg. Dec.	Dec.–Aug.
peak of pupping	end Oct.	Sept. & Oct.	–	Mar.–May
length lactation	6–7 wks	? 6 wks	–	? 6 wks
oestrous*	6–7 wks	–	–	–
delay in attach.	–	–	–	–
total gestation period	*c.* 10·5 mths	–	–	–
active gestation period	–	–	–	–

* time between parturition and oestrous

	L. weddelli	*O. rossi*	*L. carcinophagus*	*H. leptonyx*
nose to tail length ad. ♂	2·25m	3m	2·6m	3m
nose to tail length ad. ♀	2·25m	2·5m	2·6m	3·6m
nose to tail length pup	1·2–1·3m	*c.* 96cm	1·5m	*c.* 1·5m
weight pup	30–40kg	17kg	–	*c.* 30kg
pups born	Apr.	mid Nov.	Sept.–mid Oct.	? Sept.–Jan.
peak of pupping	20 Apr.	3–18 Nov.	–	–
length lactation	12–18 days	? 4 wks	? 4 wks	? 4 wks
oestrous*	12–18 days	4 wks	? 4 wks	? 4 wks
delay in attach.	2 mths	2·5–3 mths	–	2·5–3 mths
total gestation period	*c.* 11·5 mths	11 mths	? 11 mths	11 mths
active gestation period	*c.* 9 mths	8–8·5 mths	–	8–8·5 mths

* time between parturition and oestrous

	M. leonina	*M. angustirostris*
nose to tail length ad. ♂	4–5m	4–5m
nose to tail length ad. ♀	2–3m	2–3m
nose to tail length pup	1·2m	1·2m
weight pup	40kg	40kg
pups born	Oct.	Jan.
peak of pupping	Oct.	end Jan.
length lactation	23 days	27 days
oestrous*	18 days	24 days
delay in attach.	4 mths	4 mths
total gestation period	*c.* 11·5 mths	*c.* 11 mths
active gestation period	*c.* 7 mths	*c.* 7 mths

* time between parturition and oestrous

Chapter 12 Nervous system, endocrine glands

Nervous system

Brain

The pinniped brain is more spherical than that of a terrestrial carnivore, and is more highly convoluted, especially in phocids. The brain is reasonably large, representative brain weights being c. 250–300g for *P. vitulina*, 375g for *Zalophus* and up to 1kg for *Odobenus* (Harrison & Kooyman, 1968). The brain of adult Southern Elephant Seals is about 0·35 per cent of the total body weight, and in this respect falls within the normal range for terrestrial mammals on the 'mouse–elephant curve'. The growth coefficient of the brain is also 'normal' (Bryden, 1971b).

In general form the otariid brain has resemblances to the brain of bears, while the phocid brain is more dog-like. The cerebellum is large, probably because of the necessity for the fine co-ordination of movements while swimming. When compared with terrestrial carnivores, the olfactory area in pinnipeds is reduced, more so in phocids than in otariids or odobenids. The auditory area is large, and the trigeminal sensory area well developed.

Spinal cord

The spinal cord is particularly short in pinnipeds, reaching only to points between the eighth thoracic vertebra (*Phocarctos*) and the twelfth thoracic (*P. groenlandica*), the cauda equina occupying the neural canal posterior to this. For comparison the spinal cord reaches to the first lumbar vertebra in man, and to the last (seventh) lumbar vertebra in dogs.

There are few general accounts of the pinniped brain and nervous system. Detailed studies of brain anatomy, and drawings of the various brains have been published by Vrolik, 1822 (*P. vitulina*), Murie, 1874 (*O. byronia*), Turner, 1888 (*M. leonina, O. rosmarus*), Kukenthal, 1899 (*E. barbatus, P. groenlandica, P. vitulina*), Fish, 1899 (*P. vitulina, M. tropicalis, C. ursinus, Z. californianus*) and Fish, 1903 (*O. rosmarus*), and further references on this subject are to be found in Harrison & Tomlinson (1963), Harrison & Kooyman (1968) and Flanigan (1972).

Endocrine glands

Pineal

The pineal body in pinnipeds is very large. It is probably the largest among mammals, and is particularly big in those seals of polar regions. The length of the pineal body of *Callorhinus* varies between 7–14mm, and its weight between 0·11–0·61g (Elden *et al.*, 1971), while that of *L. weddelli* is c. 25×10mm and weighs approximately 1g (Cuello & Tramezzani, 1969). The pineal of an average human is about 8mm long.

The pineal appears histologically to be more active than that of man. In *L. weddelli* both cortical and medullary regions have been described, in which respect it is different from other mammals. The gland is covered with a capsule with a pigmented layer on its dorsal and distal parts. Large amounts of fat are present in the cortical cells in May and June (the Antarctic winter), and only small amounts in December and January (Cuello & Tramezzani, 1969, Cuello, 1970). A rhythmic activity of the pineal is suggested, having some correlation with the amount of light available, and probably connected with the reproductive cycle. Northern Fur Seals for instance have a very precisely timed arrival and departure from their breeding grounds on the Pribilof Islands (Elden *et al.*, 1971).

In the Southern Elephant Seal there is some evidence that the amount of daylight controls the breeding period in the males through the pineal gland (Griffiths & Bryden, 1981). The pineal weight and plasma melatonin levels vary inversely with day length, the weight of the gland being heaviest in midwinter and lightest in midsummer. In spring the pineal gland function is inhibited and the breeding season initiated.

Pituitary

The pituitary gland in newborn Harbour Seals (*P. vitulina*) shows as much secretory activity as in a two-month-old dog (Harrison, 1969). The secretion, containing a large amount of cystine, may be concerned with the production of anti-diuretic hormone and the need for water conservation by pups who feed on very rich milk with a high fat content. Hormones from the foetal pituitary may be causing the increase in size of the gonads of newly born phocids (Amoroso *et al.*, 1965) (see chapter on reproduction). Various aspects of the anatomy of the pinniped pituitary have been studied by Cuello (1968), Cannata & Tramezzani (1971) and Leatherland & Ronald (1976).

Thyroid

The thyroid gland increases in weight very rapidly in late foetal life (in *P. vitulina*), and, probably stimulated by oestrogens and progestogens of the placenta, starts to function about three months before birth (Harrison *et al.*, 1962, Harrison & Kooyman, 1968). A second period of intense thyroid activity is in the lactating female. The association between the thyroid gland and the fat content of milk is known and the active adult thyroid may be necessary to achieve the high fat content of seal milk, though little is known about this.

Thyroid serum levels in the adult Harp Seal are lower than in many other mammals, but are higher in the suckling pup than in the lactating female. The pup possibly gets some of its thyroid hormones in the milk. There is no hyperactivity of the thyroid gland in pups or adults (Leatherland & Ronald, 1979).

Adrenal

This gland is lobulated (in *P. vitulina*), and the ratio between cortical and medullary tissue is greater than in most mammals. The gland reaches its greatest weight in young pups, correlated with the general adjustment of the growing pup (Amoroso *et al.*, 1965, Bourne, 1949).

Carotid body

The carotid body partially embraces the dorsal wall of the external carotid artery at about the point where this artery leaves the common carotid. It is regarded as an endocrine gland, responding to changes in the oxygen and carbon dioxide composition of the blood, and thus affecting respiration. It has been investigated only in the Weddell Seal, where it is about 4mm long and richly innervated. Histological examination reveals evidence of secretory activities similar to those known to occur in stimulated carotid bodies. The diving habits of the Weddell Seal could well lead to stimulation of this gland but little is known about this (Morita *et al.*, 1970).

Sleep

Seals may sleep on land or in the water. In the water they usually sleep with the nostrils at or close to the surface, and the body sloping downwards at an angle, rising to the surface to breathe. In captivity Grey Seals have also been seen to sleep resting comfortably with their backs against a bottom corner of the tank, with their heads thrown back.

By implanting radiotelemetry devices in Grey Seals the activities of brain, heart and eye have been recorded while the animal is sleeping. After a period of quiet wakefulness, the seal sleeps, the heart beat becomes regular and fast (110–170 per min.) and the respiration is increased to 12–16 breaths a minute from its slower rate during the quiescent period (3–7 per min.). This is the 'rapid eye movement' sleep, which may last up to about an hour and during which the seal is difficult to arouse. After this there is a transition to 'slow wave sleep' which may last for four hours and during which the animal can be aroused.

In the way in which the rapid heart beat and regular respiration is maintained during the 'rapid eye movement' sleep which itself occurs before the 'slow wave sleep', the sleep of the Grey Seal and perhaps of other pinnipeds, is different from that of terrestrial mammals (Ridgway *et al.*, 1975b).

Chapter 13   *Senses*

Touch

Seals vary a great deal in their sensitiveness to touch. Elephant seals, walruses and sea lions are strongly thigmotactic and will crowd together so that their bodies are touching even when there is plenty of space available. Fig 13.1. This probably has at least a certain amount of thermoregulatory function. Californian Sea Lions that frequently share the rookeries with Northern Elephant Seals appear to take advantage of this habit, and lie on top of, or in the crevices between the sleeping elephant seals, who remain indifferent even when the sea lions jump on their backs. The sea lions probably get a certain amount of warmth and shelter this way. Hawaiian Monk Seals on the other hand, have never been seen lying in contact with one another, and the lightest touch will quickly wake one that is sleeping.

Southern Elephant Seals, and indeed, other seals too, are frequently seen throwing sand up on to their backs with their foreflippers. This activity may be partly thermoregulatory, but as it happens most when the humidity of the air is low, it also probably alleviates irritation of the skin caused by drying.

All seals possess relatively abundant mystacial whiskers. There are tactile organs and are mentioned in detail in the chapter on the skin and fur. The movement and the appreciation of vibrations could play an important part in the life of an aquatic animal. In otariids and walrus the mystacial whiskers are used to indicate a variety of emotions – being brought forward during tactile investigation of the environment or a neighbour, and also being used together with movements of lips and eyes to indicate submission or threat (Miller, 1975c).

Smell

The olfactory lobes of the seal brain are small, but in spite of this it is likely that this sense of smell plays quite an important part in the life of the animal. Immediately after the birth of the pup the female will spend much time touching it with her nose (Burton *et al.*, 1975, Marlow, 1975). She is presumably getting to know the particular smell of her pup, and on a crowded beach the final recognition of her pup from amongst a hundred others is almost certainly olfactory. The mother will sniff at the noses of hungry pups that come up to her, but will accept only her own. Nose to nose nuzzling, which has been recorded in all three families of pinnipeds, may be an important method for transmitting olfactory information. It also occurs in captivity in times of disturbance, when subordinate animals will nuzzle the dominant animal (Ross, 1972).

The odour of the sweat glands is presumably appreciated in the breeding season, and the bull also makes a thorough, and probably olfactory, investigation of the vulva of a cow to determine her breeding condition. Hunting Eskimos are careful to approach a walrus herd from upwind, and at a distance of 200 metres Southern Sea Lions will panic from the scent of man.

Taste

The tongues of seals are usually rather short, wide at

Fig. 13.1 A mass of Phocarctos *cows exhibiting their strongly thigmotactic tendencies.*

the back and tapering to a notched tip, except in the walrus in which the tip is rounded. The habit of swallowing food whole would not suggest any great development of taste buds, and those few tongues that have been examined do not show many of them (Sonntag, 1923, Kubota, 1968, Eastman & Coalson, 1974). *P. vitulina* is said to have a moderately good supply of gustatory organs, but the few taste buds that are present in the newborn Weddell Seal have vanished by the time the animal is adult.

Sight

Enlarged orbits are very characteristic of the pinniped skull, right back to the earliest members of the group (Repenning, 1976). Seals are agile animals that feed on relatively small prey, so they have need of the large eyes that are housed in large orbits for good visual acuity. As well as being absolutely large, the eyes are also relatively large with relation to the size of the body. From the measurements that are available, it seems that the external horizontal diameter of the eye of many seals is about 40mm. Thus the little Baikal Seal (Fig. 6.9) and also the larger Crabeater Seal and Californian Sea Lion have eyes of about the same size as those of a domestic ox. The Ross Seal and the Southern Elephant Seal appear to have the largest eyes, with the external horizontal diameter about 60mm but it is difficult to be sure that all the measurements are comparable. The eyes are directed laterally at an angle of 15° to the body axis, and are also tilted slightly upwards, which may help when searching for a landing place. Although the eyes are large, the opening of the lids is shorter than the diameter of the cornea, but the animals are so mobile in the water that this does not restrict their range of vision at all.

The nictitating membrane is present, but its supporting cartilage varies in size. In *M. angustirostris* the cartilage has been found to be particularly well developed for supporting the leading edge of the membrane. Elephant seals frequently receive a face full of sand from their sand-flipping neighbours, and the eyes have been seen to be quickly cleared of sand by a windscreen wiper effect of the nictitating membrane, frequently while the eye lids remain open (Morejohn & Briggs, 1973).

The exposed part of the eye is protected by the corneal epithelium being strongly keratinized, and the cornea has different curvatures in the vertical and horizontal planes, being slightly flattened in the vertical plane. Internal to the reasonably thick sclera is the heavily vascularized choroid which contains a particularly large tapetum cellulosum lining the whole of the vitreous cavity. This tapetum is said to reach its greatest development in seals – an indication of an eye

that is very light sensitive (Jamieson & Fisher, 1972). There is obviously restricted light in the ocean where the seals hunt their food, particularly in the deeper areas, but it seems that the ocean is not quite as dark as we think. Surface light has been noted in the Mediterranean down to 600m, and there is probably enough light for predator and prey to see each other down to 1000m, not counting the help given by bioluminescence.

The retina is rod dominated, which gives the animal increased sensitivity in dim light, and is especially useful when feeding at night. There is, however, considerable evidence for the presence of both rods and cones in both phocid and otariid retinas (Lavigne *et al.*, 1977). The cones function best in bright light and would give the seal effective vision in good illumination during the day. There is virtually no evidence of colour vision in seals, although a female *P. largha* was trained to discriminate between two differently coloured targets when it was established that the colours were the only variables (Wartzog & McCormick, 1978).

The lens is approximately spherical, the iris is very muscular and well-vascularized, and the contracted pupil is a vertical slit, except in the Bearded Seal in which it is diagonal, and the walrus in which it is broad horizontal oval. The Harderian glands are large, but the lacrimal glands are small. There is no naso-lacrimal duct, so when the seal is on land the tears can be seen running down the face.

Vision on land

In air a seal suffers from astigmatism because of the two curvatures of the cornea. A clear image will be received in bright light when the pupil is contracted. The position of the pupil then, being a vertical slit lying parallel to the astigmatic axis helps to minimize the astigmatism and focuses a point as a point on the retina. If the light is poor, the pupil will be wide open and the image will not be clear.

Thus on land a seal will see clearly only in bright light. Seals seem to respond mainly to large objects appearing on the skyline – which could represent a threatening bull, or a human being – or to rapid movements. By crawling slowly it is possible to approach many seals without causing alarm, and sometimes almost without being noticed.

Vision in water

In water – which has the same refractive index as the cornea – the astigmatic effect of the cornea is lost. With decreasing illumination the pupil becomes wide open, and the spherical lens focuses the image. The tapetum and the photoreceptors in the retina give the animal good sensitivity in dim light, and there is also increased sensitivity to green wavelengths – useful in an underwater environment (Lavigne & Ronald, 1972).

Seal eyes are thus particularly adapted for functioning in water where their visual acuity is said to be as great as that of a cat on land.

Seals, blind in one or both eyes, are not infrequently reported. As they are usually in a well-fed condition, the lack of sight is presumably no handicap in obtaining food.

Hearing

Phocid ear structure

Phocids have no external ear pinna. The opening of the external auditory meatus lies flush with the side of the head and is a small oval hole *c*. 1cm long in Northern phocids, but much smaller and only about 2mm in diameter in Southern phocids, hidden under the hair. The auditory meatus runs between the external aperture and the tympanic membrane. The tube has a lumen of *c*. 2mm, is lined with hairs and glands, and contains blood sinuses in its walls. The part of the tube immediately behind the meatus is unsupported and seems to be fairly short in Northern phocids (*c*. 5mm, and 8 per cent of length) and much longer in Southern phocids (i.e. 45mm and *c*. 55 per cent of length in a Ross Seal) (King, 1969b). The rest of the tube is supported by the auricular cartilage which is attached posteriorly to the bony rim of the external auditory meatus on the skull. Extrinsic and intrinsic muscles help in voluntarily closing the external meatus and constricting the lumen of the tube, preventing entry of water (Ramprashad *et al.*, 1971).

In the middle ear the tympanic membrane is about the same size as that of many terrestrial carnivores (*c*. 11×9mm). The auditory ossicles are dense, and particularly large, the incus with a large heavy head. The middle ear is lined with a layer of cavernous tissue containing large venous sinuses and capable of being engorged with blood. The Eustachian (or auditory) tube is supported by thick fibrous tissue and some cartilage (Ramprashad *et al.*, 1973). The bulla of phocids is considerably inflated. The cochlea has 2·5 turns, and the basal whorl is greatly enlarged in both calibre and length. The apical part of the cochlea has the same orientation as in other carnivores, but the basal whorl is flexed to lie transversely – an arrangement which is different from other carnivores, and also from otariids and odobenids. The fenestra vestibuli for the foot of the stapes, and also the fenestra cochleae are very large, the latter about three times that of the fenestra vestibuli.

Otariid ear structure

Otariids have a small external ear pinna, a scroll-like

structure about 6cm long. The pinna and the tube of the meatus are supported by cartilage and possess intrinsic muscles. The extrinsic muscles seem to be the same as those of the phocids. The tympanic membrane is small (*c*. 6×4·5mm), and the fenestra vestibuli and ossides are also smaller than in phocids. The difference in the size of the ossicles can be shown by their weight – in *P. groenlandica* they weigh 227mg, but in *Zalophus* only 18mg. Cavernous tissue lines the middle ear cavity – which is smaller than in phocids as the bulla is not inflated to such an extent. The cochlea is of fissiped form, except that the basal whorl is large.

Walrus ear structure

There is no external ear pinna, and the tube of the auditory meatus is surrounded by cartilage. The tympanic membrane is large and of phocid dimensions. The ossicles are otariid-like in shape, but larger. The middle ear cavity is large, but the amount of cavernous tissue is small. The basal whorl of the cochlea is enlarged.

Hearing in air

A measure of the efficiency of hearing can be made by comparing the area of the fenestra vestibuli with that of the tympanic membrane. When the fenestra vestibuli is relatively small the action of the ossicles is magnified on reaching the cochlea, and hearing is relatively better. The ratio tympanic membrane to fenestra vestibuli in fissipeds with good hearing is in the region of 35–50:1 (humans 22:1). The ratio in otariids and walrus is 7–71:1 and 20–24:1 respectively, in some southern phocids it is *c*. 10:1, and in some northern phocids it tends to be higher, with 22:1 in *P. hispida* and 38:1 in *Erignathus*. This would suggest that possibly northern phocids have the most efficient hearing of the seals in air, but even so they are not quite as good as most fissipeds, though a good many of them can hear at least as well as humans in air. The lack of a large pinna and the presence of a long fleshy external auditory meatus would also reduce the sensitivity of hearing in air.

Hearing in water

Under water, however, seals hear better than fissipeds and have good directional hearing. This is important for hearing the whereabouts of their prey and would allow the seal to approach its prey before vision in the low light conditions could take over, or even in total darkness. It would also help in allowing them to keep track of other members of their own species.

The modifications of the pinniped ear have obviously arisen twice – once in the phocid line and once in the otariid – odobenid line. But being adapted to the same medium, the basic changes from the fissiped arrangement are similar in all seals and may be summarized as follows:

1. The enlarged fenestra cochleae.
2. The round window fossula.
3. The presence of cavernous tissue.
4. The reduction of the difference in area between the tympanic membrane and the fenestra vestibuli.
5. The temporal unit remains unfused from the other skull bones until rather late in maturity.
6. Enlargement of the petrosal near its apex and opposite its point of attachment to the mastoid.

It should be noted that these are characters of all pinnipeds, even though they may not have been mentioned earlier in the text. These characters confer the following advantages:

1. The enlarged fenestra cochleae amplifies sound reception.
2. In pinnipeds the fenestra cochleae (also called the round window) is at the bottom of a considerable depression (the fossula fenestra cochleae or round window fossula as used by Repenning, 1972). This depression, which is not present in other carnivores, shields the membrane of the round window from contact with the expanded cavernous tissue.
3. When the seal first goes under water the air in the external auditory meatus acts as a barrier to sound. As the animal goes deeper it is necessary to equalize the pressure in the air spaces of the ear. The cavernous tissue of the middle ear and external auditory meatus is involuntarily filled with blood by increasing hydrostatic pressure, occluding the latter and reducing the air-filled part of the middle ear cavity – thus maintaining pressure equal to that outside. At great depths, where compression of air has progressed to the point where distended cavernous tissue is in contact with both sides of the tympanic membrane, this membrane then functions as though it had water on both sides of it. However, at lesser depths the membrane does not function as it remains insulated from sound by residual pockets of air.
4. Sound waves have greater pressure in water, and the reduction of the difference in area between the tympanic membrane and the fenestra vestibuli would reduce overstimulation of the cochlea by the increased sounds when the distended cavernous tissue contacts both sides of the membrane at great depths (see 3 above).
5. In seals the parts of the temporal unit (i.e. the petrosal+mastoid, tympanic and squamosal) fuse very early in foetal life. At birth this area is much more mature than in a fissiped of similar age. This possibly has some relevance when one considers that frequently very little time may elapse between birth and the need for underwater hearing. The temporal unit remains unfused from the rest of the skull for some considerable period. This isolation restricts the amount of extraneous sound reaching the cochlea through bone conduction. Sound then reaches the cochlea mainly via the mastoid, which gives it a directional quality, and adaptations to directional sensitivity are developed on the mastoid and other parts of the temporal. When, with increasing age, the temporal unit eventually fuses with the surrounding skull bones, the 'directional' surfaces of the skull have by then fully developed.
6. The enlargement of the apex of the petrosal magnifies the inertial effects and increases the distortion of the cochlea.

Under water, seals hear by several means. They may use resonant reaction – where the inertia of the ossicles conveys the sound impulses as the whole skull vibrates with the surrounding water in response to passing sound waves. They may hear by conducive reaction – where sound is transmitted into and through bones of the skull to the ear from those parts approximately normal to the direction of the sound, but is reflected from those skull parts whose surface is markedly oblique to the direction of sound. And finally, in deep dives, they may also hear by the conventional method of vibration transference across the ossicles.

The enlarged mastoid processes of otariids and walrus, and the enlarged squamosal of phocids, have large surface areas facing specific directions, and are believed to be so designed for sound reception from these directions. In approximate terms the antero-dorsal facing surfaces of the phocid squamosal parallel their antero-dorsal orbit orientation, and the more lateral orientation of the otariid mastoid parallels its more lateral orbit orientation.

Phocids differ from otariids and odobenids in being slightly more modified for the resonant reaction. The heavy ossicles are useful to create impulses from inertia, and the alignment of the basal whorl of the cochlea is effective in increasing the reception of sound from the side of the head and is correlated with the alignment of the ossicles.

Phocids seem to hear higher frequencies underwater than do otariids, and the opposite is true for airborne sounds, but all seals are more sensitive to underwater sounds than they are to airborne sounds (Schusterman, 1981).

The subject of hearing is a complicated one, and much of this account has been taken from Repenning,

1972, to which paper the reader is referred for a detailed account.

Sounds

After considering hearing, it is obviously necessary to consider also the sounds that seals make. There is variation in these of course – *Zalophus* is a noisy animal, barking almost continually during the breeding season, while *P. vitulina* and *P. hispida* are normally very quiet (Newby, 1973). There is a general tendency for polygynous species to be vocal, and monogynous species to be quiet, which is reasonable if one considers the need for communication within a colony (Evans & Bastian, 1969).

A querulous bleating sound is produced by most pups, and it eventually matures into a bark. The mother seal can recognize the individual cries of her pup. A female *Phocarctos* for instance, searching for her pup which may be a hundred metres or more deep in the forest, advances calling. The pup replies, and through the exchange of calls the female gradually moves towards the pup and contact is made. Northern Elephant Seal cows could distinguish recordings of the distress calls of their own from those of other pups (Petrinovich, 1974).

Barking, snorting, roaring, a rattling sound like a distant motor bike, a lowing, or an explosive cough-grunt are the most usual sounds produced by seals in air, and they use these in various ways to indicate alarm, aggression, or to locate pups (Peterson & Bartholomew, 1969, Brown, 1974). The roaring of elephant seals and the 'dialects' of the Northern Elephant Seal have already been mentioned, as have the bell tone of the walrus, the musical chirrups and trills of the Antarctic phocids, and the warble of the Bearded Seal, many of the sounds being produced under water (Poulter, 1968).

The sounds produced under water are not always the same as those made in air. Clicks, trains of pulses, whistles and warbles are all produced by different animals. There is some indication that male *Callorhinus* produce clicks of a lower frequency (8–32 per second) than females (38–60 per second), a difference that is detectable even in pups (Poulter, 1968). Weddell Seals produce an unusual sound – long trains of resonant pulses up to 42 seconds long and descending in pitch so that a sound like a descending whistle is made. It is

possible that these underwater sounds are not produced by the vocal cords but from the dorso-medial pouch of the larynx when it exists (otariids) or possibly by movement of air between larynx and trachea causing a vibration of the tracheal membrane. Positive information on this subject is not available.

Recording of calls, and the subsequent analysis of the resulting sonogram can give interesting results. Comparison of sonograms of various species of the genus *Arctocephalus* has shown that there are two vocal types within the genus. Where two species overlap in their distribution, as with *A.p. doriferus* and *A. forsteri* for example, it was found that the members of each pair belonged to different vocal types. This would help in recognition of each other by members of one species, and also help in keeping the two species discrete (Stirling & Warneke, 1971).

Echolocation

'If an animal listens to the echo of its own sounds, and then uses that echo to determine a bearing, range, or the characteristics of the echoing object, it is echolocating' (Norris, 1969).

The production of clicks and similar sounds under water has suggested that echolocation or sonar is being used by seals to locate food or to avoid obstacles. There have, apparently, been no definitive experiments yet which prove beyond doubt that sonar is being used, though some experiments are suggestive (Schusterman, 1968, Norris, 1969).

A sound beam that is possibly directional has been detected in recordings of *L. weddelli* and this would be useful if echolocation were being used (Schevill & Watkins, 1971). *Zalophus* has been used for many experiments in this field. An animal in an anechoic tank in total darkness had little difficulty in locating food fish, and could distinguish between the preferred fish and pieces of horsemeat of the same size (Poulter, 1966). Blind seals seem to have no difficulty in obtaining food. Blind *Zalophus* in air could follow an irregular wall by barking at it as they went, and the tempo of barking would change if a person stood in the way (Poulter, 1966). These and other experiments suggest that something else is being used besides vision, but as mentioned above, it seems that totally convincing experiments are yet to come (Schusterman, 1981).

The depths to which seals, under natural conditions, will dive, are difficult, though not impossible to measure accurately. The information from animals caught in nets or on hooks, also gives some indication of the depths they are able to achieve, although there always remains the possibility that the seal was trapped nearer the surface, and carried down. The following list gives figures that are as accurate as possible.

P. vitulina	90m	experimental conditions
H. grypus	146m	caught on hook
E. jubatus	146m	fishermen's evidence
Z. californianus	170m	trained animals
M. angustirostris	183m	caught on line
C. ursinus	190m	free swimming and depth recorder
P. groenlandica	273m	caught on line
M. angustirostris	300m	experimental conditions
L. weddelli	600m	free swimming and depth recorder

One of the most competent divers of all species – the Weddell Seal – rarely dives for longer than 30 minutes (Kooyman *et al.*, 1980). Most pinniped dives are probably of much shorter duration. As with depth, the Weddell Seal holds the record for time, and is able to stay under water for at least 73 minutes (Kooyman, 1969, 1981, Kooyman *et al.*, 1980, 1981). Young animals do not dive so deeply, or for so long as adults, but their performance increases with age.

Less work has been done on otariids and walruses, but their diving capabilities seem to be less than those of phocids. Lactating female *Callorhinus* equipped with depth recorders have been as deep as 190m for a dive that lasted 5·4 minutes. Most dives last for less than 5 minutes (Kooyman *et al.*, 1976). Walruses dive for about 40 minutes (Nyholm, 1975).

Much work on diving has been carried out on the Weddell Seal. This has been found to be a convenient animal to work on as it lacks flight response to man and lives in areas that not only force the animals to spend much time under the ice, but the ice itself has given rise almost to laboratory conditions. Research workers have made an artificial breathing hole at a suitable place in the ice, and then put a building over the top of the hole to act as a laboratory. They then capture a seal several kilometres away from this hole and bring it back. A package of recording instruments is attached to its back and the seal is released down the new hole. Almost all seals have returned to this hole, and the package of instruments such as maximum depth indicators and depth-time indicators, is removed, and the information collected. It has been found that these seals do not make just one sort of dive (Kooyman, 1968). A short dive of 15 minutes or less and down to about 100m is used to explore the neighbourhood in the vicinity of the breathing hole. Another type of 15-minute dive is used for hunting, but there the seal will go almost vertically downwards and may reach its maximum depth of 600m (*c.* 328 fathoms), although most dives are between 200 and 400m. Diving vertically in this way reduces the chances of getting lost when the seal is 'concentrating' on getting food. The longest type of dive which may last up to an hour, though seldom deeper than 200m, is when the seal is presumably searching for other breathing holes. It then travels long distances under the ice. The seals are obviously able to navigate precisely at this time, and would appear to have some appreciation of the points of the compass. During the 60 minutes that the seal may hold its breath, it may swim up to about 8km under water. It is interesting to speculate how it 'knows' when it has reached about half the capabilities of its body and should start returning to the 'home' breathing hole. Weddell Seals dive more often, and also longer and deeper during daylight hours, but because of the latitude where they live they must obviously be able to dive successfully during the long polar night (Kooyman, 1975a).

Conditions for diving

To dive successfully seals:
1. must keep the water from entering the body.
2. must adjust to the water pressure on the compressible parts of the body.
3. must not suffer from decompression sickness.
4. must use what oxygen the body contains to the best advantage.

Water occlusion

The nostrils are shut when relaxed, and increasing water pressure would close them even more firmly. Strong laryngeal muscles prevent water entering the trachea when the seal opens its mouth under water. Small muscles along the external auditory meatus probably contract and help to close this passage.

Water pressure

Under water a seal is a solid incompressible body except for those spaces normally filled with air – the middle ear and the respiratory system. The function of the cavernous tissue in the middle ear and external auditory meatus, where it equalizes pressure between these inside air spaces and the outside environment, is mentioned in the section on hearing. Water pressure on the flexible rib cage with its very oblique diaphragm pushes much air out of the lungs, the absorptive alveoli collapsing first, but the cartilaginous support of the bronchi and bronchioles keeps them open and they hold whatever air remains in the respiratory system. While diving, the trachea becomes compressed, even in those seals with circular tracheal cartilages, and at water pressures equal to 300m of sea water the trachea is compressed to less than half of its original dorso-ventral diameter (Kooyman *et al.*, 1970, Boyd, 1975, Kooyman & Andersen, 1969). It is thought necessary that a little air remains in the system so that it can be passed over the vocal cords to produce sound. There are no cranial sinuses in the pinnipeds and thus no air spaces here to be compressed.

The 'bends'

Decompression sickness, caisson sickness or the 'bends' in man is the very painful, sometimes fatal result of a diver returning to the surface too quickly from depths greater than about 14m. Under pressure, the gaseous nitrogen goes into solution in the blood, and if the pressure is released too suddenly, the nitrogen comes out of solution as bubbles which may obstruct the smaller blood vessels and interfere with the functioning of the joints, liver and central nervous system. Seals, of course, do not have a continuous air supply as does a human diver, and obviously take down a very limited supply of nitrogen. Phocids at least, exhale before diving, and as just mentioned, what little air remains does so in the cartilage-supported and therefore non absorptive parts of the lung, and is exhaled when the animal surfaces.

Oxygen conservation

As already mentioned some seals at least are capable of remaining under water for over an hour and for this length of time they are swimming about without access to further supplies of oxygen. A human pearl diver can only stay under water for about 2·5 minutes before the increased carbon dioxide in the blood stimulates the breathing reflex and the person must come to the surface before the brain is damaged from lack of oxygen. How then has a seal access to so much more oxygen?

Oxygen is carried by the oxygen binding molecule haemoglobin (Hb) which in mammals is found only in the red blood corpuscles. Another molecule closely related in structure to haemoglobin, called myoglobin is in the muscles. It gives muscle its red colour, thus in seals the muscle is very dark red because of its high concentration compared with pigs for example in which the concentration is low. Anyone that has dissected a seal will know of the very large amounts of blood these animals have – about 12 per cent of the body weight, compared with 7 per cent in man. A Weddell Seal has a blood volume as much as 14 per cent of the body weight, which means that the animal has about 55 litres (i.e. *c*. 12 gallons) of blood, and a female Southern Elephant Seal has 59 litres of blood. (The blood volume of man is 4·5–6 litres.)

Seals then, have a large quantity of blood in which the red corpuscles are fewer in number, but larger in size than in terrestrial mammals generally. Both the Hb content and the oxygen capacity are high. In a Weddell Seal, for instance, the Hb concentration and oxygen capacity are among the highest in mammals, and the oxygen can be more completely used than in terrestrial mammals. The oxygen stored in a given volume of its blood is nearly twice as much as in the Harbour Seal, and five times as much as in humans (Lenfant *et al.*, 1970b).

The most important factor in conserving the amount of oxygen used while diving is the constriction of the arteries to the peripheral parts of the body. The blood flow to the brain remains nearly constant, but the flow to the viscera, skeletal muscle, skin and flippers is reduced by about 90 per cent. However, a reduced, though still considerable blood supply is maintained to the adrenal glands (Zapol *et al.*, 1979). A rich autonomic innervation controls the blood flow and prevents a dangerous increase in intracranial pressure (Dormer *et al.*, 1977). There is some evidence that the heart and brain of at least some seals have a certain tolerance for oxygen lack (Kerem & Elsner, 1973, Elsner, Shurley *et al.*, 1970).

It was noted at the beginning of this chapter that although Weddell Seals are capable of remaining under water for over an hour, they rarely dive for longer than 30 minutes. In fact, observations on Weddell Seals have shown that 97 per cent of their free-ranging dives were shorter than 26 minutes (Kooyman *et al.*, 1980). These short dives result in little increase in the lactate concentrations in the blood, and it is concluded from this that the Weddell Seal, at least, relies mainly on aerobic metabolism for its energy while on short dives.

The oxygen store in a resting animal has been calculated to last 16 minutes. For short dives up to 26 minutes the extra oxygen needed is obtained by the reduction of the blood flow to the non-essential parts of the body, those parts then functioning at a lower metabolic rate for a while. After 26 minutes, if the dive is a long one, the animal becomes a heart-brain-lung machine, the outlying parts of the body start functioning anaerobically, and the amount of lactate in the blood increases considerably (Kooyman *et al.*, 1981).

This system is of advantage to the seal which then does not have to spend long periods recovering from its dive. The oxygen store can be replenished quickly and the animal can dive again. It has been calculated that in a series of several short aerobic dives the seal actually spends a proportionally longer period under water than in a few long dives that require a much longer period of recuperation.

Heart rate

Bradycardia – the slowing of the heart rate on diving, was one of the early diving phenomena to be observed. The normal heart rate, mentioned in the section on the vascular system, is variable, but beats up to 120 a minute have been recorded. On diving the heart rate falls to as low as 4 or 6 beats a minute, remaining more or less at this level until the seal comes to the surface again when it is increased above its normal level for a while (tachycardia) before going back to normal. It is thought that the function of bradycardia is to reduce the amount of blood that is flowing, still at normal pressure, through the restricted area of the body. Work on freely diving Weddell Seals has shown that the degree of bradycardia varies according to the type of dive, being much more pronounced in the longer dives, and on this basis it is suggested that possibly the animal somehow anticipates the nature of its intended dive. In this way experimental dives may be regarded by the seal as an emergency and full bradycardia is assumed.

Blood vessels

The anatomy of the vascular system is described separately. Modifications concerned with diving are the large extradural veins, caval sphincter, pericardial plexus and the large size of the abdominal veins for storing blood.

In phocids the venous drainage from the brain is via the extradural vein, and as this vein is connected with almost all the abdominal veins the blood may have a devious course before it is stored in the large hepatic sinus. The vessels of the cervical venous system become occluded early in forced dives so that blood from the brain cannot reach the heart this way, and has to pass through the body to reach the hepatic sinus (Ronald *et al.*, 1977). This sinus and the posterior vena cava immediately posterior to it become dilated during the first minute of diving as they take up the blood

which cannot now circulate in the peripheral parts of the body. They empty almost immediately on recovery. The caval sphincter is believed to control the flow of blood into the heart from the hepatic sinus. The dark fibres in the muscle of the caval sphincter of the Harp Seal have aggregations of mitochondria, lipid droplets and glycogen granules which would suggest the production of heat by non-shivering thermogenesis (George & Ronald, 1975).

Plexuses

The function of the venous plexuses has been questioned for some time, and it is now thought that some at least are concerned in thermoregulation. Plexuses in the Harp and Hooded Seals have been shown to be embedded in brown adipose tissue which is capable of producing heat by the oxidation of its lipid stores (Blix *et al.*, 1975). During diving, much of the blood is passing through the brain in the poorly insulated head, and the animal is presumably thus losing heat. If, as has been shown, the blood may return through the pericardial plexus, this may prevent too much loss of heat.

Pregnancy and diving

There appears to be no or little restriction of diving capabilities in the pregnant Weddell Seal, as seals bearing almost full-term pups are capable of diving for an hour, and to depths of about 300m (Elsner *et al.*, 1970a). In fact one of the longest dives recorded (55+ minutes) was made by a pregnant Weddell Seal (Lenfant *et al.*, 1969). As in all mammals the blood of the foetal Weddell Seal has a higher affinity for oxygen than that of its mother. Contrary to most mammals, however, the blood of the foetal Weddell Seal (and probably of other phocids also) has a lower Hb concentration and thus a lower oxygen capacity compared with the blood of its mother. This is advantageous for a diving mammal as it reduces the necessity for blood to go to the uterus during a dive (Lenfant *et al.*, 1969, 1970a).

Comparison

As most of the work on pinniped diving has been done on phocids, much of the subject of this chapter refers almost entirely to phocids. It will be useful to summarize the differences between phocids and otariids in this respect.

Phocids are deep divers, with a higher concentration of Hb in the blood (i.e. higher than otariids), and can thus store larger amounts of oxygen in the blood. They exhale before diving and thus have little oxygen in the lungs. They spend more time breathing between dives so it is not necessary to have the cartilaginous support right up to the alveoli. They have larger bullae, and therefore a larger middle ear cavity and cavernous sinus which can compensate for the deeper dives. The Weddell Seal has the largest bullae and it is the deepest known diver. It also has the highest Hb concentration and oxygen capacity of the blood. As mentioned in the section on the vascular system, only in the Phocidae are combined the greatest number of venous modifications thought to be of significance in diving.

Otariids (and probably odobenids also) have shallower and shorter dives. The lower Hb levels and oxygen carrying capacity of the blood indicate no adaptations for long periods of submergence (Wells, 1978). Otariids exhale before diving, but still store proportionately more oxygen in the lungs than do phocids. During swimming they surface frequently for a quick breath. The cartilaginous support extending right up to the alveoli keeps the bronchi wide open and facilitates this quick exchange of air. They have smaller bullae, thus smaller amounts of cavernous tissue, and may not be able to cope with exceptionally deep dives.

Chapter 15   *Parasites*

Parasites occurring in seals belong to the following invertebrate phyla:

Phylum Platyhelminthes
Class Cestoda – tapeworms
Class Trematoda – flukes
Phylum Aschelminthes
Class Nematoda – roundworms, lungworms, hookworms
Phylum Acanthocephala – thorny headed worms
Phylum Arthropoda
Class Insecta
Order Anoplura – sucking lice
Class Arachnida
Order Acarina – mites and ticks

Platyhelminthes

Cestoda – tapeworms

Cestodes are common in most seals, and may be present throughout the gut. Large areas of the walls of stomach, large intestine or bile duct may be occupied by mass infestations of these worms, involving large numbers of parasites. In spite of this they seem to cause very little disturbance to the seal. At least two hosts are necessary for the tapeworm life cycle and many teleost fish and pelagic crustaceans form the intermediate hosts. In fact the wide range of many of these animals at least in the northern nemisphere results in many seals sharing the same range of tapeworms. Larval cestodes are frequently found embedded in the blubber of seals, and large masses of cestodes have been seen protruding from the anus of *Neophoca* (Marlow, 1975).

Approximately 14 genera and 49 species of cestodes have been recorded from seals, and of these the greatest number (31 species) belong to the genus *Diphyllobothrium*. Many cestodes found in pinnipeds are not exclusive to them, and some are found mainly in other vertebrates, for example in toothed whales, sharks, birds and sea otters. In fact only two of the eight genera of cestodes found in the northern phocids are actually exclusive to seals. These are *Anophryocephalus* and *Pyramicocephalus*.

A few seals have cestode parasites unique to them, for example *Baylisia baylisi* is found only in *Lobodon*; *Diphyllobothrium scotti* is found only in *Ommatophoca*, *Baylisiella tecta* only in *Mirounga leonina*, and *Diphyllobothrium romeri* only in *Odobenus*. *Taenia solium* is not normally found in marine mammals, but it has been recorded from a Mediterranean Monk Seal, and also from a South African Fur Seal, where the presence of numerous cysts in the brain was causing convulsions (De Graaf *et al.*, 1980).

Trematoda – flukes

Trematodes may be found in the liver, gall bladder, bile and pancreatic ducts, and small intestine of seals. Large numbers – 1–10 000 parasites may occur in one seal. The higher numbers of parasites may cause inflammation or even death, but it is not known if the smaller numbers have any serious effect on the host. A Bearded Seal, killed by a Polar bear, was found to have, amongst other parasites, trematode eggs in numerous inflamed nodules in the pancreas, and fibrosis and inflammation of the bile ducts, probably caused by trematodes (Bishop, 1979).

Some 19 genera and 24 species of trematode are found in seals, but probably only two genera – *Pricetrema* and *Zalophotrema* are unique to pinnipeds. *Pricetrema* is a small, but heavily spined fluke which can do considerable mechanical damage to the stomach lining. Most trematodes seem to be typically found in other mammals and marine birds, and only accidentally found in seals. Amongst the sea lions it is interesting that trematodes have been recorded only from *Zalophus* and *Eumetopias*.

Aschelminthes

Nematoda – roundworms, hookworms, etc.

Nematodes are generally called roundworms, but those that live mainly in the heart or lungs are also called heartworms and lungworms respectively. About 15 genera and 46 species of parasitic nematodes have been recorded from seals. Amongst these genera, the roundworms *Anisakis, Contracaecum* and *Terranova* are known from all seals and are frequently present in large numbers. They inhabit the stomach where they are attached to the wall and cause inflamed and ulcerous areas. Peritonitis and eventual death result from perforated gastric ulcers caused by nematodes (Wilson *et al.*, 1970). Weddell Seals at least, have been seen to regurgitate a mass of nematodes (*C. osculatum*) (Johnstone, 1975).

Seals are the primary host of *Terranova decipiens*, also known as the codworm. Seals such as the Grey, Harbour and Harp for instance, that are abundant in areas that are much used by human fishermen are implicated in the spread of the codworm in the fish that are used for human food. The larvae of the codworm have been found in the isopod *Idothea neglecta* which is a significant food of the cod (Bjørge, 1979). Codworm larvae are then found in the muscles of such commercial fish as cod, where their presence is not always obvious until the fish is prepared for the table. The main problem is that the larvae make the fish look unattractive, and thus affect the sales. *Terranova* is not known to affect man and would, of course, be com-

pletely destroyed by normal cooking. Nevertheless, its presence is not welcomed in commercial fish and there have been proposals to try and control it by culling the seals involved (Bonner, 1972).

The flesh of some walruses, Ringed and Bearded Seals have been found to be infected with *Trichinella spiralis* – the nematode that causes trichinosis in man. The infected carcass of a sledgedog could be eaten by amphipods, who are devoured by fish, or directly by Ringed or Bearded Seals. The occasional walrus will eat the flesh of other seals and get infected in this way, and Polar bears, which feed on seals, are usually heavily infected (Fay, 1968).

Hookworms of the genus *Uncinaria* are a major cause of death of pups of *Callorhinus*. The larvae live in the blubber of the female seal and are then transmitted to the pup in the milk when it suckles for the first time (Keyes, 1965). In the pup the worms live in the small intestine where they feed on blood, and thus large numbers of these parasites will cause severe anaemia and death.

Lungworms occur as adult worms in the lungs, bronchi and bronchioles of seals, where lung congestion (verminous pneumonia) and abscesses may result (Migaki *et al.*, 1971). In young Californian Sea Lions in the wild, pneumonia and obstruction of the bronchi by mucus are consequent on infestation by lungworms (*P. decorus*), and are a very common cause of death (Sweeney & Gilmartin, 1974). The life history of one worm, *Parafilaroides decorus* has been worked out. The adult worm occurs in the sea lions *Zalophus* and *Eumetopias*, and the faeces, containing larvae, are eaten by the coprophagous fish *Girella nigricans* – the opal-eye. The fish, containing the developed larvae are later eaten by the sea lions (Dailey, 1970).

The only lungworm in the Antarctic seals is *Parafilaroides hydrurgae* – found only in the Leopard Seal. As a coprophagous tidepool fish may be necessary as an intermediate host, possibly this seal in its movements to the warmer shores of Australia, say, may eat such a fish and get infected, but there is no proof of this yet.

The heartworm *Skrjabinaria* seems to be unique to pinnipeds, occurring in the heart, main arteries and venous sinuses of the liver. Other heartworms are *Dirofilaria immitis*, usually reported only from *Zalophus*, *Dipetalonema odendhali* from otariids, and *Dipetalonema spirocauda* from some of the northern phocids. The life history of the latter nematode at least is not yet known. There is no good evidence that lice are implicated, and it has been suggested that the role of such blood-sucking insects as mosquitoes and simulids should be looked at more closely in this regard (Dunn & Wolke, 1976).

In *Phoca vitulina concolor*, *D. spirocauda* has been

found in the heart, great vessels, lungs and liver (Dunn & Wolke, 1976). Between 5 and 86 nematodes have been found per heart, and large numbers of the parasites were completely blocking the pulmonary artery. The presence of the adult parasites in the arteries resulted in the thickening of the innermost layer of the arterial wall (the intima) and the wall then became thrown into folds rather like villi. The liver suffers from areas of necrosis and inflammation, and a similar pathological state in the lungs results in acute interstitial pneumonia.

Acanthocephala – thorny headed worms

Acanthocephalans have an eversible proboscis provided with hooks with which they attach themselves to the wall of the small intestine of their host. The larvae pass through intermediate hosts – amphipods and fish – so it is easy to see why these parasites are so widespread in seals. In spite of this, little is apparently known of the biology of the parasites, nor of their effect on the seals.

Only two genera of acanthocephalans are found in seals – *Corynosoma* with some 20 species, and *Bolbosoma* with only two species. *Bolbosoma* is usually a cetacean parasite and seems to be endemic to the Bering Sea area, occurring in the seals of this area, though not in the walrus. The various species of *Corynosoma* are widely distributed in seals and usually it seems to be geographical boundaries that determine their distribution.

Arthropoda

Insecta

ANOPLURA – SUCKING LICE

Sucking lice are ectoparasites whose mouthparts are adapted for piercing the skin and sucking blood. Sucking lice of the family Echinophthiridae are obligate parasites, exclusive to marine Carnivora, mainly pinnipeds. They depend entirely on their host, and as they are primarily terrestrial animals, have adapted to the essentially aquatic life of their seal host. Their claws may be modified for clinging to fur or skin, and the spiracular apparatus may also be modified with an arrangement of plates and knobs that can close the tracheal system against entry of water. Some lice cling to the fur, others prefer the more naked areas of the body and are to be found on eyelids, nostrils, anus etc. (Kim, 1971, 1975). Pups are more highly infested than adults and are infected very soon after birth. Over 250 lice can be found per seal. They do not apparently cause much trouble to their hosts, though they may possibly carry *Salmonella*, or microfilaria of nematodes.

Four genera of louse are known from seals:

1 *Proechinophthirus* is known only from fur seals, where a single species *P. fluctus* lives on *Callorhinus*, and a second species *P. zumpti* is known only from *A. pusillus pusillus*. Although this genus of louse is to be expected from other fur seals, it has not yet been recorded (Kim, 1979).
2 *Echinophthirus* is known only from the northern phocids, where a single species, *E. horridus*, is common to all.
3 *Lepidophthirus* is known only from two of the southern phocids – *L. macrorhini* being found only on *Mirounga leonina* and *L. piriformis* only on *Monachus monachus*.
4 *Antarctophthirus* is the most widely distributed louse, but its six species are restricted in their distribution:
   A. callorhini occurs only on *Callorhinus*
   A. microchir occurs only on sea lions
   A. trichechi occurs only on walrus
   A. ogmorhini occurs only on *Hydrurga* and *Leptonychotes*
   A. lobodontis occurs only on *Lobodon*
   A. mawsoni occurs only on *Ommatophoca*

Thus the only seal to have two different lice is *Callorhinus* (Kim, 1972). *Proechinophthirus fluctus* is found on the fur, particularly of neck, belly and hip, and has long claws that are suitable for grasping hair. *Antarctophthirus callorhini* has pointed claws suitable for digging into and holding on to naked skin. Both lice have three nymphal stages which require blood, and the life cycle of both lice is believed to be about 17 days, so while the fur seals are on land, four full generations of lice are produced. During the time the seals are at sea no new generations of lice are produced, there is a certain amount of maturation of nymphal stages, and the life cycle is continued when the seals return to land again for their next breeding season.

Lepidophthirus macrorhini is known only from the Southern Elephant Seal, where it makes burrows in the skin of the hind flippers (Murray & Nicholls, 1965). The lice do not reproduce while the seal is at sea as they need higher temperatures (over 25°C) for reproduction. Thus they reproduce rapidly while the elephant seal is on land, and the vasodilatory capabilities of the hind flippers mean that the temperature is frequently correct. While the seal is at sea the skin temperature is much lower, and the metabolic rate of the louse falls, so that it can survive in its cutaneous burrow on the dissolved oxygen in the water. Moulting by the elephant seal removes only the top of the louse's burrow, but the older seals that tend to spend much time in a mud wallow at this time of year have fewer lice. The louse has adapted its life cycle to that of its host, and breeds rapidly twice a year at relatively high temperatures, and can survive long periods without food while its host is at sea.

Antarctophthirus ogmorhini lives in the pelage of *Leptonychotes weddelli* and *Hydrurga leptonyx*, clinging to the hairs of the hind end, usually the hip and hind flipper area. Weddell Seals live in colder places than Elephant Seals, but haul out on the ice much more often. These lice can breed at much lower temperatures than those on the Elephant Seal, but are unable to go without food for so long. They will feed within about a minute of the seal emerging from the water. Infestations of lice are heaviest on yearling and immature seals, and lightest on adult seals (Murray *et al.*, 1965).

Affinities of the lice The affinities of these highly specialized sucking lice of the pinnipeds are obscure, and it has been suggested that they have evolved with their hosts since the time when the ancestral pinnipeds started to become aquatic. Of the four echinophthiriid genera, *Proechinophthirus* is the most generalized, with many primitive characters, and is considered to be close to the ancestral echinophthiriid. It is significant that *Proechinophthirus* occurs only on fur seals – the most primitive of the otariids, and the fact that *Callorhinus* and *Arctocephalus* have different species of *Proechinophthirus* seems to confirm the fossil record that the two fur seals have evolved separately for some considerable time (Kim *et al.*, 1975).

All sea lions, both in northern and southern hemispheres, but no other pinnipeds, harbour *Antarctophthirus microchir*. This suggests that this louse was living on the ancestral sea lions before the various sea lion genera evolved and dispersed to different hemispheres (Kim *et al.*, 1975).

Of the four genera of lice, *Proechinophthirus* and *Echinophthirus* are closely related, with the former being the more generalized. *Antarctophthirus* shares some similarities with *Proechinophthirus* and some with *Lepidophthirus*, and the latter is regarded as the most specialized. Little has been done on the evolution of phocids and their lice, but it is obviously appropriate that *Lepidophthirus* occurs only on the southern phocids. Three species of *Antarctophthirus* occur only on the southern phocids also and must presumably have evolved since these seals have been in the southern hemisphere.

Arachnida

ACARINA – MITES

Three genera of mites of the family Halarachnidae are endoparasitic in mammals, and two of these – *Halarachne* and *Orthohalarachne* – are found in pinnipeds. *Halarachne* has four species, and is found only in phocids (Furman & Dailey, 1980), while *Orthohalarachne* has five species and is found only in otariids. No mites have been recorded from the walrus.

The mites occur in the nasal passages where they are attached to the mucosa of the turbinals and nasopharynx. They also occur in the trachea and bronchioles, and very small specimens less than 1mm long may be in the lungs. Larvae frequently occur on the turbinals where 670 have been recorded from one animal. Adult mites occur in much smaller numbers (less than 100 per seal) and are usually in the nasopharynx. The front end of the mite's body is buried in the mucosa and the animal feeds on the lymph. They cause some erosion, inflammation and irritation of the nasal passages. Heavy infestation will affect the respiration, cause lung lesions and more serious problems which could lead to death (Kim *et al.*, 1980, Seawright, 1964). It is possible that pathological nodules on the turbinals may be the result of such irritation. *Hydrurga* is very prone to such nodules, though whether caused by mites is not known.

The frequent contact between noses that occurs between one seal and another serves to spread the larval stages of the mites.

Another family of mites – Demodicidae – is represented by a single species known only from the Californian Sea Lion. *Demodex zalophi* lives in the hair follicles, mainly of the flippers, and causes a mange-like condition (Nutting & Dailey, 1980).

ACARINA – TICKS

A single genus and species of tick – *Dermacentor rosmari* – has been recorded from the walrus, but from no other pinniped. The tick is usually to be found attached to the skin between the digits of the hind flippers (Ass, 1935).

References

Parasitology is a specialized subject, and for those who are particularly interested the following references would be useful – Dailey & Hill, 1970, Dailey & Brownell, 1972, Dailey, 1975, Kurochkin, 1975, Margolis & Dailey, 1972, Markowski, 1952, Stroud & Dailey, 1978.

Pathology

It would hardly be possible to give a complete account of all the diseases and pathological conditions from which seals may suffer. This is primarily a veterinarians job, and even then more is known about seals in captivity, and considerably less about sick seals in the wild, most of which die without their condition ever being seen. Much information, and many references are available in a paper by Ridgway *et al.* (1975a) and in another by the same author (1972) where he lists the diseases known in pinnipeds. This latter includes references to everything from abortion, hepatitis, aortitis, lymphosarcoma and blastomycosis to tuberculosis, osteomalacia and jaundice.

In captivity, conjunctivitis and corneal cloudiness may be induced by too much chlorine in the water; in the wild similar conditions occur as the result of a scratch while in a dirty rookery. Bacteria of many kinds exacerbate the pneumonia caused by lung parasites, or are responsible for skin lesions (Anderson *et al.*, 1974); septicaemia is known; leptospirosis caused by the protozoan spirochaete *Leptospira* has caused fever, nephritis, abortion and death in *Zalophus* (Vedros *et al.*, 1971) and other seals (Smith *et al.*, 1977), and another protozoan, *Sarcocystis*, has been found encysted in the tongue muscles of a Bearded Seal (Bishop, 1979), and in the masseter muscle of a Northern Fur Seal (Brown *et al.*, 1974). The seal pox virus attacks phocids and otariids both in the wild and in captivity, showing itself as nodules over the skin (Wilson & Poglayen-Neuwall, 1971, Hadlow *et al.*, 1980). A possible contagious virus disease that killed large numbers of Crabeater Seals has been mentioned in the section on that animal.

In captivity, lack of exercise may lead to fatty degeneration of the heart. While exposed to the public there is always the danger of seals swallowing foreign bodies – a ball that may block the intestine, or a stick that may perforate the stomach. Tuberculosis and fungal infections are believed to originate with the zoo visitor (Williamson *et al.*, 1959). An infection in reverse is 'seal finger' an infection of erysipelas regarded as an occupational disease by sealing men. Similar 'whale fingers' occur among whaling men (Skinner, 1957).

Tumours in various parts of the seal body have been recorded – on the adrenal, lymph and mammary glands, in the uterus (Mawdesley-Thomas, 1971, Mawdesley-Thomas & Bonner, 1971), on the pancreatic duct, a particularly large one that occupied most of the abdominal cavity in a Californian Sea Lion, and two other Californian Sea Lions with widespread visceral tumours (Brown *et al.*, 1980). Gastric ulcers are caused by nematodes as already mentioned, and thrombosis of pulmonary vessels is attributed to such parasites.

The excessive wear of Weddell Seal upper canines and incisors due to ice sawing, which may lead to abscesses, is mentioned in the section on that seal. In captivity, walruses will 'dig' in their concrete pools and thus wear their tusks down till they too may expose the pulp cavity to infection. The continued shocks of trying to dig in concrete, when translated to the root of the canine, may wear a hole in the maxilla and eventually lead to a sore on the outside of the head. Diseases and fusions of the vertebral column have been mentioned under the separate animals involved (*C. cristata*, *H. grypus*, *M. monachus*, *Z. californianus*).

Wounds may occur as the result of a number of circumstances, apart from such man-induced things as damage from ships propellors. Adult male otariids, for instance, challenge and fight each other, resulting in many lacerations to head, chest and flippers. *Neophoca* bulls are particularly intolerant of their young pups and will pick them up and toss them. A pup landing with the thin cranial region of its skull against a hard rock will frequently suffer lethal damage, and a bull's canines will easily puncture through to the lungs of a small pup (Marlow, 1975).

Killer whales will attack seals and may be responsible for some of the missing flippers. A pack of killer whales has been seen to attack a mass of Southern Sea Lions, but in this instance they concentrated on taking the ten-week-old pups – and more than twenty of them were taken in an hour (Wilson, 1975a). These whales are also capable of tossing a full grown *Otaria* into the air – possibly thus stunning it before killing it (Bartlett, 1976).

In appropriately warm waters sharks also take their toll of seals and a huge semicircular row of jagged tooth marks on the side of a seal leaves no doubt as to the attacker (Fig 16.1). Certainly many seals recover from some horrific wounds. Attacks by Leopard Seals are thought to cause many of the long parallel scars on Crabeater Seals (Siniff & Bengtson, 1977) and Leopard

Fig. 16.1 Sleeping male Zalophus *on the Ano Nuevo Island, California. The semicicular wound is probably due to shark attack. Photo. J. E. King.*

Seals themselves have succumbed to the barbed needlelike spines of stingrays which have worked themselves into a vital spot or set up abscesses. Such a stingray spine perforated the oesophagus of an Australian Fur Seal. It then went through the pericardium and the wall of the right ventricle, coming to lodge in the ventricle, setting up septicaemia and causing death (Obendorf & Presidente, 1978).

Crater wounds, sometimes with the central plug still in position, are known on fishes and cetaceans (Jones, 1971), and these wounds are believed to be made by the shark *Isistius brasiliensis*, which is known, for obvious reasons, as the cookie-cutter shark. Similar wounds, presumably inflicted by this shark, have been seen on a Leopard Seal and a Southern Elephant Seal that stranded in Sydney (E. A. Smith, pers. comm.).

Barnacles are sometimes found as passengers, usually on the dorsal surface of the seal. Three species of the goose barnacle *Lepas* have been recorded from seals – *L. pacifica* on *Mirounga angustirostris* (Baldridge, 1977), *L. australis* on *Arctocephalus gazella* (Bonner, 1968) and *Mirounga leonina* (Laws, 1953), and *L. hilli* on *Callorhinus ursinus* (Scheffer, 1962). A second genus *Conchoderma auritum* is also known from *M. leonina* (Best, 1971) and 'barnacles ?*Balanus*' are known from *P.v. richardsi* (Scheffer & Slipp, 1944). The barnacles must become attached while the seals are at sea for a considerable period, but will soon die when the seal spends a short time ashore. Green algae have been found on the hairs of the West Indian Monk Seal (Ward, 1887), Hawaiian Monk Seal (Kenyon & Rice, 1959), Northern Elephant Seal (Baldridge, 1977) and Grey Seal (Mackenzie, 1954); green and blue-green algae on the Harbour Seal (Scheffer & Slipp, 1944) and green, brown and red algae on the Northern Fur Seal (Scheffer, 1962). Both barnacles and algae are doubtless found on far more seals than those recorded here.

Pollution

Seals are not excluded from the pollution that man has inflicted on the world. Tissues of seals from the Arctic to the Antarctic have been found to contain organochlorine pesticides such as dieldrin and DDT. These substances are, of course, found in other members of the food chain – in amphipods and fish for example, and their concentrations are probably increased in the seals at the top of the food chain.

These organochlorine compounds, and also the industrial polychlorinated biphenyls (PCB) that are used in such things as paint and plastics, are soluble in fats, and are thus found in the fatty tissues of seals. High concentrations are found in the blubber, and low concentrations in the brain.

The physiological effects of these poisons in seals is not yet known – one cannot always be sure of the cause of death of a stranded animal, and one cannot easily watch for behavioural and other abnormalities in a wild animal. Studies on other animals suggest that there may be a toxic effect on the central nervous system and decreased resistance to infection. Some correlation between the increased production of premature pups by *Zalophus* has been suggested. Pesticide concentration was considerably higher in the cows and their dead premature pups than in the cows that produced normal pups, and reproductive failure from pesticides has been found in experimental rabbits (DeLong *et al.*, 1973).

Mercury has also been found in seal tissues, but as mercury is not soluble in fat the blubber has relatively little, and the highest concentrations are found in the liver kidney, spleen and brain. Toxic levels have been found, but again it is difficult to estimate effects. A Ringed Seal with high levels of mercury was found to have obvious symptoms of poisoning and had difficulty in co-ordinating its movements (Holden, 1975). It is interesting that the mercury in seals is not methylated to any great extent as it is in fish, and there is the possibility that the seals may be able to demethylate the mercury in the fish they eat (Holden, 1978). Other heavy metals such as zinc, copper, lead and cadmium have also been found in seal tissues, but there seems to be no evidence of abnormal levels.

The levels of pollutants in seals appears to bear a correlation with the degree of industrial development of the shores of the seas where the seals are found. The Baltic and the Gulf of St Lawrence, for instance, have particularly high levels of pollution, reflected in the tissues of the seals.

Oil spills from such notorious sources as the *Torrey Canyon* disaster off Land's End, England in March, 1967 (Bonner, 1972) and the Santa Barbara Channel (California) oil blow out in January 1969 (LeBouef,

1971a), have been blamed for causing much damage to seals, although the seals may have been as much affected by the detergent used to disperse the oil. Although many seals were coated in oil at these times, luckily no obvious serious effect has been proved. Oil-covered Grey Seal pups have been seen to continue suckling (Davies, 1949), but the oil could well impede swimming and indirectly increase mortality. Oily Grey Seal pups were significantly lower in weight than clean pups, but it is possible that this difference was due to disturbance by attempts to clean the animals (Davis & Anderson, 1976).

A review of pollutants in seals (Holden, 1978) concludes that although the presence of PCBs may be implicated in the high abortion rate of Baltic *P. hispida* and also *Zalophus,* there is still no conclusive evidence that pollutants, with the possible exception of mercury, are having an adverse effect on seals, though not enough is known about the subject yet.

SECTION III

Appendices

Origins of scientific names

It is interesting to know the derivations of scientific names. Most come from either Greek or Latin roots, and have frequently some reference to a distinguishing character, the locality, or perhaps the discoverer or collector of the animal. The list includes the names of recent seals, in alphabetical order.

ARCTOCEPHALUS
Greek αρκτος (arktos), a bear, and κεφαλή (kephale), head, from the bear-like appearance of the head.

A. australis
Latin australis, southern.

A. forsteri
Named after George Forster, assistant naturalist to his father on Cook's second voyage round the world 1772–75 in HMS Resolution. Seals were killed at Dusky Sound, South Island, New Zealand, a drawing made, and an account given of them in the narrative of the voyage, from which the animal was eventually named.

A. galapagoensis
A geographical reference.

A. gazella
First described from a specimen brought back from Kerguelen by the German vessel SMS Gazelle, which went out to observe the transit of Venus in 1874.

A. philippii
Named after Dr Rodolfo Amando Philippi, who, in 1864 when he was Director of the Natural History Museum in Santiago, Chile, collected the skull from which the animal was described.

A. pusillus
Latin pusillus, little. The first description was based on the picture of a young pup.

A.p. doriferus
Greek δόρα (dora), a skin or hide – the word applying only when the skin is removed for the fur. Latin fero I bear. A name applicable to a fur seal (Gotch, 1979).

A. townsendi
Named after Dr C. H. Townsend of the American Museum of Natural History, who collected the original specimen on a trip to Guadalupe in 1892.

A. tropicalis
A reference to the locality, as the first specimen given this name was said, mistakenly, to have come from the north coast of Australia. Greek tropikos – tropical.

CALLORHINUS
Greek καλός (kalos), beautiful, and ρινός (rhinos), the skin or hide (Gotch, 1979). This fur seal produces the best known commercial seal fur.

C. ursinus
Latin ursinus, bear-like. This, and all the fur seals were previously known as 'sea bears'.

CYSTOPHORA
Greek κύστις (kustis), a bladder, and φορος (phoros), carrying, referring to the inflatable nasal appendage.

C. cristata
Latin crista, a crest, also referring to the nasal appendage.

ERIGNATHUS
Greek ερι (eri), intensive prefix, + γνάθος (gnathos), jaw, referring to the rather deep jaw.

E. barbatus
Latin barba, a beard, referring to the abundant moustachial whiskers of this seal.

E.b. nauticus
Greek ναυτικός (nautikos), nautical.

EUMETOPIAS*
Greek ευ (eu), well, typical, and μετωπίας (metopias), having a broad forehead.

E. jubatus
Latin jubatus, having a mane, referring to the well-developed mane of the adult male.

HALICHOERUS
Greek αλιος (halios), of the sea, and χοιρος (choiros), a pig.

H. grypus
Greek γρυπός (grupos), hook-nosed, referring to the high 'Roman' nose of this seal.

---

* According to Liddell & Scott's Greek–English Lexicon, 'metopias' is masculine.

HISTRIOPHOCA

Latin *histrio*, a stage-player, + phoca. An allusion to the strikingly coloured coat.

HYDRURGA

Greek υδωǫ (udor), water, and (possibly) suffix ουǫγός (ourgos), a worker in, a description of its aquatic habits.

H. leptonyx

Greek λεπτός (leptos), small, slender, and ονυξ (onux), claw, as in *Leptonychotes*.

LEPTONYCHOTES

Greek λεπτός (leptos), small, slender, with ονυξ (onux), claw, + suffix οτης (otes), denoting possession, referring to the small size of the claws on the hind digits.

L. weddelli

Named after James Weddell who commanded the British sealing expedition 1822–24, during which trip he penetrated far south into the Weddell Sea, also named after him. The description of this seal was based on a drawing by Weddell, and skeletons from the South Orkney Islands were brought home from this voyage.

LOBODON

Greek λοβός (lobos), lobe, + οδούς (odous), tooth, referring to the extremely lobed character of the cheek teeth.

L. carcinophagus

Greek καǫκίνος (karkinos), a crab, + φαγειν (phagein), to eat, a mistaken reference to its diet.

MIROUNGA

From Miouroung, the Australian native name for the Elephant Seal.

M. angustirostris

Latin *angustus*, narrow, and *rostrum*, snout. This seal is said to differ from the Southern Elephant Seal in having a narrower snout.

M. leonina

Latin *leoninus*, lion-like, referring probably partly to size and partly to the roaring. Earlier, most large seals were called Sea Lions, but now the term is more restricted to the Otariid hair seals, the large males of which are also lion-like in having a mane.

MONACHUS

Greek μοναχός (monachos), a monk, probably referring to the cowl-like effect of the rolls of fat of the neck, seen particularly when the head is drawn back.

M. schauinslandi

Dr H. Schauinsland brought back from Laysan Island in 1899 the skull on which the original description of this seal was based in 1905.

M. tropicalis

A reference to its habitat.

NEOPHOCA

Greek νέος (neos), new, + phoca.

N. cinerea

Latin *cinereus*, ash coloured, referring to the greyish colour of the animal.

ODOBENUS

Greek οδους (odous), tooth, and βαίνω (baino), I walk, a reference to the supposed use of the long tusks in helping the animal to walk.

O. rosmarus

From the Norwegian *rossmaal* or *rossmaar*, based on various forms of earlier Scandinavian words for the walrus, e.g. rosval, rosm hval. These, and the word 'walrus' itself are connected with the Old English horsch-wael, the Swedish hvalross, the Icelandic hross-hvalr etc., all indicating 'whale-horse'. The walrus was frequently known as the 'seahorse' or 'morse', the latter name coming from the Lapland morssa, a descriptive word for the noise made by the animal.

O.r. divergens

Latin *di* + *vergo*, to turn in different directions, referring to the supposedly greater inward curvature of the tusks.

OMMATOPHOCA

Greek ομμα (omma), eye, + phoca, referring to the enormous orbits.

O. rossi

Named after Sir James Clark Ross, Commander of HMS *Erebus*, which together with HMS *Terror* composed the British Expedition to the Antarctic 1839–43. These ships were the first to force a way through the pack ice of the Ross Sea, and this seal was first described from two skeletons brought home from this voyage.

OTARIA

Greek ωτάǫιον (otarion), a little ear, referring to the small external ear.

O. byronia

Named after Commodore John Byron, in command of HMS *Dolphin*, which, with HMS *Tamar* went on a voyage of discovery to the South Seas in 1764, returning in 1766. The animal named from a skull brought home by Byron, probably from the Straits of Magellan.

PAGOPHILUS
Greek πάγος (pagos), ice, and φίλος (philos), loving, referring to the habits of the animal.

PHOCA
Greek φώκη (phoce), a seal, which is connected with the Sanskrit root sphâ – to swell up, and refers to the plumpness of the animal.

P. caspica
Comes from the Caspian Sea.

P. fasciata
Latin *fascia*, a band or ribbon, referring to the white band-like markings on the coat.

P. groenlandica
A geographical reference.

P. hispida
Latin *hispidus*, rough, bristly, referring to the coat of the adult.

P.h. botnica
A geographical reference to the Gulf of Bothnia.

P.h. krascheninikovi
A Russian surname.

P.h. ladogensis
Referring to Lake Ladoga.

P.h. ochotensis
The seal comes from the Okhotsk Sea.

P.h. saimensis
A geographical reference to Lake Saimaa.

P. largha
Larga is the Tungus vernacular for Harbour Seals of the western Okhotsk Sea.

P. sibirica
A geographical reference. The seal is from Lake Baikal.

P. vitulina
Latin *vitulus*, a calf. The Common Seal was often known as the sea-dog, or sea-calf.

P.v. concolor
Latin *concolor*, of the same colour. A reference to the fact that the original specimen was grey all over, without spots.

P.v. richardsi
Named after Captain G. H. Richards, Hydrographer to the Admiralty, and captain of HMS *Hecate*, on whose surveying voyage, 1859–64, the type skull was collected.

P.v. stejnegeri
Named after Dr L. Stejneger who collected the original specimens on the Commander Islands in April, 1883.

PHOCARCTOS
From Phoca, and the Greek αρκτος (arktos), a bear, referring to the slightly bear-like skull.

P. hookeri
Named after Sir Joseph Hooker, the botanist with the British Expedition to the Antarctic 1839–43 – the ships *Erebus* and *Terror* by which the original specimens were brought home.

PINNIPED
Latin *pinna* (or *penna*), a feather, possibly from the root of the Greek πέτομαι, I fly, + ped, Latin *pes* (genit. *pedis*), the foot; winged, or finny-footed.

PUSA
Derived from the general term used for seals by the Greenlanders, or probably from the Greenland name – puirse – for the Harp Seal.

SEAL
From the Old English *seolh*.

ZALOPHUS
Greek ζά- (za), intensive prefix, and λόφος (lophos), crest, because of the extremely high sagittal crest on skulls of adult males.

Z. californianus
A geographical reference to the locality.

Z.c. japonicus
The Japanese race.

Z.c. wollebaeki
Named after Dr Alf Wollebaek who collected, in 1925, the skull from which the animal was described.

Geographical index

The numbers following the geographical locations correspond to those seals listed below that are found in these regions. The text should, however, be consulted for details of seal distribution, as some seals are found only rarely in these locations. e.g. Crozet Is, 10, 34, indicates that 10–*Arctocephalus tropicalis* and 34–*Mirounga leonina* occur on the Crozet Is.

KEY

 1 *Eumetopias*
 2 *Zalophus*
 3 *Otaria*
 4 *Neophoca*
 5 *Phocarctos*
 6 *Arctocephalus townsendi*
 7 *A. galapagoensis*
 8 *A. philippii*
 9 *A. australis*
10 *A. tropicalis*
11 *A. gazella*
12 *A. pusillus pusillus*
13 *A. p. doriferus*
14 *A. forsteri*
15 *Callorhinus*
16 *Odobenus*
17 *Halichoerus*
18 *Phoca vitulina*
19 *P. largha*
20 *P. hispida*
21 *P. caspica*
22 *P. sibirica*
23 *P. groenlandica*
24 *P. fasciata*
25 *Cystophora*
26 *Erignathus*
27 *Monachus monachus*
28 *M. tropicalis*
29 *M. schauinslandi*
30 *Leptonychotes*
31 *Ommatophoca*
32 *Lobodon*
33 *Hydrurga*
34 *Mirounga leonina*
35 *M. angustirostris*

Aegean Sea, 27
Alaska, 1, 20, 26, 18, 19, 15, 35, 16
Aleutian Is., 1, 18, 19, 24, 15, 16
Algeria, 27
Algoa Bay, 12
Amirantes, 34
Amsterdam I., 10, 34
Anacapa, 35
Angola, 34
Ano Nuevo I., 1, 2, 18, 35
Antarctic, 30, 32, 33, 31, 34
Antipodes Is., 14
Arctic, 25, 26, 20, 16
Arctic coasts, 20, 16
Argentina, 3, 9, 34
Auckland Is., 5, 33, 34, 14
Australia, 4, 33, 13, 14, 34
Azores, 27

Baffin I., 18, 20, 23, 25, 16
Bahama I., 28
Baja California, 18, 2, 6, 35
Balearic Is., 27
Ballestas Is., 3
Baltic, 17, 18, 20
Baranof Id., 1
Barents Sea, 26, 16
Basque Is., 17
Bass Strait, 4, 13, 14, 34
Bay of Fundy, 17, 18
Beaufort Sea, 26
Bering Sea, 1, 18, 19, 20, 24, 26, 15, 16
Bird I., 11
Black Sea, 27
Bounty Is., 14
Bouvet I., 11, 34
Bras d'Or Lakes, 17, 18
Brazil, 10
Bristol Bay, 18, 19, 16
British Columbia, 1, 18, 15, 35
Brittany, 17

California, 1, 2, 18, 6, 15, 35
Campbell I., 5, 33, 34, 14
Campeche, 28
Canada, 1, 16, 17, 18, 20, 23, 25, 2, 26, 15, 35
Canadian Arctic, 26, 16
Canary Is., 27
Cap Blanc, 27
Cape Breton I., 17

Cape Chelyuskin, 23
Cape Cross, 12
Caribbean Sea, 28
Caspian Sea, 21
Cedros I., 2, 18, 6, 35
Chagos Archipel., 34
Chatham I., 14, 34
Chicacof I., 1
Chile, 3, 9
Chincha Is., 3
Chukchi Sea, 19, 24, 26, 16
Coetivy, 34
Commander Is., 1, 18, 19, 20, 24, 15, 16
Cornwall, 17
Corsica, 27
Crete, 27
Crozet Is., 10, 34
Cyprus, 27

Dangerous Reef, 4
Davis Strait, 25
Denmark, 17, 18
Denmark Strait, 25
Desertas Is., 27
Dodecanese Is., 27
Dundas I., 5
Dyer I., 12

Egypt, 27
Enderby Id., 5
Europe, 20, 16
Elephant Rock, 12
Escondida Id., 9

Faeroes, 17
Falkland Is., 3, 9, 34
False Bay, 12
Farne Is., 17
Florida, 18
Foxe Basin, 26, 16
France, 17, 18, 25
Franz Josef Land, 23, 16
French Frigates Shoal, 29

Galapagos Is., 2, 3, 7
Germany, 18, 16
Graham Land, 32
Great Britain, 17, 18, 20, 23, 25, 26, 16
Greece, 27
Greenland, 18, 20, 23, 25, 26, 16

References

ABBOTT, G. J. & NAIRN, N. B. (eds) 1969. *Economic growth of Australia 1788–1821*. Melbourne University Press, Melbourne.

ABBOTT, I. 1979. The past and present distribution and status of sea lions and fur seals in Western Australia. *Rec. West. Aust. Mus.* **7** (4): 375–390.

ABE, H., HASEGAWA, Y. & WADA, K. 1977. A note on the air sac of ribbon seal. *Sci. Rep. Whales Res. Inst. Tokyo* **29**: 129–135.

AGUAYO, A., 1970. Census of Pinnipedia in the South Shetland Islands. *Antarctic ecology*. Ed. M. W. Holdgate. **1**:395–397.

AGUAYO, A. 1973. The Juan Fernandez fur seal. pp. 140–143 in *Seals. IUCN Suppl. Paper* **No. 39**.

AGUAYO, A. 1978. The present status of the Antarctic fur seal *Arctocephalus gazella* at South Shetland Islands. *Polar Record* **19** (119): 167–176.

AGUAYO, A., MATURANA, R. & TORRES, D. 1971. El lobo fino de Juan Fernandez. *Rev. Biol. Mar. Valparaiso* **14** (3): 135–149.

ALBRECHT, C. B. 1950. Toxicity of sea water in mammals. *Amer. J. Physiol.* **163**: 370–385.

ALLEN, G. M. 1928. The walrus in New England. *Bull. Boston Soc. Nat. Hist.* **47**: 10–12.

AMOROSO, E. C., BOURNE, G. H., HARRISON, R. J., MATTHEWS, L. H., ROWLANDS, I. W. & SLOPER, J. C. 1965. Reproductive and endocrine organs of foetal, newborn and adult seals, *J. Zool. Lond.* **147**: 430–486.

AMOROSO, E. C. & MATHEWS, J. H. 1951. The growth of the grey seal (*Halichoerus grypus* (Fabricius)) from birth to weaning. *J. Anat.* **85**: 427–428.

ANDERSON, S. S. 1977. The grey seal in Wales. *Nature in Wales* **15** (3): 114–123.

ANDERSON, S. S. 1981. Seals in Shetland waters. *Proc. Roy. Soc. Edinburgh.* **80B**: 181–188.

ANDERSON, S. & KNOX JONES, J. (eds) 1967. *Recent mammals of the world. A synopsis of families.* Ronald Press Co., New York.

ANDERSON, S. S., BAKER, J. R., PRIME, J. H. & BAIRD, A. 1979. Mortality in grey seal pups: incidence and causes. *J. Zool., Lond.* **189**: 407–417.

ANDERSON, S. S., BONNER, W. N., BAKER, J. R. & RICHARDS, R. 1974. Grey seals, *Halichoerus grypus* of the Dee Estuary, and observations and characteristic skin lesion in British seals. *J. Zool., Lond.* **174** (3): 429–440.

ANDERSON, S. S., BURTON, R. W. & SUMMERS, C. F. 1975. Behaviour of grey seals (*Halichoerus grypus*) during a breeding season at North Rona. *J. Zool., Lond.* **177**: 179–195.

ANDREWS, J. C. & MOTT, P. R. 1967.

Gray seals at Nantucket, Massachusetts. *J. Mamm.* **48** (4): 657–658.

ANGAS, G. F. 1847. *Savage life and scenes in Australia and New Zealand*. Vols. 1 & 2. London.

ANGOT, M. 1954. Observations sur les mammifères marins de l'Archipel de Kerguelen. *Mammalia,* Paris. **18(1)**: 1–111.

ANON. 1980. Notes in *Oryx* **15** (3).

ANTHONY, R., LIOUVILLE, J. 1920. Les caractères d'adaptation du rein du Phoque de Ross (*Ommatophoca rossi* Gray) aux conditions de la vie aquatique. *C. R. Acad. Sci. Paris* **171**: 318–320.

ARNASON, U. 1974. Comparative chromosome studies in Pinnipedia. *Hereditas* **76**: 179–226.

ARVY, L. & HIDDEN, G. 1973. Les caracteristiques rénales et réniculaires de *Mirounga leonina. C. R. Acad. Sci. Paris.* ser. D. **277** (24): 2713–2714.

ASHWORTH, V. S., RAMAIAH, G. D. & KEYES, M. C. 1966. Species difference in the composition of milk with special reference to the northern fur seal. *J. Dairy Sci.* **49**: 1206–1211.

ASS, M. I. 1935. Zur Kenntnis der Ektoparasiten der Flossenfüsser (Pinnipedia). Eine neue Zeckenart auf dem Walross. *Zeit f. Parasitenkunde* **7**: 601–607

AUSTIN, K. A. 1964. *The voyage of the* Investigator *1801–1803, Comdr. Matthew Flinders R.N.* Seal Books. Rigby Ltd, Adelaide.

BACKHOUSE, K. M. 1961. Locomotion of seals with particular reference to the forelimb. *Symp. Zool. Soc. Lond.* No. **5**: 59–75.

BACKHOUSE, K. M. & HEWER, H. R. 1957. A note on spring pupping in the grey seal. *Proc. Zool. Soc. Lond.* **128** (4): 593–596.

BALAZA, G. H. & WHITTOW, G. C. 1978. Bibliography of the Hawaiian monk seal *Monachus schauinslandi* Matschie 1905. *Hawaii Inst. Mar. Biol. Univ. Hawaii Tech. Rep.* **35**: 27pp.

BALDRIDGE, A. 1977. The barnacle *Lepas pacifica* and the alga *Navicula grevillei* on northern elephant seals *Mirounga angustirostris. J. Mamm.* **53** (3): 428–429.

BALL, G. H. 1930. An acanthocephalan, *Corynosoma strumosum* (Rudolphi) from the Californian harbour seal. *Univ. Calif. Pub. Zool.* **33**: 301–305.

BARNES, L. G. 1972. Miocene Desmatophocinae (Mammalia, Carnivora) from California. *Univ. Calif. Pub. in Geol. Sci.* **89**: 68pp.

BARNETT, C. H., HARRISON, R. J. & TOMLINSON, J. D. W. 1958. Variations in the venous systems of mammals. *Biol. Rev.* **33**: 442–487.

BARTHOLOMEW, G. A. 1952. Reproductive and social behaviour of the northern elephant seal. *Univ. Calif. Pub. Zool.* **47** (15): 369–472.

BARTHOLOMEW, G. A. 1959. Mother-young relations

and the maturation of pup behaviour in the Alaska fur seal. *Anim. Behav.* **7**: 163–171.

BARTHOLOMEW, G. A. 1967. Seal and sea lion populations of the California islands. *Proc. Symp. on Biol. of Calif. Islands*: 229–244.

BARTHOLOMEW, G. A. & HUBBS, C. L. 1960. Population growth and seasonal movements of the northern elephant seal, *Mirounga angustirostris. Mammalia*, Paris. **24**: 313–324.

BARTHOLOMEW, G. A. & WILKE, F. 1956. Body temperature in the northern fur seal *Callorhinus ursinus. J. Mamm.* **37**: 327–337.

BARTLETT, D. & J. 1976. Patagonia's wild shore. *Nat. Geogr. Mag.* **149** (3): 312–317.

BECK, B., SMITH, T. G. & MANSFIELD, A. W. 1970. Occurrence of the harbour seal *Phoca vitulina* Linnaeus in the Thlewiaza River, N. W. T. *Can. Field Nat.* **84**: 297–300.

BENJAMINSEN, T. 1973. Age determination and the growth and age distribution from cementum growth layers of bearded seals at Svalbard. *Fisk. Div. Skr. Ser. HavUnders.* **16**: 159–170.

BENJAMINSEN, T., BERGFLØDT, B., HUSE, I., BRODIE, P. & TOKLUM, K. 1977. Grey seal investigations on the Norwegian coast from Lofoten to Frøya, September–November 1976. *Fiskerinaeringens Forsøksfond. Rapporter* nr. **1**: 24–33.

BERLAND, B. 1966. The hood and its extrusible balloon in the hooded seal *Cystophora cristata* Erxl. *Norsk Polarinstitutt* **1965**: 95–102.

BERRY, J. A. & KING, J. E. 1970. The identity of the Pliocene seal from Cape Kidnappers, New Zealand, previously known as *Arctocephalus caninus. Tuatara* **18** (1): 13–18.

BERTRAM, G. C. L. 1940. The biology of the Weddell and Crabeater seals. *Brit. Graham Land Exped. 1934–37. Sci. Rep.* **1** (1): 1–139.

BEST, P. B. 1971. Stalked barnacles *Conchoderma auritum* on an elephant seal: occurrence of elephant seals on South African coast. *Zool. Africana* **6** (2): 181–185.

BESTER, M. N. 1975. The functional morphology of the kidney of the Cape fur seal *Arctocephalus pusillus* (Schreber). *Madoqua ser. II* **4** (74–80): 69–92.

BESTER, M. N. 1980. Population increase in the Amsterdam Island fur seal *Arctocephalus tropicalis* at Gough Island. *S. Afr. J. Zool.* **15** (4): 229–234.

BIGG, M. A. 1969. The harbour seal in British Columbia. *Fish Res. Bd. Canada Bull.* **172**: 1–33.

BIGG, M. A. 1973. Census of California sea lions on southern Vancouver Island, British Columbia. *J. Mamm.* **54** (1): 285–287.

BISHOP, L. 1979. Parasite-related lesions in a bearded seal, *Erignathus barbatus. J. Wildl. Dis.* **15**: 285–293.

BJØRGE, A. J. 1979. An isopod as intermediate host of codworm. *Fisk. Dir. Skr. Ser. HavUnders.* **16**: 561–565.

BLIX, A. S., GRAV, H. J. & RONALD, K. 1975. Brown adipose tissue and the significance of the venous plexuses in pinnipeds. *Acta Physiol. Scand.* **94**: 133–135.

BLIX, A. S., GRAV, H. J. & RONALD, K. 1979. Some aspects of temperature regulation in newborn harp seal pups. *Amer. J. Physiol.* **236** (3): R188–R197.

BONHAM, K. 1943. Duration of life and behaviour of Alaska fur seals in captivity. *J. Mamm.* **24**: 504.

BONNER, W. N. 1968. The fur seal of South Georgia. *Brit. Ant. Survey Sci. Rep.* **56**: 81pp.

BONNER, W. N. 1971. An aged grey seal (*Halichoerus grypus*). In Notes from the Mammal Society No. 22. *J. Zool., Lond.* **164**: 261–262.

BONNER, W. N. 1972. The grey seal and common seal in European waters. *Oceanogr. Mar. Biol. Ann. Rev.* **10**: 461–507.

BONNER, W. N. 1973. Grey seals in the Baltic. pp. 164–174 in *Seals. IUCN Suppl. Paper No. 39.*

BONNER, W. N. 1976. The stocks of grey seals (*Halichoerus grypus*) and common seals (*Phoca vitulina*) in Great Britain. 16pp. *Nat. Env. Res. Council Pub. Ser. C. No. 16.*

BONNER, W. N. 1979a. Harbour (Common) Seal. pp. 58–62 in *Mammals in the seas* 2 FAO Fisheries Series No. 5.

BONNER, W. N. 1979b. Largha Seal. pp. 63–65 in *Mammals in the seas.* 2 FAO Fisheries Series No. 5.

BONNER, W. N. & HICKLING, G. 1971. The grey seals of the Farne Islands. Report for the period October 1969 to July 1971. *Trans. Nat. Hist. Soc. Northumb.* **17** (4): 141–162.

BONNER, W. N. & HICKLING, G. 1974. The grey seals of the Farne Islands 1971 to 1973. *Trans. Nat. Hist. Soc. Northumb.* **42** (2): 65–84.

BOSHIER, D. P. 1979. Electron microscopic studies on the endometrium of the grey seal (*Halichoerus grypus*) during its preparation for nidation. *J. Anat.* **128** (4): 721–735.

BOSWALL, J. 1972. The South American sea lion *Otaria byronia* as a predator on penguins. *Bull. Brit. Ornith. Club* **92** (5): 129–131.

BOULVA, J. & MCLAREN, I. A. 1979. Biology of the harbor seal, *Phoca vitulina* in Eastern Canada. *Bull. Fish. Res. Bd. Canada Bull.* **200**: 24pp.

BOURNE, G. H. 1949. *The mammalian adrenal gland.* 239pp. Clarendon Press, Oxford.

BOYD, R. B. 1975. A gross and microscopic study of the respiratory anatomy of the Antarctic weddell seal, *Leptonychotes weddelli. J. Morph.* **147**: 309–336.

BOYD, J. M. & CAMPBELL, R. N. 1971. The grey seal (*Halichoerus grypus*) at North Rona, 1959 to 1968. *J. Zool., Lond.* **164**: 469–512.

BRAHAM, H. W., EVERITT, R. D. & RUGH, D. J. 1980. Northern sea lion population decline in the Eastern Aleutian Islands. *J. Wildl. Mgt.* **44** (1): 25–33.

BRAZIER HOWELL, A. 1929. Contribution to the comparative anatomy of the eared and earless seals (genera *Zalophus* and *Phoca*). *Proc. U.S. Nat. Mus.* **73** (15): 1–142.

BREE, P. J. H. VAN 1972. On a luxation of the skull–atlas joint in a grey seal, *Halichoerus grypus* (Fabricius 1791), with notes on other grey seals from the Netherlands. *Zool. Med.* **47**: 331–336.

BREE, P. J. H. VAN 1977. On a walrus which recently visited the coast of the Netherlands and Belgium. *De Levende Natuur.* No. 3, March 1977: 58–62.

BRIEN, Y. 1974. La reproduction du phoque gris *Halichoerus grypus* Fabricius en Bretagne. *Mammalia*, Paris. **38** (2): 346–347.

BRIGGS, K. T. 1974. Dentition of the northern elephant seal. *J. Mamm.* **55** (1): 158–171.

BRIGGS, K. T. & MOREJOHN, G. V. 1975. Sexual dimorphism in the mandibles and canine teeth of the northern elephant seal. *J. Mamm.* **56** (1): 224–231.

BRIGGS, K. T. & MOREJOHN, G. V. 1976. Dentition, cranial morphology and evolution in elephant seals. *Mammalia*, Paris. **40** (2): 199–222.

BROOKS, J. W. 1954. A contribution to the life history and ecology of the Pacific walrus. 103pp. *Special Report No. 1. Alaska Cooperative Wildlife Research Unit.*

BROSSET, A. 1963. Statut actuel des mammifères des îles Galapagos. *Mammalia*. Paris **27**: 323–338.

BROWN, D. H. & ASPER, E. D. 1966. Further observations on the Pacific walrus, *Odobenus rosmarus divergens*, in captivity. *Int. Zoo Yearbook* **6**: 78–82.

BROWN, D. L. 1974. Vocal communication of the New Zealand fur seal at Open Bay Islands, Westland. *N. Z. Min. Agric. Fish. Tech. Rep.* **130**: 1–15.

BROWN, K. G. 1952. Observations on the newly born leopard seal. *Nature* **170**: 982–983.

BROWN, R. J., SMITH, A. W. & KEYES, M. C. 1974. Sarcocystis in the northern fur seal. *J. Wildl. Dis.* **10**: 53.

BROWN, R. J., SMITH, A. W., MOREJOHN, G. V. & DELONG, R. L. 1980. Metastatic adenocarcinoma in two Californian sea lions, *Zalophus c. californianus. J. Wildl. Dis.* **16** (2): 261–266.

BROWNELL, R. L., DELONG, R. L. & SCHREIBER, R. W. 1974. Pinniped populations at Islas de Guadalupe, San Benito, Cedros and Natividad, Baja California in 1968. *J. Mamm.* **55** (2): 469–472.

BRUEMMER, F. 1977. The gregarious but contentious walrus. *Nat. Hist. N.Y.* **86** (9): 52–61.

BRUN, E., LID, G. & LUND, H. M.-K. 1968. Hvalross, *Odobenus rosmarus*, på norskekysten. *Fauna* **21**: 7–20.

BRYDEN, M. M. 1967. Testicular temperature in the southern elephant seal. *J. Reprod. Fert.* **13**: 583–584.

BRYDEN, M. M. 1968. Lactation and suckling in relation to early growth of the southern elephant seal, *Mirounga leonina* L. *Aust. J. Zool.* **16**: 739–748.

BRYDEN, M. M. 1971a. Myology of the southern elephant seal, *Mirounga leonina. Antarctic Pinnipedia. Am. Geophys Union. Ant. Res. Series* **18**: 109–140.

BRYDEN, M. M. 1971b. Size and growth of viscera in the southern elephant seal *Mirounga leonina* (L). *J. Anat.* **116**: 121–133.

BRYDEN, M. M. 1973. Growth patterns of individual muscles of the elephant seal, *Mirounga leonina* (L). *J. Anat.* **116** (1): 121–133.

BRYDEN, M. M. 1978. Arteriovenous anastomoses in the skin of seals. III. The harp seal *Pagophilus groenlandicus* and the hooded seal *Cystophora cristata. Aquatic Mammals* **6** (3): 67–75.

BRYDEN, M. M. & ERIKSON, A. W. 1976. Body size and composition of crabeater seals (*Lobodon carcinophagus*) with observations on tissue and organ size in Ross seals (*Ommatophoca rossi*). *J. Zool., Lond.* **179**: 235–247.

BRYDEN, M. M. & FELTS, W. J. L. 1974. Quantitative anatomical observations on the skeletal and muscular systems of four species of Antarctic seals. *J. Anat.* **118** (3): 589–600.

BRYDEN, M. M. & MOLYNEUX, G. S. 1978. Arteriovenous anastomoses in the skin of seals. II. The Californian sea lion *Zalophus californianus* and the northern fur seal *Callorhinus ursinus. Anat. Rec.* **191** (2): 253–260.

BUDD, G. M. 1970. Rapid population increase in the Kerguelen fur seal, *Arctocephalus tropicalis gazella* at Heard Island. *Mammalia*, Paris. **34** (3): 410–414.

BUDD, G. M. 1972. Breeding of the Fur Seal at McDonald Islands, and further population growth at Heard Island. *Mammalia*. Paris. **36** (3): 423–427.

BUDD, G. M. & DOWNES, M. C. 1969. Population increase and breeding in the Kerguelen fur seal, *Arctocephalus tropicalis gazella*, at Heard Island. *Mammalia*, Paris **33** (1): 58–67.

BURNE, R. H. 1909. Notes on the viscera of a walrus (*Odobaenus rosmarus*) *Proc. Zool. Soc. Lond.* 732–738.

BURNS, J. J. 1981. Ribbon seal *Phoca fasciata* Zimmerman, 1783 pp. 89–109 in *Handbook of marine mammals.* Eds. Ridgway, S. H. & Harrison, R. J. 1 Academic Press, London.

BURNS, J. J. & FAY, F. H. 1970. Comparative morphology of the skull of the ribbon seal, *Histriophoca fasciata*, with remarks on systematics of Phocidae. *J. Zool., Lond.* **161**: 363–394.

BURNS, J. J. & FAY, F. H. 1972. *Comparative biology of Bering Sea harbour seal populations.* Science in Alaska. Proc. 23rd Alaska Science Conf. Fairbanks, Alaska. Alaska Div. Amer. Assoc. for Adv. Sci. (abstract).

BURNS, J. J. & FROST, K. J. 1979. *The natural history and ecology of the bearded seal,* Erignathus barbatus. Final report outer continental shelf environmental assessment. 77pp. Program Contract 02.5.022.53. Alaska Dept. Fish & Game. Fairbanks.

BURNS, J. J., RAY, G. C., FAY, F. H. & SHAUGHNESSY, P. D. 1972. Adoption of a strange pup by the ice-inhabiting harbor seal *Phoca vitulina largha. J. Mamm.* **53** (3): 594–598.

BURTON, R. W., ANDERSON, S. S. & SUMMERS, C. F. 1975. Perinatal activities in the grey seal (*Halichoerus grypus*). *J. Zool., Lond.* **177**: 197–201.

BYCHKOV, V. A. 1973a. Atlantic walrus, *Odobenus rosmarus rosmarus* L. Novaya Zemlya population. pp. 56–58 in *Seals IUCN Suppl. Paper No. 39*.

BYCHKOV, V. A. 1973b. The Laptev walrus, *Odobenus rosmarus laptevi* Chapskii 1940. pp. 54–55 in *Seals IUCN Suppl. Paper No. 39*.

BYCHKOV, V. A. & ANTONIUK, A. A. 1975. Ladoga seal and the problems of its conservation. *Scientific Foundations of Nature Conservation*. **3**: 255–267.

CAMERON, A. W. 1967. Breeding behaviour in a colony of western Atlantic gray seals. *Can. J. Zool.* **45**: 161–173.

CAMERON, A. W. 1969. The behaviour of adult gray seals (*Halichoerus grypus*) in the early stages of the breeding season. *Canad. J. Zool.* **47** (2): 229–234.

CAMERON, A. W. 1970. Seasonal movements and diurnal activity rhythms of the grey seal (*Halichoerus grypus*). *J. Zool., Lond.* **161**: 15–23.

CANNATA, M. A. & TRAMEZZANI, J. H. 1971. Neurohypophysis of the Weddell seal; an electron microscope study. *J. Anat.* **108** (1): 185–195.

CARRARA, I. S. 1954. *Observaciones sobre el estado actual de las poblaciones de pinnipedos de la Argentina.* 17pp. Univ. Nac. Eva Peron. Fac. Cienc. Vet.

CARRICK, R. & INGHAM, S. E. 1962. Studies on the southern elephant seal *Mirounga leonina* (L.) II Canine tooth structure in relation to function and age determination. *CSIRO Wildl. Res.* **7** (2): 102–118.

CASTELLO, H. P. & PINEDO, M. C. 1977. *Arctocephalus tropicalis*, first record for Rio Grande do Sul coast. *Atlantica, Rio Grande* **2** (2): 111–119.

CAVE, A. J. E. & KING, J. E. 1964. The ossiculum mastoideum of the otariid skull. *Ann. Mag. N.H.* ser 13, **7**: 235–240.

CHAPSKY, K. K. 1936. The walrus of the Kara Sea. *Trans. Arctic Inst. Leningrad* **67**: 124pp.

CLARKE, R. 1954. Whales and seals as resources of the sea. *Norsk Hvalfangst-tidende* **9**: 489–508.

CLARK, T. W. 1979. Galapagos fur seal. pp. 31–33 in *Mammals in the seas* **2** FAO Fisheries Series No. 5.

CLAUSEN, G. & ERSLAND, A. 1969. The respiratory properties of the blood of the bladdernose seal (*Cystophora cristata*). *Resp. Physiol.* **7** (1): 1–6.

CLELAND, J. B. & SOUTHCOTT, R. V. 1969a. Illnesses following the eating of seal liver in Australian waters. *Med. J. Aust. 1969* **1**: 760–763.

CLELAND, J. B. & SOUTHCOTT, R. V. 1969b. Hypervitaminosis A in the Antarctic in the Australasian Antarctic expedition of 1911–1914: a possible explanation of the illnesses of Mertz and Mawson. *Med. J. Aust. 1969* **1**: 1337–1342.

COBB, W. M. 1933. The dentition of the walrus *Odobenus obesus*. *Proc. Zool. Soc. Lond.* 645–668.

CONDY, P. R. 1978. Distribution , abundance and annual cycle of fur seals (*Arctocephalus* spp.) on the Prince Edward Islands. *S. Afr. J. Wildl. Res.* **8**: 159–168.

COOK, H. W. & BAKER, B. E. 1969. Seal milk. I Harp seal (*Pagophilus groenlandicus*) milk, composition and pesticide residue content. *Can. J. Zool.* **47** (6): 1129–1132.

COTT, H. B. 1961. Scientific results of an inquiry into the ecology and economic status of the Nile crocodile (*Crocodilus niloticus*) in Uganda and Northern Rhodesia. *Trans. Zool. Soc. Lond.* **29** (4): 211–356.

CRANDALL, L. S. 1964. *The management of wild mammals in captivity*. 769pp. Univ. Chicago Press, Chicago.

CRAWLEY, M. C. 1975. Growth of New Zealand fur seal pups. *N.Z. J. Mar. F.W. Res.* **9** (4): 539–545.

CRAWLEY, M. C. & CAMERON, D. B. 1972. New Zealand sea lions *Phocarctos hookeri* on the Snares Islands. *N.Z. J. Mar. F.W. Res.* **6** (1 and 2): 127–132.

CRAWLEY, M. C. & WILSON, G. J. 1976. The natural history and behaviour of the New Zealand fur seal (*Arctocephalus forsteri*). *Tuatara* **22** (1): 1–29.

CSORDAS, S. E. 1962. The Kerguelen fur seal on Macquarie Island. *Vict. Nat.* **79**: 226–229.

CSORDAS, S. E. 1964. Wandering elephant seal. *Vict. Nat.* **80**: 336–338.

CSORDAS, S. E. & INGHAM, S. E. 1965. The New Zealand fur seal *Arctocephalus forsteri* (Lesson) at Macquarie Island, 1949–64. *CSIRO Wildl. Res.* **10**: 83–99.

CUELLO, A. C. 1968. Relationship between the pars intermedia and the pars nervosa in the hypophysis of an Antarctic seal. *Experientia (Switzerland)* **24** (4): 399–400.

CUELLO, A. C. 1970. The glandular pattern of the epiphysis cerebri of the Weddell seal. *Antarctic Ecology* **1**: 483–489.

CUELLO, A. C. & TRAMEZZANI, J. H. 1969. The epiphysis cerebri of the Weddell seal, its remarkable size and glandular pattern. *Gen. & Compar. Endocr.* **12**: 154–164.

CUMPSTON, J. S. 1969. *Macquarie Island*. 380pp. Canberra Ant. Div. Dept. Ext. Affairs.

CURRY-LINDAHL, K. 1970. Breeding biology of the Baltic grey seal (*Halichoerus grypus*). *Zool. Gart. Leipzig* **38**: 16–29.

DAILEY, M. D. 1970. The transmission of *Parafilaroides decorus* (Nematoda: Metastrongyloidea) in the California sea lion (*Zalophus californianus*). *Proc. Helminth Soc. Wash.* **37**(2): 215–222.

DAILEY, M. D. 1975. The distribution and intraspecific variation of helminth parasites in pinnipeds. *Rapp. P-v. Réun. Cons. int. Explor. Mer.* **169**: 338–352.

DAILEY, M. D. & HILL, B. L. 1970. A survey of metazoan parasites infecting the California (*Zalophus californianus*) and Steller (*Eumetopias jubatus*) sea lions. *Bull. So. Calif. Acad. Sci.* **69** (3–4): 126–132.

DAILEY, M. D. & BROWNELL, R. L. 1972. A checklist of marine mammal parasites. pp. 528–589 in *Mammals of the sea. Biology and medicine*.

Ed. S. H. Ridgway. Charles Thomas, Springfield, Ill.

DAVIES, J. L. 1949. Observations on the grey seal (*Halichoerus grypus*) at Ramsey Island, Pembrokeshire. *Proc. Zool. Soc. Lond.* **119** (3): 673–692.

DAVIS, J. E. & ANDERSON, S. S. 1976. Effects of oil pollution on breeding grey seals. *Marine Pollution Bull.* **7** (6): 115–118.

DAVYDOV, A. F. & MAKAROVA, A. R. 1965. Changes in the temperature of the skin of the harp seal during ontogenesis as related to the degree of cooling. *Morskie Mlekopitayushchie, Akademiya Nauk SSSR*: 262–265. (Translation by Trans-Bureau, Foreign Languages Division, Canada)

DE GRAAF, A. S., SHAUGHNESSY, P. D., MCCULLY, R. M. & VERSTER, A. 1980. Occurrence of *Taenia solium* in a Cape fur seal (*Arctocephalus pusillus*). *Onderstepoort J. Vet. Res.* **47**: 119–120.

DELONG, R. L., GILMARTIN, W. G. & SIMPSON, J. G. 1973. Premature births in California sea lions: association with high organochlorine pollutant residue levels. *Science* **181**: 1168–1170.

DEMASTER, D. P. 1979. Weddell seal. pp. 130–134 in *Mammals in the seas* **2** FAO Fisheries Series No. 5.

DENISON, D. M. & KOOYMAN, G. L. 1973. The structure and function of the small airways in pinniped and sea otter lungs. *Resp. Physiol.* **17** (1): 1–10.

DEPOCAS, F., HART, S. J. & FISHER, H. D. 1971. Seawater drinking and water flux in starved and in fed harbor seals, *Phoca vitulina. Can. J. Physiol. Pharmacol.* **49** (1): 53–62.

DESPIN, B., MOUGIN, J. L. & SEGONSAC, M. 1972. Oiseaux et mammifères de l'Ile de l'Est, Archipel Crozet (47°25′S, 52°12′E). *C.N.F.R.A.* **31**: 1–106.

DIVINYI, C. A. 1971. Growth and movements of a known-age harbor seal. *J. Mamm.* **52** (4): 824.

DORMER, K. J., DENN, M. J. & STONE, H. L. 1977. Cerebral blood flow in the sea lion (*Zalophus californianus*) during voluntary dives. *Comp. Biochem. Physiol.* **58** (A): 11–18.

DOUTT, J. K. 1942. A review of the genus *Phoca. Ann. Carnegie Mus.* **29**: 61–125.

DRABEK, C. M. 1975. Some anatomical aspects of the cardiovascular system of Antarctic seals and their possible functional significance in diving. *J. Morph.* **145** (1): 85–106.

DRABEK, C. M. 1977. Some anatomical and functional aspects of seal hearts and aortae. pp. 217–234 in *Functional anatomy of marine mammals* **3**. Ed. R. J. Harrison. Academic Press, London.

DRAGERT, J., CORY, S. & RONALD, K. 1975. Anatomical aspects of the kidney of the harp seal *Pagophilus groenlandicus* (Erxleben, 1777). *Rapp. P-v. Réun. Cons. int. Explor. Mer.* **169**: 133–140.

DUNN, J. L. & WOLKE, R. E. 1976. *Dipetalonema spirocauda* infection in the Atlantic harbor seal (*Phoca vitulina concolor*). *J. Wildl. Dis.* **12** (4): 531–538.

EASTMAN, J. T. & COALSON, R. E. 1974. The digestive system of the Weddell seal, *Leptonychotes weddelli* – a review. pp. 253–320 in *Functional anatomy of marine mammals* **2** Ed. R. J. Harrison. Academic Press, London.

EHLERS, K. 1957. Uber die Seelöwin (*Eumetopias californianus*) 'Inge' der Tiergrotten Bremerhaven. *Zool. Gart. Leipzig* **23**: 189–194.

EHLERS, K., SIERTS, W. & MOHR, E. 1958. Die Klappmütze *Cystophora cristata* Erxl. der Tiergrotten Bremerhaven. *Zool. Gart. Leipzig* **24**: 149–210.

ELDEN, C. A., KEYES, M. C. & MARSHALL, C. E. 1971. Pineal body of the Northern fur seal (*Callorhinus ursinus*): a model for studying the probable function of the mammalian pineal body. *Amer. J. Vet. Res.* **32** (4): 939–947.

ELSNER, R., KOOYMAN, G. L. & DRABEK, C. M. 1970a. Diving duration in pregnant Weddell seals. pp. 477–482 in *Antarctic Ecology* **1** Ed. M. W. Holdgate.

ELSNER, R., SHURLEY, J. T., HAMMOND, D. D. & BROOKS, R. E. 1970. Cerebral tolerance to hypoxemia in asphyxiated Weddell seals. *Resp. Physiol.* **9**(2): 287–297.

ENGLE, E. T. 1926. The intestinal length in Steller's sea lion. *J. Mamm.* **7**: 28–30.

ENGLISH, A. W. 1976a. Limb movements and locomotor function in the California sea lion (*Zalophus californianus*). *J. Zool., Lond.* **178**: 341–364.

ENGLISH, A. W. 1976b. Functional anatomy of the hands of fur seals and sea lions. *Am. J. Anat.* **147** (1): 1–18.

ENGLISH, A. W. 1977. Structural correlates of forelimb function in fur seals and sea lions. *J. Morph.* **151** (3): 325–352.

ERHARDT, A. 1940. Ein Walross (*Odobenus rosmarus* (L)) und eine Sattel-robbe (*Phoca groenlandica* Fabr.) für Mecklenburg nachgewiesen. *Arch. Ver. Naturg. Mecklenburg. N.F.* **15**: 9–12.

EVANS, W. E. & BASTIAN, J. 1969. Marine mammal communication: social and ecological factors. pp. 424–475 in *The biology of marine mammals* Ed. H. T. Anderson. Academic Press, London.

FALLA, R. A., TAYLOR, R. H. & BLACK, C. 1979. Survey of Dundas Island, Auckland Islands, with particular reference to Hooker's sea lion (*Phocarctos hookeri*). *N. Z. J. Zool.* **6**: 347–355.

FAVA-DE-MORAES, F., XIMENEZ, I. RADTKE, B. & JUNQUEIRA, L. C. U. 1966. Morphological and chemical studies on the salivary glands and pancreas of two species of Pinnipedia. *Ann. Histochim.* **11**: 199–212.

FAY, F. H. 1957. History and present status of the Pacific walrus population. *Trans. 22nd N. Am. Wildl. Conf.* 431–445.

FAY, F. H. 1960a. Structure and function of the pharyngeal pouches of the walrus (*Odobenus rosmarus* L.). *Mammalia* Paris **24**: 361–371.

FAY, F. H. 1960b. Carnivorous walrus and some Arctic zoonoses. *Arctic J. Arctic Inst. Am.* **13**: 111–122.

FAY, F. H. 1967. The number of ribs and thoracic vertebrae in pinnipeds. *J. Mamm.* **48** (1): 144.

FAY, F. H. 1968. Experimental transmission of *Trichinella spiralis* via marine amphipods. *Can. J. Zool.* **46**: 597–599.

FAY, F. H. 1979. Industrial utilization of marine mammals. *Proc. 29th Alaska Science Conf.* pp. 75–79.

FAY, F. H. 1981. Walrus–*Odobenus rosmarus.* pp. 1–23 in *Handbook of marine mammals.* **1** Eds. Ridgway, S. H. & Harrison, R. J. Academic Press, London.

FAY, F. H. 1982, Ecology and biology of the Pacific Walrus, *Odobenus rosmarus divergens* Illiger. 279 pp. U. S. Dept. Int. Fish Wildl. Serv. North American Fauna No. 74.

FAY, F. H. & KELLY, B. P. 1980. Mass natural mortality of walruses (*Odobenus rosmarus*) at St Lawrence Island, Bering Sea, Autumn 1978. *Arctic* **33** (2): 226–245.

FAY, F. H., RAUSCH, V. R. & FELTZ, E. T. 1967. Cytogenetic comparison of some pinnipeds (Mammalia: Eutheria). *Can. J. Zool.* **45**: 773–778.

FAY, F. H. & RAY, C. 1968. Influence of climate on the distribution of walruses, *Odobenus rosmarus* (Linnaeus). I Evidence from thermoregulatory behavior. *Zoologica. N.Y.* **53** (1): 1–14.

FAY, F. H. & RAY, G. C. 1979. Reproductive behavior of the Pacific walrus in relation to population structure. pp. 409–410 in *Alaska fisheries: 200 years and 200 miles of change* Ed. B. R. Melteff. Proc. 29th Alaska Sci. Conf. 1978.

FEDOSEEV, G. A. 1968. Determination of abundance and grounds for establishing the catch quota for ringed seals in the Sea of Okhotsk. *Trudy vses. nauchno-issled Inst. morsk. ryb. Khoz. Okeanogr.* **62** (68): 180–188.

FELTZ, E. T. & FAY, F. H. 1966. Thermal requirements in vitro of epidermal cells from seals. *Cryobiology* **3** (3): 261–264.

FISH, P. A. 1899. The brain of the fur seal, *Callorhinus ursinus*; with a comparative description of those of *Zalophus californianus, Phoca vitulina, Ursus americanus* and *Monachus tropicalis.* pp. 24–41 in *The fur seals and fur seal islands of the North Pacific Ocean.* Part 3. D. S. Jordan. Washington.

FISH, P. A. 1903. The cerebral fissures of the Atlantic walrus. *Proc. U.S. Nat. Mus.* **26**: 675–688.

FISHER, H. D. 1952. Harp seals of the Northwest Atlantic. *Fish Res. Bd. Canada. Atlantic Biol. St. Gen. Series No 20.* 4pp.

FLANIGAN, J. J. 1972. The central nervous system. pp. 215–246 in *Mammals of the sea, biology and medicine.* Ed. S. H. Ridgway. Charles C. Thomas, Springfield, Ill.

FLEMING, C. A. 1951. Sea lions as geological agents. *J. Sed. Petrology* **21** (1): 22–25.

FLINDERS, M. 1814. *A voyage to Terra Australis in the years 1801, 1802 and 1803 in HM ship* Investigator. G & W Nicol, London.

FLOWER, S. S. 1931. Contributions to our knowledge of the duration of life in vertebrate animals. V Mammals. *Proc. Zool. Soc. Lond.* 145–243.

FORBES, W. A. 1882. Notes on the external characters and anatomy of the Californian sea lion (*Otaria gillespii*). *Trans. Zool. Soc.* **11**: 225–231.

FRASER, F. C. 1935. Zoological notes from the voyage of Peter Mundy 1655–56. Sea elephant on St. Helena. *Proc. Linn. Soc. Lond.* **147** (2): 33–35.

FROST, K. J. & LOWRY, L. F. 1980. Feeding of ribbon seals (*Phoca fasciata*) in the Bering Sea in spring. *Can. J. Zool.* **58**: 1601–1607.

FURMAN, D. P. & DAILEY, M. D. 1980. The genus *Halarachne* (Acari: Halarachnidae), with the description of a new species from the Hawaiian monk seal. *J. Med. Entomol.* **17** (4): 352–359.

GEORGE, J. C. & RONALD, K. 1975. The harp seal *Pagophilus groenlandicus* (Erxleben, 1777). XVII Structure and metabolic adaptation of the caval sphincter muscle with some observations on the diaphragm. *Acta anat.* **93**: 88–99.

GENTRY, R. L. 1973. Thermoregulatory behavior of eared seals. *Behavior* **46**: 73–93.

GENTRY, R. L. 1980. Set in their ways. Survival formula of the northern fur seal. *Oceans* **13** (3): 34–37.

GENTRY, R. L. 1981. Seawater drinking in eared seals. *Comp. Biochem. Physiol.* **68A**: 81–86.

GENTRY, R. L. & JOHNSON, J. H. 1981. Predation by sea lion on northern fur seal neonates. *Mammalia,* Paris **45** (4): 423–430.

GERACI, J. R. 1975. Pinniped nutrition. *Rapp. P-v. Réun. Cons. Int. Explor. Mer.* **169**: 312–323.

GORDON, K. R. 1981. Locomotor behaviour of the walrus (*Odobenus*) *J. Zool. Lond.* **195**: 349–367.

GOTCH, A. F. 1979. *Mammals – their Latin names explained.* Blandford Press, Dorset.

GRAV, H. J., BLIX, A. S. & PÅSCHE, A. 1974. How do seal pups survive birth in Arctic winter. *Acta Physiol. Scand.* **92**: 427–429.

GRAV, H. J. & BLIX, A. S. 1979. A source of non-shivering thermogenesis in fur seal skeletal muscle. *Science* **204**: 87–89.

GRIFFITHS, D. J. & BRYDEN, M. M. 1981. The annual cycle of the pineal gland of the elephant seal (*Mirounga leonina*). pp. 57–66 in *Pineal function.* Eds C. Matthews & R. F. Seamark. Elsevier Press.

GRIGORESCU, D. 1976. Paratethyan seals. *Syst. Zool.* **25** (4): 407–419.

GRIMWOOD, I. 1968. Endangered mammals in Peru. *Oryx* **9** (6): 411–421.

GUILER, E. R. 1978. Whale strandings in Tasmania since 1945 with notes on some seal reports. *Pap. Proc. Roy. Soc. Tasmania* **112**: 189–213.

HAAFTEN, J. L. VAN 1962. Diseases of seals in the Dutch coastal waters. 4th Int. Symp. on Diseases in Zoo Animals. *Nord. Vet-med.* **14**: Suppl. 1:138–140.

HADLOW, W. C., CHEVILLE, N. F. & JELLISON, W. L. 1980. Occurrence of pox in a northern fur seal on the Pribilof Islands in 1951. *J. Wildl. Dis.* **16** (2): 305–312.

HALL-MARTIN, A. J. 1974. Observations on population density and species composition of seals in King Haakon VII Sea, Antarctica. *S. Afr. J. Antarct. Res.* No. 4: 34–39.

HAMILTON, J. E. 1934. The southern sea lion *Otaria byronia* (de Blainville). *Discovery Reports* 8: 269–318.

HAMILTON, J. E. 1939a. The leopard seal *Hydrurga leptonyx* (de Blainville). *Discovery Reports* 18: 239–264.

HAMILTON, J. E. 1939b. A second report on the southern sea lion *Otaria byronia* (de Blainville). *Discovery Reports* 19: 121–164.

HAMILTON, J. E. 1940. On the history of the elephant seal *Mirounga leonina* (Linn.). *Proc. Linn. Soc., Lond.* Session 152: 33–37.

HAMILTON, J. E. 1946. Seals preying on birds. *The Ibis* Jan. 1946: 131–132.

HARINGTON, C. R. 1966. Extralimital occurrences of walruses in the Canadian Arctic. *J. Mamm.* 47: 506–513.

HARESTAD, A. S. & FISHER, H. D. 1975. Social behavior in a non-pupping colony of Steller sea lions (*Eumetopias jubata*). *Can. J. Zool.* 53 (11): 1596–1613.

HARRISON, R. J. 1960. Reproduction and reproductive organs in common seals (*Phoca vitulina*) in the Wash, East Anglia. *Mammalia*, Paris 24: 372–385.

HARRISON, R. J. 1969. Endocrine organs: hypophysis, thyroid and adrenal. pp. 349–390 in *The biology of marine mammals*. Ed. H. T. Andersen. Academic Press, London.

HARRISON, R. J. & KOOYMAN, G. L. 1968. General physiology of the pinnipedia. pp. 211–296 in *The behavior and physiology of pinnipeds*. Eds Harrison *et al.* Appleton-Century-Crofts, NY.

HARRISON, R. J., HARRISON MATTHEWS, L. & ROBERTS, J. M. 1952. Reproduction in some pinnipedia. *Trans. Zool. Soc. Lond.* 27 (5): 437–540.

HARRISON, R. J., ROWLANDS, I. W., WHITTING, H. W. & YOUNG, B. A. 1962. Growth and structure of the thyroid gland in the common seal (*Phoca vitulina*). *J. Anat.* 96: 3–15.

HARRISON, R. J. & TOMLINSON, J. D. W. 1963. Anatomical and physiological adaptations in diving mammals. *Viewpoints in biology* 2: 115–162.

HARRISON, R. J. & YOUNG, B. A. 1966. Functional characteristics of the pinniped placenta. *Comp. Biol. of Reprod. in Mammals. Symp. Zool. Soc.* No. 15: 47–67. Academic Press.

HART, J. S. & IRVING, L. 1959. The energetics of harbor seals in air and in water with special consideration of seasonal changes. *Can. J. Zool.* 37: 447–457.

HARWOOD, J. 1978. The effect of management policies on the stability and resilience of British grey seal populations. *J. Applied Ecology* 15: 413–421.

HEATH, M. E., MCGINNIS, S. M. & AKORN, D. 1977. Comparative thermoregulation of suckling and weaned pups of the northern elephant seal, *Mirounga angustirostris*. *Comp. Biochem. Physiol.* 57A (2): 203–206.

HELLER, E. 1904. Mammals of the Galapagos Archipelago, exclusive of the Cetacea. Papers from the Hopkins Stanford Galapagos Expedition 1898–1899. *Proc. Calif. Acad. Sci.* 3rd ser. 3 (7): 233–250.

HENDEY, Q. B. 1972. The evolution and dispersal of the Monachinae. *Ann. S. Afr. Mus.* 59 (5): 99–113.

HENDEY, Q. B. & REPENNING, C. A. 1972. A Pliocene phocid from South Africa. *Ann. S. Afr. Mus.* 59 (4): 71–98.

HEWER, H. R. 1957. A Hebridean breeding colony of grey seals, *Halichoerus grypus* (Fab) with comparative notes on the grey seals of Ramsey Island, Pembrokeshire. *Proc. Zool. Soc. Lond.* 128: 23–66.

HEWER, H. R. 1964. The determination of age, sexual maturity, longevity and a life table in the grey seal (*Halichoerus grypus*). *Proc. Zool. Soc. Lond.* 142 (4): 593–624.

HEWER, H. R. 1974. *British seals*. 256pp. The New Naturalist. Collins, London.

HEWER, H. R. & BACKHOUSE, K. M. 1959. Field identification of bulls and cows of the grey seal *Halichoerus grypus* Fab. *Proc. Zool. Soc. Lond.* 132: 641–645.

HEWER, H. R. & BACKHOUSE, K. M. 1968. Embryology and foetal growth of the grey seal, *Halichoerus grypus*. *J. Zool., Lond.* 155: 507–533.

HICKLING, G. 1962. *Grey seals and the Farne Islands.* 180pp. Routledge & Kegan Paul, London.

HOFMAN, R., ERICKSON, A. & SINIFF, D. 1973. The Ross seal (*Ommatophoca rossi*). pp. 129–139 in *Seals. IUCN Suppl. Paper No. 39.*

HOL, R., BLIX, A. S. & MYHRE, H. O. 1975. Selective redistribution of the blood volume in the diving seal (*Pagophilus groenlandicus*). *Rapp. P-v Réun. Cons. int. Explor. Mer.* 169: 423–431.

HOLDEN, A. V. 1975. The accumulation of oceanic contaminants in marine mammals. *Rapp. P-v. Réun. Cons. int. Explor. Mer.* 169: 353–361.

HOLDEN, A. V. 1978. Pollutants and seals. A review. *Mammal Rev.* 8 (1 & 2): 53–66.

HOLDGATE, M. W. 1963. Fur seals in the South Sandwich Islands. *The Polar Record.* 11 (73): 1 page.

HOLDGATE, M. W. 1965. The biological report of the Royal Society expedition to Tristan da Cunha 1962. Part 3 The fauna of the Tristan da Cunha Islands. *Phil. Trans. Roy. Soc.* 249B: 361–402.

HOLDGATE, M. W., TILBROOK, P. J., & VAUGHAN, R. W. 1968. The biology of Bouvetøya. *Bull. Brit. Ant. Survey.* 15: 1–7.

HOSE, C. 1927. *Fifty years of romance and research or a jungle-wallah at large.* 301pp. Hutchinson, London.

HOWORTH, P. C. 1976. The seals of San Miguel Island. *Oceans* 9 (5): 38–43.

HUBBS, C. L. 1956. Back from oblivion. Guadalupe fur seal: still a living species. *Pacific Discovery* 9 (6): 14–21.

HUBBS, C. L. & NORRIS, K. S. 1971. Original teeming abundance, supposed extinction, and survival of the Juan Fernandez fur seal. *Antarctic Pinnipedia. Am. Geophys. Union. Ant. Res. Series* **18**: 35–52.

HUBER, E. 1934. Anatomical notes on Pinnipedia and Cetacea. *Carnegie Inst. Wash. Publ.* No. **447**: 105–136.

HUNT, R. M. 1974. The auditory bulla in Carnivora: an anatomical basis for reappraisal of carnivore evolution. *J. Morph.* **143** (1): 21–76.

HSÜ, K. J. 1978. When the Black Sea was drained. *Sci. Am.* **238** (5): 52–63.

ICHIHARA, T. & YOSHIDA, K. 1972. Diving depth of northern fur seals in the feeding time. *Sci. Rep. Whales Res. Inst.* No. **24**: 145–148.

INGHAM, S. E. 1960. The status of seals (Pinnipedia) at Australian Antarctic stations. *Mammalia*, Paris **24**: 422–430.

INNS, R. W., AITKEN, P. F. & LING, J. K. 1979. 7. Mammals. *Natural History of Kangaroo Island. Roy. Soc. S. Aust.* 91–102.

IRVING, L. & HART, J. S. 1957. The metabolism and insulation of seals as bare-skinned mammals in cold water. *Can. J. Zool.* **35**: 497–511.

IRVING, L., PEYTON, L. J., BAHN, C. H. & PETERSON, R. S. 1962. Regulation of temperature in fur seals. *Physiol. Zool.* **35**: 275–284.

JAMESON, R. J. & KENYON, K. W. 1977. Prey of sea lions in the Rogue River, Oregon. *J. Mamm.* **58** (4): 672.

JAMIESON, G. S. & FISHER, H. D. 1972. The pinniped eye: a review. pp. 245–261 in *Functional anatomy of marine mammals.* Ed. R. J. Harrison. Academic Press, London.

JENNISON, G. 1914. A hybrid sea lion. *Proc. Zool. Soc. Lond.* 219–220.

JOHNSON, A. M. 1975. The status of northern fur seal populations. *Rapp. P-v. Réun. Cons. int. Explor. Mer.* **169**: 263–266.

JOHNSTONE, G. W. 1972. A review of biological research by Australian National Antarctic Research Expeditions 1947–71. *Polar Record* **16** (102): 519–532.

JOHNSTONE, G. W. 1975. Regurgitation of nematodes by a Weddell seal. *Saugetierk. Mitt.* **23**: 159–160.

JONES, E. C. 1971. *Isistius brasiliensis*, a squaloid shark, the probable cause of crater wounds of fishes and cetaceans. *Fisheries Bull. U.S. Dept. Commerce* **69** (4): 791–798.

JONES, R. 1966. A speculative archaeological sequence for north-west Tasmania. *Rec. Queen Vict. Mus.* No. **25**: 1–12.

KARPOVICH, V. N., KOKHANOV, V. D. & TATARINKOVA, I. P. 1967. The grey seal on the Murman coast. *Trans. Polar Inst. Mar. Fish & Oceanogr.* **21**: 117–125.

KELLER, O. 1887. *Thiere des Classischer Alterthums.* 488pp. Innsbruck.

KENYON, K. W. 1973a. Hawaiian monk seal (*Monachus schauinslandi*). pp. 88–97 in *Seals. IUCN Supplementary Paper No. 39.*

KENYON, K. W. 1973b. Guadalupe fur seal (*Arctocephalus townsendi*). pp. 82–87 in *Seals. IUCN Supplementary Paper No. 39.*

KENYON, K. W. 1977. Caribbean monk seal extinct. *J. Mamm.* **58** (1): 97–98.

KENYON, K. W. 1980. No man is benign. The endangered monk seal. *Oceans* **13** (3): 48–54.

KENYON, K. W. & FISCUS, C. H. 1963. Age determination in the Hawaiian monk seal. *J. Mamm.* **44**: 280–281.

KENYON, K. W. & RICE, D. W. 1959. Life history of the Hawaiian monk seal. *Pacific Science* **13**: 215–252.

KENYON, K. W. & RICE, D. W. 1961. Abundance and distribution of the Steller sea lion. *J. Mamm.* **42** (2): 223–234.

KENYON, K. W. & SCHEFFER, V. B. 1954. A population study of the Alaska fur seal herd. *U.S. Dept. Int. Special Scientific Report, Wildlife No. 12.*

KEREM, D. & ELSNER, R. 1973. Cerebral tolerance to asphyxial hypoxia in the harbor seal. *Respir. Physiol.* **19**: 188–200.

KERRY, K. R. & MESSNER, M. 1968. Intestinal glycosidases of three species of seals. *Comp. Biochem. Physiol.* **25** (2): 437–446.

KEYES, M. C. 1965. Pathology of the northern fur seal. *J. Am. Vet. Med. Ass.* **147**: 1091–1095.

KEYES, M. C. 1968. The nutrition of pinnipeds. pp. 359–395 in *The behaviour and physiology of pinnipeds.* Eds R. J. Harrison *et al.* Appleton-Century-Crofts, NY.

KILIAAN, H. P. L. & STIRLING, I. 1978. Observations on overwintering walruses in the eastern Canadian high arctic. *J. Mamm.* **59** (1): 197–200.

KIM, K. C. 1971. The sucking lice (Anoplura: Echinophthiriidae) of the northern fur seal. Descriptions and morphological adaptation. *Ann. Entomol. Soc. Amer.* **64**: 280–292.

KIM, K. C. 1972. Louse populations of the northern fur seal (*Callorhinus ursinus*). *Am. J. Vet. Res.* **33** (10): 2027–2036.

KIM, K. C. 1975. Ecology and morphological adaptation of the sucking lice (Anoplura: Echinophthiriidae) on the northern fur seal. *Rapp. P-v. Réun. Cons. int. Explor. Mer.* **169**: 504–515.

KIM, K. C. 1979. Life stages and population of *Proechinophthirus zumpti* (Anoplura: Echinophthiriidae) from the Cape fur seal (*Arctocephalus pusillus*). *J. Med. Entomol.* **16** (6): 497–501.

KIM, K. C., HAAS, V. L. & KEYES, M. C. 1980. Populations, microhabitat preference and effects of infestation of two species of *Orthohalarachne* in the northern fur seal. *J. Wildl. Dis.* **16** (1): 45–51.

KIM, K. C., REPENNING, C. A. & MOREJOHN, G. V. 1975. Specific antiquity of the suckling lice and evolution of otariid seals. *Rapp. P-v. Réun. Cons. int. Explor. Mer.* **169**: 544–549.

KING, J. E. 1954. The Otariid seals of the Pacific coast of America. *Bull. Br. Mus. nat. Hist.* (Zool.) **2** (10): 311–337.

KING, J. E. 1956. The monk seals genus *Monachus*. *Bull. Br. Mus. nat. Hist.* (Zool.) **3** (5): 203–256.

KING, J. E. 1959a. Northern and southern populations of *Arctocephalus gazella*. *Mammalia*, Paris **23** (1): 19–40.

KING, J. E. 1959b. A note on the specific name of the Kerguelen fur seal. *Mammalia*, Paris **23**: 381.

KING, J. E. 1961. Notes on the Pinnipedes from Japan described by Temminck in 1844. *Zool. Med. Leiden* **37** (13): 211–224.

KING, J. E. 1964. *Seals of the World*. 154pp. British Museum (Nat. Hist.), London.

KING, J. E. 1966. Relationships of the hooded and elephant seals (genera Cystophora and Mirounga). *J. Zool. Lond.* (1966) **148**: 385–398.

KING, J. E. 1969a. The identity of the fur seals of Australia. *Aust. J. Zool.* **17**: 841–853.

KING, J. E. 1969b. Some aspects of the anatomy of the Ross seal, *Ommatophoca rossi* (Pinnipedia: Phocidae). 54pp. *Brit. Ant. Survey Sci. Rep. No. 63*.

KING, J. E. 1971. The lacrimal bone in the *Otariidae*. *Mammalia*, Paris **35** (3): 465–470.

KING, J. E. 1972a. Observations on phocid skulls. pp. 81–115 in *Functional anatomy of marine mammals*. Ed. R. J. Harrison. Academic Press, London.

KING, J. E. 1972b. On the laryngeal skeletons of the leopard seal, *Hydrurga leptonyx* and the Ross seal, *Ommatophoca rossi*. *Mammalia*, Paris **36** (1): 146–156.

KING, J. E. 1973. Pleistocene Ross seal (*Ommatophoca rossi*) from New Zealand. *N.Z. J. Mar. F. W. Res.* **7** (4): 391–397.

KING, J. E. 1976. On the identity of the three young fur seals (genus *Arctocephalus*) stranded in New Caledonia. *Beaufortia* **25** (324): 97–105.

KING, J. E. 1977. Comparative anatomy of the blood vessels of the sea lions *Neophoca* and *Phocarctos*; with comments on the differences between the otariid and phocid vascular systems. *J. Zool., Lond.* **181**, 69–94.

KING, J. E. 1978. On the specific name of the southern sea lion. *J. Mamm.* **59** (4): 861–863.

KING, J. E. & HARRISON, R. J. 1961. Some notes on the Hawaiian monk seal. *Pacific Science* **15**: 282–293.

KING, J. E. & MARLOW, B. J. 1974. With a thousand sea lions on the Auckland Islands. *Aust. Nat. Hist.* **18** (1): 6–11.

KIPARSKY, V. 1952. L'Histoire du Morse. *Ann. Acad. Scient. Fennicae B.* **73** (3): 53pp.

KIRCHSCHOFER, R. 1968. Notizen über zwei Bastarde zwichen *Otaria byronia* und *Zalophus californianus*. *Z. Saugetierk.* **33**: 45–49.

KOOYMAN, G. L. 1968. An analysis of some behavioral and physiological characteristics related to diving in the Weddell seal. Ant. Res. Series vol. 2. pp. 227–261 in *Biology of the Antarctic Seas III*. Eds W. L. Schmitt & G. A. Llano. Am. Geophys. Union, Wash. DC.

KOOYMAN, G. L. 1969. The Weddell seal. *Sci. Am.* Aug. 1969 **221** (2): 100–106.

KOOYMAN, G. L. 1973. Respiratory adaptations in marine mammals. *Amer. Zool.* **13**: 457–468.

KOOYMAN, G. L. 1975a. A comparison between day and night diving in the Weddell seal. *J. Mamm.* **56** (3): 563–574.

KOOYMAN, G. L. 1975b. Physiology of freely diving Weddell seals. *Rapp. P-v. Réun. Cons. int. Explor. Mer.* **169**: 441–444.

KOOYMAN, G. L. 1981. *Weddell seal: consummate diver*. 135pp. Cambridge Univ. Press, Cambridge.

KOOYMAN, G. L. & ANDERSEN, H. T. 1969. Deep diving. pp. 65–94 in *The biology of marine mammals*. Ed. H. T. Andersen. Academic Press, London.

KOOYMAN, G. L. & DRABEK, C. M. 1968. Observations on milk, blood and urine constituents of the Weddell seal. *Physiol. Zool.* **41** (2): 187–194.

KOOYMAN, G. L., GENTRY, R. L. & URQUHART, D. L. 1976. Northern fur seal diving behavior: a new approach to its study. *Science* **193** (4251): 411–412.

KOOYMAN, G. L., HAMMOND, D. D. & SCHROEDER, J. P. 1970. Bronchograms and tracheograms of seals under pressure. *Science* **169**: 82–84.

KOOYMAN, G. L., WAHRENBROCK, E. A., CASTELLINI, M. A., DAVIS, R. W. & SINNETT, F. E. 1980. Aerobic and anaerobic metabolism during voluntary diving in Weddell seals; evidence of preferred pathways from blood chemistry and behavior. *J. Comp. Physiol.* **138**: 335–346.

KOOYMAN, G. L., CASTELLINI, M. A. & DAVIS, R. W. 1981. Physiology of diving in marine mammals. *Ann. rev. Physiol.* **43**: 343–356.

KOSYGIN, G. M. 1968. Some data on morphological characteristics of bearded seal fetus. *Trudy vses nauchno-issled Inst. morsk ryb. Khoz Okeanogr.* **62** (68): 244–251.

KOSYGIN, G. M. & POTELOV, V. A. 1971. Age, sex and population variability of the craniological characters of bearded seals. *Izv. TINRO* **80**: 266–288.

KRETCHMER, N. & SUNSHINE, P. 1967. Intestinal disaccharide deficiency in the sea lion. *Gastroenterology* **53**: 123–129.

KUBOTA, K. 1968. Comparative anatomical and neurohistological observations on the tongue of the northern fur seal, *Callorhinus ursinus*. *Anat. Rec.* **161** (2): 257–266.

KUKENTHAL, W. 1899. *Vergl.-anat, und entwickelungsgeschichtliche Untersuchungen an Walthieren*. Part 1. Sect. 3. Das Centralnervensystem der Cetaceen.

KUMLIEN, L. 1879. Contributions to the natural history of Arctic America, made in connection with the

Howgate Polar Expedition 1877–78. *Bull. US Nat. Mus.* **15**: 1–179.

KUROCHKIN, Y. V. 1975. Parasites of the Caspian seal *Pusa caspica. Rapp. P-v. Réun. Cons. int. Explor. Mer.* **169**: 363–365.

LANDER, R. H. 1979. Alaskan or Northern fur seal. pp. 19–23 in *Mammals in the seas.* **2** FAO Fisheries Series. No. 5.

LANE, R. A. B., MORRIS, R. J. H. & SHEEDY, J. W. 1972. A haematological study of the southern elephant seal *Mirounga leonina* (Linn). *Comp. Biochem. Physiol.* **42**A: 841–850.

LAVIGNE, D. M. 1979. Harp seal. pp. 76–80 in *Mammals in the seas.* **2** FAO Fisheries Series. No. 5.

LAVIGNE, D. M., BERNHOLZ, C. D. & RONALD, K. 1977. Functional aspects of pinniped vision. pp. 135–173 in *Functional anatomy of marine mammals* **3**. Ed. R. J. Harrison. Academic Press, London.

LAVIGNE, D. & RONALD, K. 1972. The harp seal, *Pagophilus groenlandicus* (Erxleben, 1777). XXIII. Spectral sensitivity. *Can. J. Zool.* **50** (9): 1197–1206.

LAWS, R. M. 1953. The elephant seal (*Mirounga leonina* Linn.). I. Growth and age. *Falk. Is. Dep. Surv. Sci. Rep.* **8**: 62pp.

LAWS, R. M. 1956. The elephant seal (*Mirounga leonina* Linn.). 2. General, social and reproductive behaviour. 88pp. *Falk. Is. Dep. Surv. Sci. Rep.* **13**.

LAWS, R. M. 1957. On the growth rates of the leopard seal, *Hydrurga leptonyx* (de Blainville, 1820). *Saugetierk. Mitt.* **5** (2): 49–55.

LAWS, R. M. 1958. Growth rates and ages of crabeater seals, *Lobodon carcinophagus* Jacquinot & Pucheran. *Proc. Zool. Soc. Lond.* **130** (2): 275–288.

LAWS, R. M. 1962. Age determination of pinnipeds with special reference to growth layers in the teeth. *Zeitschrift f. Säugetierk.* **27** (3): 129–146.

LAWS, R. M. 1973. The current status of seals in the southern hemisphere. pp. 144–161 in *Seals. IUCN Suppl. Paper No. 39.*

LAWS, R. M. 1979. Southern elephant seal. pp. 106–109 in *Mammals in the seas.* **2** FAO Fisheries Series No. 5.

LAWS, R. M. & TAYLOR, R. J. F. 1957. A mass dying of crabeater seals *Lobodon carcinophagus* (Gray). *Proc. Zool. Soc. Lond.* **129**: 315–324.

LEATHERLAND, J. F. & RONALD, K. 1976. Structure of the adenohypophysis in juvenile harp seal, *Pagophilus groenlandicus. Cell Tiss. Res.* **173**: 367–382.

LEATHERLAND, J. F. & RONALD, K. 1979. Thyroid activity in adult and neonate harp seals *Pagophilus groenlandicus. J. Zool., Lond.* **189**: 399–405.

LE BOUEF, B. J. 1971a. Oil contamination and elephant seal mortality: a 'negative' finding. pp. 277–285 in *Biological and oceanographic survey of Santa Barbara Channel oil spill 1969–70.* **1** Ed. D. Straughan. Allan Hancock Foundation.

LE BOUEF, B. J. 1971b. The aggression of the breeding bulls. *Nat. Hist. N.Y.* **80** (2): 83–94.

LE BOUEF, B. J. 1977. Back from extation? *Pacific Discovery* **30** (5): 1–9.

LE BOUEF, B. J., AINLEY, D. G. & LEWIS, T. J. 1974. Elephant seals on the Farallones: population structure of an incipient breeding colony. *J. Mamm.* **55** (2): 370–385.

LE BOUEF, B. J. & MATE, B. R. 1978. Elephant seals colonize additional Mexican and Californian islands. *J. Mamm.* **59** (3): 621–622.

LE BOUEF, B. J. & ORTIZ, C. L. 1977. Composition of elephant seal milk. *J. Mamm.* **58** (4): 683–685.

LE BOUEF, B. J. & PETERSON, R. S. 1969. Dialects in elephant seals. *Science* **166** (3913): 1654–1656.

LE BOUEF, B. J. & PETRINOVICH, L. F. 1975. Elephant seal dialects: are they reliable? *Rapp. P-v. Réun. Cons. int. Explor Mer.* **169**: 213–218.

LE BOUEF, B. J., WHITING, R. J. & GANTT, R. F. 1972. Perinatal behaviour of northern elephant seal females and their young. *Behaviour* **43** (1–4): 121–156.

LENFANT, C. 1969. Physiological properties of blood of marine mammals. pp. 95–116 in *The biology of marine mammals.* Ed. H. T. Andersen. Academic Press, London.

LENFANT, C., ELSNER, R., KOOYMAN, G. L. & DRABEK, C. M. 1969. Respiratory function of blood of the adult and fetus Weddell seal *Leptonychotes weddelli. Am. J. Physiol.* **216** (6): 1595–1597.

LENFANT, C., ELSNER, R., KOOYMAN, G. L. & DRABEK, C. 1970a. Tolerance to sustained hypoxia in the Weddell seal *Leptonychotes weddelli. Antarctic Ecology* **1**: 471–476.

LENFANT, C., JOHANSEN, K. & TORRANCE, J. D. 1970b. Gas transport and oxygen storage capacity in some pinnipeds and the sea otter. *Resp. Physiol.* **9** (2): 277–286.

LING, J. K. 1965. Functional significance of sweat glands and sebaceous glands in seals. *Nature* **208**: 560–562.

LING, J. K. 1966. The skin and hair of the southern elephant seal *Mirounga leonina* (Linn.) I. The facial vibrissae. *Aust. J. Zool.* **14**: 855–866.

LING, J. K. 1968. The skin and hair of the southern elephant seal *Mirounga leonina* (L.). III. Morphology of the adult integument. *Aust. J. Zool.* **16**: 629–645.

LING, J. K. 1974. The integument of marine mammals. pp. 1–44 in *Functional anatomy of marine mammals* **2**. Ed. R. J. Harrison. Academic Press, London.

LING, J. K. & BUTTON, C. E. 1975. The skin and pelage of grey seal pups (*Halichoerus grypus* Fabricius): with a comparative study of foetal and neonatal mouting in the Pinnipedia. *Rapp. P-v. Réun. Cons. int. Explor. Mer.* **169**: 112–132.

LING, J. K. & THOMAS, C. D. B. 1967. The skin and hair of the southern elephant seal, *Mirounga leonina* (L.). II. Prenatal and early post-natal development and moulting. *Aust. J. Zool.* **15**: 349–365.

LING, J. K. & WALKER, G. E. 1976.
Seal studies in South Australia: Progress report for the year 1975. *South Aust. Nat.* **50** (4): 59–68.

LING, J. K. & WALKER, G. E. 1977.
Seal studies in South Australia: Progress report for the period January 1976 to March 1977. *South Aust. Nat.* **52** (2): 18–30.

LING, J. K. & WALKER, G. E. 1978.
An 18-month breeding cycle in the Australian sea lion? *Search* **9** (12): 464–465.

LIPPS, J. H. 1980. Hunters among the ice floes. *Oceans* **13** (3): 45–47.

LOCKLEY, R. M. 1966. The distribution of grey and common seals on the coasts of Ireland. *Irish Nat. J.* **15** (5): 136–143.

LÖNNBERG, E. 1929. A hybrid between grey seal, *Halichoerus grypus* Nils, and Baltic ringed seal, *Phoca hispida annellata* Nils. *Arkiv für Zoologi* **21A** (5): 1–8.

LOUGHREY, A. G. 1959. Preliminary investigation of the Atlantic walrus, *Odobenus rosmarus rosmarus* (Linnaeus). *Wildl. Mgt. Bull.* ser. 1. No. **14**: 123pp.

MACKENZIE, B. A. 1954. Green algal growth on gray seals. *J. Mamm.* **35**: 595–596.

MADDEN, F. 1832. Historical remarks on the introduction of the game of chess into Europe, and on the ancient chess-men discovered in the Isle of Lewis. *Archaeologia*. London **24**: 203–291.

MANNING, T. H. 1974. Variations in the skull of the bearded seal. *Biol. Pap. Univ. of Alaska, Fairbanks* **16**: 1–21.

MANSFIELD, A. W. 1966a. The walrus in Canada's Arctic. *Canad. Geogr. J.* **72** (3): 88–95.

MANSFIELD, A. W. 1966b. The Grey seal in Eastern Canadian waters. *Can. Audubon Mag.* Nov–Dec 1966. 160–166.

MANSFIELD, A. W. 1967a. Seals of arctic and eastern Canada. 2nd ed. *Fish. Res. Bd. Can.* Bulletin No. 137.

MANSFIELD, A. W. 1967b. Distribution of the harbor seal, *Phoca vitulina* Linnaeus, in Canadian arctic waters. *J. Mamm.* **48** (2): 249–257.

MANSFIELD, A. W. 1973. The atlantic walrus, *Odobenus rosmarus* in Canada and Greenland. pp. 69–79 in *Seals. IUCN Suppl. Paper No. 39.*

MANSFIELD, A. W. & BECK, B. 1977.
The grey seal in eastern Canada. Tech. Rep. No. **704**: 81pp.

MANSFIELD, A. W. & FISHER, H. D. 1960.
Age determination in the harbour seal, *Phoca vitulina* L. *Nature* **186**: 92–93.

MARGOLIS, L. & DAILEY, M. D. 1972.
Revised annotated list of parasites from sea mammals caught off the west coast of North America. NOAA Tech. Rep. NMFS. SSRF-647.

MARKOWSKI, S. 1952. The cestodes of seals from the Antarctic. *Bull. Br. Mus. nat. Hist.* (Zool.) **1** (7): 123–150.

MARLOW, B. J. 1967. Mating behaviour in the leopard seal, *Hydrurga leptonyx* in captivity. *Aust. J. Zool.* **15**: 1–5.

MARLOW, B. J. 1974. Ingestion of placenta in Hooker's sea lion. *N.Z. J. Mar. F.W. Res.* **8** (1): 233–238.

MARLOW, B. J. 1975. The comparative behaviour of the Australasian sea lions *Neophoca cinerea* and *Phocarctos hookeri. Mammalia* Paris **39** (2): 159–230.

MARLOW, B. J. & KING, J. E. 1974.
Sea lions and fur seals of Australia and New Zealand – the growth of knowledge. *J. Aust. Mamm. Soc.* **1** (2): 117–136.

MARSHALL, F. H. A. 1922. *Physiology of reproduction.* 770pp. London.

MATHESON, C. 1950. Longevity in the grey seal. *Nature* **166**: 73–74.

MATSUURA, D. T. & WHITTOW, G. C. 1974.
Evaporative heat loss in the California sea lion and harbor seal. *Comp. Biochem. Physiol.* **48A** (1): 9–20.

MATTHEWS, L. H. 1950. The natural history of the grey seal, including lactation. *Proc. Zool. Soc. Lond.* **120**: 763.

MAWDESLEY-THOMAS, L. E. 1971. An ovarian tumour in a southern elephant seal (*Mirounga leonina*). *Vet. Path.* **8**: 9–15.

MAWDESLEY-THOMAS, L. E. & BONNER, W. N. 1971.
Uterine tumours in a grey seal (*Halichoerus grypus*). *J. Path.* **103**: 205–208.

MCCANN, T. S. 1980. Territoriality and breeding behaviour of adult male Antarctic fur seal, *Arctocephalus gazella. J. Zool. Lond.* **192** (3): 295–310.

MCCANN, T. S. 1981. Aggression and sexual activity of male southern elephant seals, *Mirounga leonina. J. Zool. Lond.* **195**: 295–310.

MCDERMID, E. M. & BONNER, W. N. 1975.
Red cell and serum protein systems of grey seals and harbour seals. *Comp. Biochem. Physiol.* **50B** (1): 97–101.

MCLAREN, I. A. 1958a. Some aspects of growth and reproduction of the bearded seal, *Erignathus barbatus* (Erxleben). *J. Fish. Res. Bd. Canada* **15** (2): 219–227.

MCLAREN, I. A. 1958b. The biology of the ringed seal (*Phoca hispida* Schreber) in the eastern Canadian Arctic. *Bull. Fish. Res. Bd. Canada* **118**: 1–97.

MCLAREN, I. A. 1960. Are the Pinnipedia biphyletic? *Syst. Zool.* **9** (1): 18–28.

MCLAREN, I. A. 1962. Population dynamics and exploitation of seals in the eastern Canadian Arctic. pp. 168–183 in *The exploitation of natural animal populations.* Eds E. D. LeCren & M. W. Holdgate. Blackwell Scientific Publications, Oxford.

MCNAB, A. G. & CRAWLEY, M. C. 1975.
Mother and pup behaviour of the New Zealand fur seal, *Arctocephalus forsteri* (Lesson). *Mauri Ora* **3**: 77–88.

MERCER, M. C. 1967. Records of the Atlantic walrus, *Odobenus rosmarus rosmarus* from Newfoundland. *J. Fish. Res. Bd. Can.* **24** (12): 2631–2635.

MERCER, M. C. 1976. *The seal hunt. 25pp. Fisheries and Marine Service, Ottawa.*

MIGAKI, G., VAN DYKE, D. & HUBBARD, R. C. 1971. Some histopathological lesions caused by helminths in marine mammals. *J. Wildl. Dis.* **7** (4): 281–289.

MILLER, E. H. 1974. Social behaviour between adult male and female New Zealand fur seals, *Arctocephalus forsteri* (Lesson) during the breeding season. *Aust. J. Zool.* **22**: 155–173.

MILLER, E. H. 1975a. Walrus ethology. I The social role of tusks and applications of multidimensional scaling. *Can. J. Zool.* **53** (5): 590–613.

MILLER, E. H. 1975b. Body and organ measurements of fur seals, *Arctocephalus forsteri* (Lesson) from New Zealand. *J. Mamm.* **56** (2): 511–513.

MILLER, E. H. 1975c. Comparative study of facial expressions of two species of pinnipeds. *Behaviour* **53**: 268–284.

MILLER, E. H. 1976. Walrus ethology. II Herd structure and activity budgets of summering males. *Can. J. Zool.* **54** (5): 704–715.

MILLER, W. C. S. 1888. The myology of the Pinnipedia. pp. 139–234 in Report on the seals. *The zoology of the voyage of HMS* Challenger. W. Turner **26** (68).

MITCHELL, E. 1966. The Miocene pinniped *Allodesmus*. *Univ. Calif. Publ. in Geol. Sci.* **61**: 1–105.

MITCHELL, E. D. 1968. The Mio-Pliocene pinniped *Imagotaria*. *J. Fish. Res. Bd. Can.* **25** (9): 1843–1900.

MITCHELL, E. & TEDFORD, R. H. 1973. The Enaliarctidae, a new group of extinct aquatic Carnivora and a consideration of the origin of the Otariidae. *Bull. Am. Mus. Nat. Hist.* **151** (art. 3): 203–284.

MOLYNEUX, G. S. & BRYDEN, M. M. 1978. Arteriovenous anastomoses in the skin of seals. I The weddell seal *Leptonychotes weddelli* and the elephant seal *Mirounga leonina*. *Anat. Rec.* **191** (2): 239–252.

MONTAGNA, W. & HARRISON, R. J. 1957. Specializations in the skin of the seal (*Phoca vitulina*). *Am. J. Anat.* **100**: 81–114.

MOREJOHN, G. V. 1969. Vertebral column deformity and osteonecrosis of pelvis and femur in the California sea lion. *Calif. Fish and Game* **55** (4): 323–326.

MOREJOHN, G. V. 1975. A phylogeny of otariid seals based on morphology of the baculum. *Rapp. P-v. Réun. Cons. int. Explor. Mer.* **169**: 49–56.

MOREJOHN, G. V. & BRIGGS, K. T. 1973. Post mortem studies of northern elephant seal pups. *J. Zool., Lond.* **171** (1): 67–78.

MORI, M. 1958. The skeleton and musculature of *Zalophus*. *Okajimas Folia anat. Jap.* **31**: 203–284.

MORITA, E., CHIOCCHIO, S. R. & TRAMEZZANI, J. H. 1970. The carotid body of the Weddell seal (*Leptonychotes weddelli*). *Anat. Rec.* **167** (3): 309–328.

MUIZON, C. de 1978. *Arctocephalus (Hydrarctos) lomasiensis*, subgen. nov. et nov. sp., un nouvel Otariidae du Mio-Pliocene de Sacaco (Pérou). *Bull. Inst. Fr. Et. And.* **7** (3–4): 169–188.

MUIZON, C. de 1981 *Les vertébrés fossiles de la formation Pisco (Pérou)*. Institut Français d'études Andines. Recherche sur les grandes civilisations. Mém. No. 6. 150 pp. Paris.

MUIZON, C. de & HENDEY, Q. B. 1980. Late Tertiary seals of the South Atlantic Ocean. *Ann. South Afr. Mus.* **82** (3): 91–128.

MURIE, J. 1872a. Researches upon the anatomy of the Pinnipedia. Part 1. On the walrus (*Trichechus rosmarus* Linn.). *Trans. Zool. Soc. Lond.* **7**: 411–464.

MURIE, J. 1872b. Researches upon the anatomy of the Pinnipedia. Part 2. Descriptive anatomy of the sea lion (*Otaria jubata*). *Trans. Zool. Soc. Lond.* **7**: 527–596.

MURIE, J. 1874. Researches upon the anatomy of the Pinnipedia. Part 3. Descriptive anatomy of the sea lion (*Otaria jubata*). *Trans. Zool. Soc. Lond.* **8**: 501–582.

MURRAY, M. D. & NICHOLLS, D. G. 1965. Studies on the ectoparasites of seals and penguins. I. The ecology of the louse *Lepidophthirus macrorhini* Enderlein on the southern elephant seal, *Mirounga leonina* (L). *Aust. J. Zool.* **13**: 437–454.

MURRAY, M. D., SMITH, M. S. R. & SOUCEK, Z. 1965. Studies on the ectoparasites of seals and penguins. II The ecology of the louse *Antarctophthirus ogmorhini* Enderlein on the Weddell seal *Leptonychotes weddelli* Lesson. *Aust. J. Zool.* **13**: 761–771.

MUSGRAVE, T. 1866. *Castaway on the Auckland Isles*. 174pp. London.

NAEVDAL, G. 1965. Protein polymorphism used for identifications of harp seal populations. *Årbok Univ. Bergen Mat.-Naturv. Serie* **1965** (9): 3–20.

NAEVDAL, G. 1969. Blood protein polymorphism in harp seals off eastern Canada. *J. Fish. Res. Bd. Can.* **26** (5): 1397–1399.

NAIRN, R. G. W. 1979. The status and conservation of the common seal *Phoca vitulina* in Northern Ireland. *Irish Natur J.* **19** (10): 360–363.

NAITO, Y. 1974. The hyoid bones of two kinds of harbour seals in the adjacent waters of Hokkaido. *Scient. Rep. Whales Res. Inst. Tokyo* **25**: 301–310.

NAITO, Y. & OSHIMA, M. 1976. The variation in the development of pelage of the ribbon seal with reference to the systematics. *Sci. Rep. Whales Res. Inst. Tokyo* **28**: 187–197.

NAUMOV, S. P. 1933. The seals of the USSR. The raw material of the marine mammal fishery. Series: *Economically exploited animals of the USSR*. 105pp. Gen. ed. N. A. Bobrinskii. Moscow.

NEILAND, K. A. 1961. Suspected role of parasites in non-rookery mortality of fur seals (*Callorhinus ursinus*). *J. Parasit.* **47**: 732.

NERC. 1981. Annual assessment of the stocks of grey seals and common seals in Great Britain 1981. *NERC Newsletter* **2** (11): 10–11.

NEWBY, T. C. 1973. Observations on the breeding behaviour of the harbour seal in the State of Washington. *J. Mamm.* 54 (2): 540–543.

NISHIWAKI, M. 1972. General Biology. pp. 3–204 in *Mammals of the sea. Biology and medicine.* Ed. S. H. Ridgway. Charles Thomas, Springfield, Ill.

NISHIWAKI, M. 1973. Status of the Japanese sea lion. pp. 80–81 in *Seals. IUCN Suppl. Paper No. 39.*

NISHIWAKI, M. & NAGASAKI, F. 1960. Seals of the Japanese coastal waters. *Mammalia,* Paris 24: 459–467.

NOAA 1977. *The story of the Pribilof fur seals.* 13pp. US Dept. Commerce. National Oceanic & Atmospheric Administration.

NORRIS, K. S. 1969. The echolocation of marine mammals. pp. 391–423 in *The biology of marine mammals.* Ed. H. T. Andersen. Academic Press, London.

NUTTING, W. B. & DAILEY, M. D. 1980. Demodicosis (Acari: Demodicidae) in the Californian sea lion. *J. Med. Entomol.* 17 (4): 344–347.

NYHOLM, E. S. 1975. Observations on the walrus (*Odobenus rosmarus* L.) in Spitsbergen in 1971–1972. *Ann. Zool. Fennici* 12: 193–196.

OBENDORF, D. L. & PRESIDENTE, P. J. A. 1978. Foreign body perforation of the esophagus initiating traumatic pericarditis in an Australian fur seal. *J. Wildl. Dis.* 14: 451–454.

ODELL, D. K. 1971. Censuses of pinnipeds breeding on the California Channel Islands. *J. Mamm.* 52 (1): 187–190.

OGNEV, S. I. 1935. *Mammals of USSR and adjacent countries.* 3: Carnivora. Israel Program for Sci. Transl. Jerusalem 1962.

O'GORMAN, F. 1961. Fur seals breeding in the Falkland Islands Dependencies. *Nature* 192 (4806): 914–916.

O'GORMAN, F. 1963. Observations on terrestrial locomotion in Antarctic seals. *Proc. Zool. Soc. Lond.* 141: 837–850.

ORR, M. F. 1971. Survival of histological structure and biochemical constituents in an ancient mummified Weddell seal. Part II. Survival of histological structure. *Antarctic Pinnipedia. Am. Geophys. Union. Ant. Res. Series* 18: 197–206.

ORR, R. T. 1967. The Galapagos sea lion. *J. Mamm.* 48 (1): 62–69.

ORR, R. T. 1973. Galapagos fur seal (*Arctocephalus galapagoensis*). pp. 124–128 in *Seals. IUCN Suppl. Paper No. 39.*

ORR, R. T. & POULTER, T. C. 1965. The pinniped population of Ano Nuevo Island, California. *Proc. Calif. Acad. Sci.* 4th ser. 32 (13): 377–404.

ORR, R. T. & POULTER, T. C. 1967. Some observations on reproduction, growth, and social behavior in the Steller sea lion. *Proc. Calif. Acad. Sci.* 4th ser. 35 (10): 193–226.

ORR, R. T., SCHONEWALD, J. & KENYON, K. W. 1970. The California sea lion: skull growth and a comparison of two populations. *Proc. Calif. Acad. Sci.* 4th ser. 37 (11): 381–394.

ORTIZ, C. L., COSTA, D. & LEBOEUF, B. J. 1978. Water and energy flux in elephant seal pups fasting under natural conditions. *Physiol. Zool.* 51 (2): 166–178.

OSGOOD, W. H., PREBLE, E. A. & PARKER, G. H. 1916. The fur seals and other life of the Pribilof Islands, Alaska in 1914. *Bull. US Bur. Fish.* 34: 1–172.

OWEN, R. 1853. On the anatomy of the walrus. *Proc. Zool. Soc. Lond.* 103–106.

PAULIAN, P. 1953. Pinnipèdes, cétacés, oiseaux des Îles Kerguelen et Amsterdam. Mission Kerguelen 1951. *Mém. Inst. Sci. Madagascar* Ser. A 8: 111–234.

PAULIAN, P. 1964. Contribution a l'étude de l'otarie de l'Ile Amsterdam. *Mammalia,* Paris 28: suppl. 1, 1–146.

PAYNE, M. R. 1977. Growth of a fur seal population. *Phil. Trans. R. Soc. Lond.* B. 279: 67–79.

PAYNE, M. R. 1978. Population size and age determination in the Antarctic fur seal *Arctocephalus gazella. Mammal. Rev.* 8 (1 and 2): 67–73.

PAYNE, M. R. 1979a. Growth in the Antarctic fur seal *Arctocephalus gazella. J. Zool. Lond.* 187: 1–20.

PAYNE, M. R. 1979b. Fur seals *Arctocephalus tropicalis* and *A. gazella* crossing the Antarctic Convergence at South Georgia. *Mammalia,* Paris 43: 93–98.

PEAKER, M. & GOODE, J. A. 1978. The milk of the fur seal *Arctocephalus tropicalis gazella*; in particular the composition of the aqueous phase. *J. Zool., Lond.* 185 (4): 469–476.

PEARSE, R. J. 1979. Distribution and conservation of the Australian fur seal in Tasmania. *Vict. Nat.* 96 (2): 48–53.

PENNEY, R. L. 1969. The leopard seal – south polar predator. *Animal Kingdom.* NY. 72 (5): 2–7.

PENNEY, R. L. & LOWRY, G. 1967. Leopard seal predation on Adelie penguins. *Ecology* 48: 878–882.

PETERS, W. 1875. Uber eine neue Art von Seebaren, *Arctophoca gazella* von den Kerguelen Inseln. *Monatsb. Akad. Berlin.* 393–399.

PETERS, W. 1866. Uber die Ohrenrobben (Seelöwen un Seebären) *Otariae,* insbesondere über die in dem Sammlungen zu Berlin befindlichen Arten. *Monatsb. Akad. Berlin* 261–281, 665–672.

PETERSON, R. S. 1968. Social behaviour in pinnipeds, with particular reference to the northern fur seal. pp. 3–53 in *The behavior and physiology of pinnipeds.* Eds R. J. Harrison *et al.* Appleton-Century-Crofts, NY.

PETERSON, R. S. & BARTHOLOMEW, G. A. 1967. The natural history and behavior of the California sea lion. 79pp. *Am. Soc. Mamm. Spec. Pub. No. 1.*

PETERSON, R. S. & BARTHOLOMEW, G. A. 1969. Airborne vocal communication in the Californian sea lion, *Zalophus californianus. Anim. Behaviour* (Lond.) 17 (1): 17–24.

PETERSON, R. S., HUBBS, C. L., GENTRY, R. L. & DELONG, R. L. 1968a.

The Guadalupe fur seal: habitat, behaviour, population size and field identification. *J. Mamm.* **49** (4): 665–675.

PETERSON, R. S., LEBOEUF, B. J. & DELONG, R. L. 1968b. Fur seals from the Bering Sea breeding in California. *Nature.* Lond. **219** (5157): 899–901.

PETERSON, R. S. & REEDER, W. G. 1966. Multiple births in the northern fur seal. *Z. Saugetierk.* **31**: 52–56.

PETRINOVICH, L. 1974. Individual recognition of pup vocalization by northern elephant seal mothers. *Z. Tierpsychol.* **34** (3): 308–312.

PIÉRARD, j. 1966. Sexual dimorphism in laryngeal size of the northern fur seal. *J. Mamm.* **47**: 143–145.

PIÉRARD, J. 1969. Le larynx du phoque de Weddell (*Leptonychotes weddelli*, Lesson 1826). *Canad. J. Zool.* **47** (1): 77–87.

PIÉRARD, J. 1971. Osteology and myology of the Weddell seal *Leptonychotes weddelli* (Lesson 1826). *Antarctic Pinnipedia. Am. Geophys. Union Ant. Res. Series 18: 53–108.*

PIÉRARD, J. & BISAILLON, A. 1978. Osteology of the Ross seal *Ommatophoca rossi* Gray 1844. *Biol. of Antarctic seas* **9**: 1–24. *Ant. Res. Series 31: Am. Geophys. Union.*

PILLERI, G. & GIHR, M. 1977. *Radical extermination of the South American sea lion* Otaria byronia (*Pinnipedia, Otariidae*) *from Isla Verde, Uruquay.* 15pp. Verlag des Hirnanatomischen Institutes. Ostermundigen (Bern).

PILSON, M. E. Q. 1965. Absence of lactose from the milk of the Otarioidea, a superfamily of marine mammals. *Am. Zool.* **5**: 220.

PILSON, M. E. Q. & KELLY, A. L. 1962. Composition of the milk from *Zalophus californianus*, the California sea lion. *Science* **135** (3498): 104–105.

PITCHER, K. W. & CALKINS, D. G. 1981. Reproductive biology of Steller sea lions in the Gulf of Alaska. *J. Mamm.* **62** (3): 599–605.

PLATT, N. E., PRIME, J. H. & WITTHAMES, S. R. 1975. The age of the grey seal at the Farne Islands. *Trans. Nat. Hist. Soc.* Northumbria **42** (4): 99–106.

POPOV, L. 1979a. Ladoga seal. pp. 70–71 in *Mammals in the seas* **2** FAO Fisheries Series No. 5.

POPOV, L. 1979b. Baikal seal. pp. 72–73 in *Mammals in the seas.* **2** FAO Fisheries Series No. 5.

POTELOV, V. A. 1975. Reproduction of the bearded seal (*Erignathus barbatus*) in the Barents Sea. *Rapp. P-v. Réun. Cons. int. Explor. Mer.* **169**: 554.

POULTER, T. C. 1966. The use of active sonar by the California sea lion. *J. Auditory Res.* **6**: 165–173.

POULTER, T. C. 1968. Underwater vocalisation and behavior of pinnipeds. pp. 69–84 in *The behavior and physiology of pinnipeds.* Eds R. J. Harrison *et al.* Appleton-Century-Crofts, NY.

POUVREAU, B., DUGUY, R., ALZIEU, C. & BABIN, P. 1980. Capture d'un phoque à crête, *Cystophora cristata* (Erxleben, 1777) sur la côte française atlantique et recherches sur sa pathologie. *Bull Cent. Étud. Rech. sci. Biarritz.* **13** (1): 7–12.

PRIEUR, D. & DUGUY, R. 1981. Les phoques des côtes de France. III Le phoque gris *Halichoerus grypus* (Fabricius, 1791). *Mammalia*, Paris. **45** (1): 83–98.

PRIME, J. H. 1981. Breeding grey seals on the Isle of May 1980. *Trans. Nat. Hist. Soc. Northumb.* **47**: 13–16.

RADDE, G. 1862. *Reisen im Süden von Ost-Siberien in den Jahren 1855–1859 incl.* I: *Mammalia.*

RAE, B. B. 1969. Twin seals in Scotland. *J. Zool., Lond.* **158**: 243–245.

RAE, B. B. 1973. Further observations on the food of seals. *J. Zool., Lond.* **169** (3): 287–297.

RAMPRASHAD, F., COREY, S. & RONALD, K. 1971. The harp seal *Pagophilus groenlandicus* (Erxleben, 1777). XIII. The gross and microscopic structure of the auditory meatus. *Can. J. Zool.* **49** (2): 241–248.

RAMPRASHAD, F., COREY, S. & RONALD, K. 1972. Anatomy of the seal's ear (*Pagophilus groenlandicus* Erxleben, 1977). pp. 263–306 in *Functional anatomy of marine mammals* **1**. Ed. R. J. Harrison. Academic Press, London.

RAMPRASHAD, F., COREY, S. & RONALD, K. 1973. The harp seal, *Pagophilus groenlandicus* (Erxleben, 1777). XIV. The gross and microscopic structure of the middle ear. *Can. J. Zool.* **51** (6): 589–600.

RANCUREL, P. 1975. Échouages d'otaries à fourrure dans le sud de de la Nouvelle-Calédonie. *Mammalia*, Paris **39**(3): 499–504.

RAND, R. W. 1956a. The Cape fur seal, *Arctocephalus pusillus* (Schreber), its general characteristics and moult. 52pp. *Union S. Africa Dept. Commerce and Industries. Div. Fisheries Invest. Rep. No. 21.*

RAND, R. W. 1956b. Notes on the Marion Island fur seal. *Proc. Zool. Soc. Lond.* **126**: 65–82.

RAND, R. W. 1967. The Cape fur seal (*Arctocephalus pusillus*). 3. General behavior on land and at sea. 40pp. *Rep. S. Africa Dept. Commerce and Industries. Div. Sea Fish. Invest. Rep. No. 60.*

RAND, R. W. 1972. The Cape fur seal *Arctocephalus pusillus*. 4. Estimates of population size. *Rep. S. Africa Dept. Indust. Div. Sea Fisheries. Invest. Report.* **89**: 1–28.

RAY, C., 1963. Locomotion in pinnipeds. *Nat. Hist.* NY. **72**: 10–21.

RAY, C. 1966. Snooping on seals for science in Antarctica. *Animal Kingdom* **69** (3): 66–75.

RAY, C. & FAY, F. H. 1968. Influence of climate on the distribution of walruses *Odobenus rosmarus* (Linnaeus). II. Evidence from physiological characteristics. *Zoologica* NY. **53** (1): 19–32.

RAY, C. & SMITH, M. S. R. 1968. Thermoregulation of the pup and adult Weddell seal, *Leptonychotes weddelli* (Lesson) in Antarctica. *Zoologica* NY. **53** (1): 33–46.

RAY, C., WATKINS, W. A. & BURNS, J. J. 1969. The underwater song of *Erignathus* (bearded seal). *Zoologica* NY. **54** (2): 79–83.

RAY, C. E. 1976. Geography of phocid evolution. *Syst. Zool.* **25** (4): 391–406.

RAY, C. E. & LING, J. K. 1981. A well-documented early record of the Australian sea lion. *Archives of Natural History.* **10** (1): 155–171.

RAY, G. C. & WATKINS, W. A. 1975. Social function of underwater sounds in the walrus *Odobenus rosmarus. Rapp. P-v. Réun. Cons. int. Explor. Mer.* **169**: 524–526.

REEVES, R. R. 1978. Atlantic walrus (*Odobenus rosmarus rosmarus*): a literature survey and status report. *US Dept. Int. Fish Wildl. Serv. Wildl. Res. Rep.* **10**: 1–41.

REITER, J., STINSON, N. L. & LEBOEUF, B. J. 1978. Northern elephant seal development: the transition from weaning to nutritional independence. *Behav. Ecol. Sociobiol.* **3**: 337–367.

REPENNING, C. A. 1972. Underwater hearing in seals: functional morphology. pp. 307–331 in *Functional anatomy of marine mammals* **1** Ed. R. J. Harrison. Academic Press, London.

REPENNING, C. A. 1976. Adaptive evolution of sea lions and walruses. *Syst. Zool.* **25** (4): 375–390.

REPENNING, C. A., PETERSON, R. S. & HUBBS, C. L. 1971. Contributions to the systematics of the southern fur seals, with particular reference to the Juan Fernandez and Guadalupe species. *Antarctic Pinnipedia. Ant. Res. Series* **18**: 1–34. Am. Geophys. Union.

REPENNING, C. A. & RAY, C. E. 1977. The origin of the Hawaiian monk seal. *Proc. Biol. Soc. Wash.* **89** (58): 667–688.

REPENNING, C. A., RAY, C. E. & GRIGORESCU, D. 1979. Pinniped biogeography. pp. 357–369 in *Historical biogeography, plate tectonics and the changing environment*. Eds J. Gray & A. J. Boucot. Oregon State Univ. Press.

REPENNING, C. A. & TEDFORD, R. H. 1977. Otarioid seals of the Neogene. 93pp. *U.S. Dept. Int. Geol. Surv. Professional Paper 992.*

REVENTLOW, A. 1951. Observations on the walrus (*Odobenus rosmarus*) in captivity. *Der. Zool. Gart.* (NF) **18**: 227–234.

RICE, D. W. 1973. Caribbean monk seal (*Monachus tropicalis*). pp. 98–112 in *Seals. IUCN Suppl. Paper No. 39.*

RIEDMAN, M. & ORTIZ, C. L. 1979. Changes in milk composition during lactation in the northern elephant seal. *Physiol. Zool.* **52** (2): 240–249.

RIDGWAY, S. H. 1972. Homeostasis in the aquatic environment. pp. 590–747 in *Mammals of the sea. Biology and medicine*. Ed. S. H. Ridgway. Charles C. Thomas, Illinois.

RIDGWAY, S. H., GERACI, J. R. & MEDWAY, W. 1975a. Diseases of pinnipeds. *Rapp. P-v. Réun. Cons. int. Explor. Mer.* **169**: 327–337.

RIDGWAY, S. H., HARRISON, R. J. & JOYCE, P. L. 1975b. Sleep and cardiac rhythm in the gray seal. *Science* **187** (4176): 553–555.

RIGDON, R. H. & DRAGER, G. A. 1955. Thiamine deficiency in sea lions (*Otaria californiana*) fed only frozen fish. *J. Amer. Vet. Med. Ass.* **127**: 453–455.

RITCHIE, J. 1921. The walrus in British waters. *Scot. Nat.* **5–9**, 77–86.

ROBINETTE, H. R. & STAINS, H. J. 1970. Comparative study of the calcanea of the Pinnipedia. *J. Mamm.* **51** (3): 527–541.

RODAHL, K. 1949. The toxic effect of polar bear liver. 90pp. *Norsk Polarinst. Skrifter* No. 92.

RONALD, K. 1980. Newsletter of the league for the conservation of the monk seal. 44pp. *IUCN* No. 5.

RONALD, K., JOHNSON, E., FOSTER, M. & VANDER POL, D. 1970. The harp seal, *Pagophilus groenlandicus* (Erxleben 1777). I. Methods of handling, molt and diseases in captivity. *Can. J. Zool.* **48** (5): 1035–1040.

RONALD, K., MCCARTNER, R. & SELLEY, L. J. 1977. Venous circulation in the harp seal *Pagophilus groenlandicus. Funct. Anat. Marine Mammals.* Ed. R. J. Harrison **3**: 235–270.

ROSS, G. J. B. 1972. Nuzzling behaviour in captive Cape fur seals. *Int. Zoo Yearbook* **12**: 183–184.

ROSS, G. J. B., RYAN, F., SAAYMAN, G. S. & SKINNER, J. 1976. Observations on two captive crabeater seals at the Port Elizabeth Oceanarium. *Int. Zoo Yearbook* **16**: 160–164.

ROUNSEVELL, D. & EBERHARD, I. 1980. Leopard seals, *Hydrurga leptonyx* (Pinnipedia) at Macquarie Island from 1949 to 1979. *Aust. Wildl. Res.* **7**: 403–415.

SANDEGREN, F. E. 1970. Breeding and maternal behavior of the Steller sea lion (*Eumetopias jubata*) in Alaska. 138pp. Thesis pres. to Univ. Alaska.

SARICH, V. M. 1969. Pinniped origins and the rate of evolution of carnivore albumins. *Syst. Zool.* **18** (3): 286–295.

SAVAGE, R. J. G. 1957. The anatomy of *Potamotherium*, an Oligocene lutrine. *Proc. Zool. Soc. Lond.* **129**: 151–244.

SCHEFFER, V. B. 1951. Cryptorchid fur seals. *Amer. Mid. Nat.* **46** (3): 646–648.

SCHEFFER, V. B. 1958. *Seals, sea lions and walruses. A review of the Pinnipedia.* 179pp. Stanford Univ. Press.

SCHEFFER, V. B. 1962. Pelage and surface topography of the northern fur seal. 92pp. *US Dept. Int. N. Am. Fauna.* No. 64.

SCHEFFER, V. B. 1964. Hair patterns in seals (Pinnipedia). *J. Morph.* **115**: 291–304.

SCHEFFER, V. B. 1967. Standard measurements of seals. *J. Mamm.* **48** (3): 459–462.

SCHEFFER, V. B. & JOHNSON, A. M. 1963. Molt in the northern fur seal. *US Fish & Wildl. Service. Special Scientific Report. Fisheries* No. 450.

SCHEFFER, V. B. & KENYON, K. W. 1963. Baculum size in pinnipeds. *Zeit. für Säugetierk.* **28** (1): 38–41.

SCHEFFER, V. B. & KRAUS, B. S. 1964. Dentition of the northern fur seal. *Fishery Bulletin* **63**: 293–342.

SCHEFFER, V. B. & PETERSON, R. S. 1967.
Growth layers in teeth of suckling fur seals.
Growth **31**: 35–38.

SCHEFFER, V. B. & SLIPP, J. W. 1944.
The harbor seal in Washington State. *Am. Midl.
Nat.* **32** (2): 373–416.

SCHEFFER, V. B. & WILKE, F. 1953.
Relative growth in the northern fur seal. *Growth*
17: 129–145.

SCHEVILL, W. E., WATKINS, W. A. & RAY, C. 1966.
Analysis of underwater *Odobenus* calls with
remarks on the development and function of the
pharyngeal pouches. *Zoologica Sci. Contr. NY Zool.
Soc.* **51** (3): 103–106.

SCHEVILL, W. E. & WATKINS, W. A. 1971.
Directionality of the sound beam in *Leptonychotes
weddelli. Antarctic Pinnipedia. Am. Geophys. Union.
Ant. Res. Series* **18**: 163–168.

SCHLIEMANN, H. 1968. Notiz über einen Bastard
zwischen *Arctocephalus pusillus* und *Zalophus
californianus. Z. Saugetierk.* **33** (1): 42–45.

SCHNAPP, B., HELLWING, S. & GHIZELEA, G. 1962.
Contributions to the knowledge of the Black Sea
seal (*Monachus monachus*). *Trav. Mus. Hist. Nat.
'Gr. Antipa'.* Bucarest **3**: 382–400.

SCHNEIDER, R. 1962. Vergleichende Untersuchungen
am Kehlkopf der Robben (Mammalia, Carnivora,
Pinnipedia). *Morph. Jb.* **103**: 177–262.

SCHNEIDER, R. 1963. Morphologische
Anpassungserscheinungen am Kehlkopf einiger
aquatiler Säugetiere. *Zeit. f. Saugetierk.* **28**: 257–267.

SCHREIBER, R. W. & KRIDLER, E. 1969.
Occurrence of an Hawaiian monk seal (*Monachus
schauinslandi*) on Johnston Atoll, Pacific Ocean. *J.
Mamm.* **50** (4): 841–842.

SCHROEDER, C. R. 1933. Cow's milk protein
hypersensitivity in a walrus. *J. Am. Vet. Med. Ass.*
83: 810–815.

SCHUSTERMAN, R. J. 1968. Experimental laboratory
studies of pinniped behavior. pp. 87–171 in *The
behavior and physiology of pinnipeds.* Eds R. J.
Harrison *et al.* Appleton-Century-Crofts, NY.

SCHUSTERMAN, R. J. 1981. Behavioral capabilities of
seals and sea lions: a review of their hearing,
visual, learning and diving skills. *The Psychological
Record.* **31**: 1251–143.

SEAWRIGHT, A. A. 1964. Pulmonary acariasis in a
Tasmanian fur seal. *J. Comp. Path. Therapeutics* **74**
(1): 97–100.

SERGEANT, D. E. 1965. Exploitation and conservation
of harp and hooded seals. *Polar Record* **12** (80):
541–551.

SERGEANT, D. E. 1973a. Environment and
reproduction in seals. *J. Reprod. Fert. Suppl.* **19**:
555–561.

SERGEANT, D. E. 1973b. Transatlantic migration of a
harp seal, *Pagophilus groenlandicus. J. Fish. Res. Bd.
Can.* **30** (1): 124–125.

SERGEANT, D. E. 1974. A rediscovered whelping
population of hooded seals *Cystophora cristata*

Erxleben and its possible relationship to other
populations. *Polarforschung* **44** (1): 1–7.

SERGEANT, D. E. 1976. History and present status of
populations of harp and hooded seals. *Biol.
Conserv.* **10**: 95–118.

SERGEANT, D., RONALD, K., BOULVA, J. & BERKES, F.
1978.
The recent status of *Monachus monachus*, the
Mediterranean monk seal. *Biol. Conserv.* **14**:
259–287.

SHAUGHNESSY, P. D. 1970. Serum protein variation in
southern fur seals, *Arctocephalus* spp in relation to
their taxonomy. *Aust. J. Zool.* **18**: 331–343.

SHAUGHNESSY, P. D. 1975. Observations on the seals of
Gough Island. *S. Afr. J. Antarct. Res.* No. **5**: 42–44.

SHAUGHNESSY, P. D. 1976. Controversial harvest.
African Wildlife **30** (6): 26–31.

SHAUGHNESSY, P. D. 1979a Cape (South African) fur
seal. pp. 37–40 in *Mammals in the seas.* **2** FAO
Fisheries Series No. 5.

SHAUGHNESSY, P. D. 1979b. *Cape fur seals in Southwest
Africa.* S.W.A. Annual. pp. 101–105.

SHAUGHNESSY, P. D. & FAY, F. H. 1977.
A review of the taxonomy and nomenclature of
North Pacific harbour seals. *J. Zool., Lond.* **182**:
385–419.

SHAUGHNESSY, P. D. & ROSS, G. J. B. 1980.
Records of the subantarctic fur seal (*Arctocephalus
tropicalis*) from South Africa with notes on its
biology and some observations on captive
animals. *Ann. S. Afr. Mus.* **82** (2): 71–89.

SHULDHAM, M. 1775. Account of the sea-cow, and the
use made of it. *Phil. Trans.* Lond. **65**: 249–251.

SINIFF, D. B. & BENGTSON, J. L. 1977.
Observations and hypotheses concerning the
interactions among crabeater seals, leopard seals
and killer whales. *J. Mamm.* **58** (3): 414–416.

SINIFF, D. B., STIRLING, I., BENGTSON, I. L. & REICHLE,
R. A. 1979.
Social and reproductive behaviour of crabeater
seals (*Lobodon carcinophagus*) during the austral
spring. *Can. J. Zool.* **57** (11): 2243–2255.

SIVERTSON, E. 1941. On the biology of the harp seal.
Hvalrådets Skrifter Nr. **26**: 1–166.

SIVERTSEN, E. 1953. A new species of sea lion,
Zalophus wollebaeki from the Galapagos Islands. *Det
Kong. Norske Vid. Selsk. Forh.* **26** (1): 1–3.

SIVERTSEN, E. 1954. A survery of the eared seals
(Family Otariidae) with remarks on the Antarctic
seals collected by M/K *Norvegia* in 1928–1929 Det
Norske Videnskaps-Akademi i Oslo. *Sci. Res.
Norw. Ant. Exp.* 1927–1928. No. 36. 76pp.

SKINNER, J. S. 1957. Seal finger. *Amer. Med. Assoc.
Arch. Dermatol.* **75**: 559–561.

SLIJPER, E. J. 1961. Foramen ovale and ductus
arteriosus Botalli in aquatic mammals. *Mammalia,*
Paris. **25**: 528–570.

SLIJPER, E. J. 1968. On the heart of temporary aquatic
mammals with special reference to the embryonic
pathways. *Bijdragen Dierk.* **38**: 75–84.

SMITH, A. W., BROWN, R. J., SKILLING, D. E., BRAY, H. L. & KEYES, M. C. 1977.
Naturally occurring leptospirosis in northern fur seals (*Callorhinus ursinus*). *J. Wildl. Dis.* **13** (2): 144–148.

SMITH, E. A. 1966. A review of the world's grey seal population. *J. Zool., Lond.* **150**: 463–489.

SMITH, I. W. G. 1978. Seasonal sea mammal exploitation and butchering patterns in an archaic site (Tairua N44/2) on the Coromandel Peninsula. *Rec. Auckland Inst. Mus.* **15**: 17–26.

SMITH, I. W. G. 1979. Prehistoric sea mammal hunting in Palliser Bay. *Prehistoric Man in Palliser Bay, National Museum Bulletin* **21**: 215–224.

SMITH, T. G. 1973a. Population dynamics of the ringed seal in the Canadian eastern Arctic. *Fish. Res. Bd. Can. Bull.* **181**: 55pp.

SMITH, T. G. 1973b. Management research on the eskimo's ringed seal. *Can. Geog. J.* **86** (4): 118–125.

SMITH, T. G. 1976. Predation of ringed seal pups (*Phoca hispida*) by the arctic fox (*Alopex lagopus*). *Can. J. Zool.* **54**: 1610–1616.

SMITH, T. G. 1980. Polar bear predation of ringed and bearded seals in the land-fast sea ice habitat. *Can. J. Zool.* **58** (12): 2201–2209.

SMITH, T. G. & STIRLING, I. 1975. The breeding habitat of the ringed seal (*Phoca hispida*). The birth lair and associated structures. *Can. J. Zool.* **53** (9): 1297–1305.

SOKOLOV, A. S., KOSYGIN, G. M. & SHUSTOV, A. P. 1968. Lungs and trachea structure of the Bering Sea pinnipeds. *Trudy vses. nauchno-issled. Inst. morsk ryb. Khoz. Okeanogr.* **62** (68): 252–263.

SONNTAG, C. F. 1923. The comparative anatomy of the tongues of the Mammalia. VIII. Carnivora. *Proc. Zool. Soc. Lond.* 129–153.

SOUTHCOTT, R. V., CHESTERFIELD, N. J. & LUGG, D. J. 1971. Vitamin A content of the livers of huskies and some seals from Antarctic and subantarctic regions. *Med. J. Aust.* **1971** (1): 311–313.

SOUTHCOTT, R. V., CHESTERFIELD, N. J. & WARNEKE, R. M. 1974. The vitamin A content of the liver of the Australian fur seal, *Arctocephalus pusillus doriferus*. *Aust. Wildl. Res.* **1** (2): 145–148.

SPALDING, D. J. 1964. Comparative feeding habits of the fur seal, sea lion and harbour seal on the British Columbia coast. *Fish. Res. Bd. Canada. Bull. No. 146.*

SPALDING, D. 1966. Eruption of permanent canine teeth in the northern sea lion. *J. Mamm.* **47**: 157–158.

STEPHENS, R. J., BEEBE, I. J. & POULTER, T. C. 1973. Innervation of the vibrissae of the Californian sea lion, *Zalophus californianus*. *Anat. Rec.* **176** (4): 421–442.

STEWART, R. E. A. & LAVIGNE, D. M. 1980. Neonatal growth of northwest Atlantic harp seals, *Pagophilus groenlandicus*. *J. Mamm.* **61** (4): 670–680.

STIRLING, I. 1969. Tooth wear as a mortality factor in the Weddell seal, *Leptonychotes weddelli*. *J. Mamm.* **50** (3): 559–565.

STIRLING, I. 1970. Observations on the behaviour of the New Zealand fur seal (*Arctocephalus forsteri*). *J. Mamm.* **51** (4): 766–778.

STIRLING, I. 1971. Population dynamics of the Weddell seal (*Leptonychotes weddelli*) in McMurdo Sound, Antarctica, 1966–1968. *Antarctic Pinnipedia, Ant. Res. Series.* **18**: 141–161. Am. Geophys. Union.

STIRLING, I. 1971a. Studies on the behaviour of the South Australian fur seal, *Arctocephalus forsteri* (Lesson). I. Annual cycle, posture and calls, and adult males during the breeding season. *Aust. J. Zool.* **19**: 243–266.

STIRLING, I. 1972a. Observations on the Australian sealion *Neophoca cinerea* (Péron). *Aust. J. Zool.* **20**: 271–279

STIRLING, I. 1972b. The economic value and management of seals in South Australia. *Dept. Fisheries Pub.* No. 2: 11pp.

STIRLING, I. & ARCHIBALD, R. 1979. Bearded seal. pp. 83–85 in *Mammals in the seas*. **2** FAO Fisheries Series No. 5.

STIRLING, I. & CALVERT, W. 1979. Ringed seal. pp. 66–69 in *Mammals in the seas*. **2** FAO Fisheries Series. No. 5.

STIRLING, I. & GENTRY, R. L. 1972. An elimination posture in the Pinnipedia. *J. Mamm.* **53** (1): 191–192.

STIRLING, I. & MCEWAN, E. H. 1975. The caloric value of whole ringed seals (*Phoca hispida*) in relation to polar bear (*Ursus maritimus*) ecology and hunting behaviour. *Can. J. Zool.* **53** (8): 1021–1027.

STIRLING, I. & WARNEKE, R. M. 1971. Implications of a comparison of the airborne vocalizations and some aspects of the behaviour of the two Australian fur seals, *Arctocephalus* spp, on the evolution and present taxonomy of the genus. *Aust. J. Zool.* **19**: 227–241.

STODDART, D. R. 1972. Pinnipeds or sirenians at western Indian Ocean Islands? *J. Zool., Lond.* **167**: 207–217.

STRANGE, I. 1972. Wildlife in the Falklands. *Oryx* **11** (4): 240–257.

STRANGE, I. 1979. Sea lion survey in the Falklands. *Oryx* **15** (2): 175–184.

STREET, R. J. 1964. Feeding habits of the New Zealand fur seal *Arctocephalus forsteri*. *N.Z. Mar. Dept. Fish. Tech. Rep.* No. 9: 20pp.

STROUD, R. K. & DAILEY, M. D. 1978. Parasites and associated pathology observed in pinnipeds stranded along the Oregon coast. *J. Wildl. Dis.* **14**: 292–298.

STROUD, R. K. & ROFFE, T. J. 1979. Causes of death in marine mammals stranded along the Oregon coast. *J. Wildl. Dis.* **15** (1): 91–97.

SUMMERS, C. F. 1974. The grey seal (*Halichoerus grypus*) in Cornwall and the Isles of Scilly. *Biol. Conserv.* **6** (4): 285–291.

SUMMERS, C. F. 1979. The scientific background to seal stock management in Great Britain. *Nat. Env. Res. Council, Publ. Ser.* C. No. 21. 14 pp.

SUMMERS, C. F., BONNER, W. N. & VAN HAAFTEN, J. 1978.
Changes in the seal populations of the North Sea. *Rapp. P-v. Réun. Cons. int. Explor. Mer.* **172**: 278–285.

SUMMERS, C. F., BURTON, R. W. & ANDERSON, S. S. 1975.
Grey seal (*Halichoerus grypus*) pup production at North Rona: A study of birth and survival statistics collected in 1972. *J. Zool. Lond.* **175**: 439–451.

SUMMERS, C. F., WARNER, P. J., NAIRN, R. G. W., CURREY, M. G. & FLYNN, J. 1980.
An assessment of the status of the common seal *Phoca vitulina vitulina* in Ireland. *Biol. Conserv.* **17** (2): 115–123.

SWALES, M. K. 1956. The fur seals of Gough Island. Unpublished report. Colonial Office London.

SWEENEY, J. C. & GILMARTIN, W. G. 1974.
Survey of diseases in freeliving California sea lions. *J. Wildl. Dis.* **10** (4): 370–376.

TALENT, L. G. & TALENT, C. L. 1975.
An extrauterine fetus in the Steller sea lion *Eumetopias jubata. Calif. Fish & Game* **61** (4): 233–234.

TARASOFF, F. J. 1974. Anatomical adaptations in the river otter, sea otter and harp seal with reference to thermal regulation. pp. 111–141 in *Functional anatomy of marine mammals* 2 Ed. R. J. Harrison. Academic Press, London.

TARASOFF, F. J., BISAILLON, A., PIÉRARD, J. & WHITT, A. P. 1972.
Locomotory patterns and external morphology of the river otter, sea otter, and harp seal (Mammalia). *Can. J. Zool.* **50** (7): 915–929.

TARASOFF, F. J. & FISHER, H. D. 1970.
Anatomy of the hind flippers of two species of seals with reference to thermoregulation. *Can. J. Zool.* **48** (4): 821–829.

TARASOFF, F. J. & KOOYMAN, G. L. 1973a.
Observations on the anatomy of the respiratory system of the river otter, sea otter and harp seal. I. The topography, weight, and measurements of the lungs. *Can. J. Zool.* **51** (2): 163–170.

TARASOFF, F. J. & KOOYMAN, G. L. 1973b.
Observations on the anatomy of the respiratory system of the river otter, sea otter and harp seal. II. The trachea and bronchial tree. *Can. J. Zool.* **51** (2): 171–177.

TAYLOR, F. H. C., FUJINAGA, M. & WILKE, F. 1955.
Distribution and food habits of the fur seals of the North Pacific Ocean. 86pp. US Dept. Int. Fish & Wildl. Service. Washington.

TAYLOR, M. 1978. *The Lewis Chessmen.* 16pp. British Museum Publications Ltd.

TAYLOR, R. H. 1971. Influence of man on vegetation and wildlife of Enderby and Rose Is., Auckland Is. *NZ J. Bot.* **9** (2): 225–268.

TEDFORD, R. H. 1976. Relationship of pinnipeds to other carnivores. *Syst. Zool.* **25** (4): 363–374.

TEDMAN, R. A. & BRYDEN, M. M. 1979.
Cow-pup behaviour of the Weddell seal, *Leptonychotes weddelli* in McMurdo Sound, Antarctica. *Aust. Wildl. Res.* **6**: 19–37.

TEDMAN, R. A. & BRYDEN, M. M. 1981.
The mammary gland of the Weddell seal, *Leptonychotes weddelli* (Pinnipedia). 1. Gross and microscopic anatomy. *Anat. Rec.* **199** (4): 519–529.

THOMAS, J., DEMASTER, D., STONE, S. & ANDRIOSHEK, D. 1980.
Observations of a newborn Ross seal pup (*Ommatophoca rossi*) near the Antarctic Peninsula. *Can. J. Zool.* **58** (11): 2156–2158.

TIKHOMIROV, E. A. 1964. Some data on the distribution and biology of pinnipeds in the Bering Sea. *Trudy vses. nauch-iss. Inst. morsk. ryb. Khoz. Okeano.* **53**: 277–285. (English abstract in *Polar Record* 1965 **12**: 748–749.

TIKHOMIROV, E. A. 1975. Biology of the ice forms of seals in the Pacific section of the Antarctic. *Rapp. P-v Réun. Cons. int. Explor. Mer.* **169**: 409–412.

TIMOSHENKO, YU. K. 1975. Craniometric features of seals of the genus *Pusa. Rapp. P-v. Réun. Cons. int. Explor. Mer.* **169**: 161–164.

TOWNSEND, C. H. 1934. The fur seal of the Galapagos Islands. *Zoologica* NY **18** (2): 43–56.

TURNER, W 1888. Report on the seals. *Zoology of the voyage of HMS* Challenger. **26** (68).

TYSON, R. M. 1977. Birth of an elephant seal on Tasmania's east coast. *Vict. Nat.* **94** (5): 212–213.

UCHIYAMA, K. 1965. Californian sea lion twins at Tokuyama Zoo. *Int. Zoo Year Book* **5**: 111.

VARDY, P. H. & BRYDEN, M. M. 1981. The kidney of *Leptonychotes weddelli* (Pinnipedia: Phocidae) with some observations on the kidneys of two other southern phocid seals. *J. Morph.* **167**: 13–34.

VAUGHAN, R. W. 1975. Seals in Orkney. pp. 95–97 in *The Natural Environment of Orkney.* Ed. R. Goodier. Proc. NCC Symp. Edinburgh. 1975.

VAUGHAN, R. W. 1978. A study of common seals in the Wash. *Mammal. Rev.* **8** (1–2): 25–34.

VAZ-FERREIRA, R. 1950. Observaciones sobre la Isla de Lobos. *Revista Fac. Hum. y Ciencias Montevideo* No. **5**: 145–176.

VAZ-FERREIRA, R. 1965. Comportamiento antisocial en machos subadultos de *Otaria byronia. Revista Fac. Hum. y Ciencias* **22**: 203–207.

VAZ-FERREIRA, R. 1979. South American fur seal. pp. 34–36 in *Mammals in the seas.* 2 FAO Fisheries Series. No. 5.

VEDROS, N. A., SMITH, A. W., SCHONEWALD, J., MIGAKI, G. & HUBBARD, R. C. 1971.
Leptospirosis epizootic among California sea lions. *Science* **172** (3989): 1250–1251.

VINSON, J. 1956. Sur la présence de l'élèphant de mer

aux Mascareignes. *Proc. Roy. Soc. Arts & Sci.* Mauritius **1** (4): 313–318.

VROLIK, W. 1822. *Specimen Anatomica – Zoologicum de Phocis speciatim de* Phoca vitulina. 138pp. Trajecti ad Rhenum.

WACE, N. M. & HOLDGATE, M. W. 1976. Man and nature in the Tristan da Cunha Islands. *IUCN Monograph* No. **6**: 1–114.

WARD, H. A. 1887. Notes on the life history of *Monachus tropicalis*, the West Indian seal. *Amer. Nat.* **21**: 257–264.

WARNEKE, R. M. 1975. Dispersal and mortality of juvenile fur seals *Arctocephalus pusillus doriferus* in Bass Strait, southeastern Australia. *Rapp. P-v. Réun. Cons. int. Explor. Mer.* **169**: 296–302.

WARNEKE, R. M. 1979. Australian fur seal. p. 41–44 in *Mammals in the seas.* **2** FAO Fisheries Series. No. 5.

WARTZOG, D. & MCCORMICK, M. G. 1978. Colour discrimination by a Bering Sea spotted seal *Phoca largha. Vision Research* **18** (7): 781–784.

WELLINGTON, G. M. & DE VRIES, T. J. 1976. The South American sea lion *Otaria byronia* in the Galapagos Islands. *J. Mamm.* **57** (1): 166–167.

WELLS, R. M. G. 1978. Observations on the haematology and oxygen transport of the New Zealand fur seal *Arctocephalus forsteri. NZ J. Zool.* **5** (2): 421–424.

WESTON, R. J., REPENNING, C. A. & FLEMING, C. A. 1973. Modern age of supposed Pliocene seal, *Arctocephalus caninus* Berry (*Phocarctos hookeri* Gray) from New Zealand. *NZ J. Sci.* **16**: 591–598.

WHITTOW, G. C. 1974. Sun, sand and sea lions. *Natur. Hist.* **83** (7): 56–63.

WHITTOW, G. C. 1978. Thermoregulatory behavior of the Hawaiian monk seal (*Monachus schauinslandi*). *Pacific Science* **32** (1): 47–60.

WHITTOW, G. C., OHATA, C. A. & MATSUURA, D. T. 1971. Behavioural control of body temperature in the unrestrained California sea lion. *Commun. Behav. Biol.* **6** (2)A: 87–91.

WHITTOW, G. C., MATSUURA, D. T. & OHATA, C. A. 1975. Physiological and behavioural temperature regulation in the California sea lion (*Zalophus californianus*). *Rapp. P-v. Réun. Cons. int. Explor. Mer.* **169**: 479–480.

WHITTOW, G. C., MATSUURA, D. T. & LIN, Y. C. 1972. Temperature regulation in the California sea lion (*Zalophus californianus*). *Physiol. Zool.* **45** (1): 68–77.

WILKE, F. 1942. Large walrus tusk from St. Paul Island, Alaska. *The Murrelet* **23**: 17.

WILLIAMSON, W. M., LOMBARD, L. S. & GETTY, R. E. 1959. North American blastomycosis in a northern sea lion. *J. Am. Vet. Med. Ass.* **135**: 513–515.

WILSON, G. J. 1974. A preliminary report on the distribution and abundance of the New Zealand fur seal (*Arctocephalus forsteri*) at the Auckland Islands December 8 1972 to January 15 1973. *NZ Min. Ag. & Fish. Fisheries Tech. Rep.* No. **133**: 10pp.

WILSON, G. J. 1979. Hooker's sea lions in southern New Zealand. *NZ J. Mar. F.W. Res.* **13** (3): 373–375.

WILSON, J. 1975a. Killers in the surf. *Audubon* Sept. 1975 **77** (5): 2–5.

WILSON, S. C. 1975b. Attempted mating between a male grey seal and female harbor seals. *J. Mamm.* **56** (2): 531–534.

WILSON, T. M. & POGLAYEN-NEUWALL, I. 1971. Pox in South American sea lions (*Otaria byronia*). *Canad. J. Comp. Med.* **35**: 174–177.

WILSON, T. M. & STOCKDALE, P. H. 1970. The harp seal *Pagophilus groenlandicus* (Erxleben 1777). XI. *Contracaecum* sp. infestation in a harp seal. *J. Wildl. Dis.* **6** (3): 152–154.

WILSON, V. J. 1975c. A second survey of seals in the King Haakon VII Sea, Antarctica. *S. Afr. J. Antarct. Res.* No. **5**: 31–36.

WIRTZ, W. O. 1968. Reproduction, growth and development, and juvenile mortality in the Hawaiian monk seal. *J. Mamm.* **49** (2): 229–238.

XIMENEZ, I. 1976. Dinámica de la Población de *Otaria flavescens* (Shaw) en el area de Peninsula Valdés y zonas adyacentes Provincia del Chubut, Republica Argentina. *Publ. Centro. Nacional Patagonico Informes Tecnicos* No. **1.4.1**: 52pp.

YALDWYN, J. C. 1958. Decapod crustacea from subantarctic seal and shag stomachs. *Rec. Dominion Mus.* **3** (2): 121–127.

YALDWYN, J. C. (ed.) 1975. *Preliminary results of the Auckland Islands Expedition 1972–73.* Dept. Lands & Survey. Wellington. NZ.

YABLOKOV, A. V. & SERGEANT, D. E. 1963. Cranial variation in the harp seal (*Pagophilus groenlandicus* Erxleben 1777). *Zool. Zh.* **42**: 1857–1865.

ZAPOL, W. M., LIGGINS, G. C., SCHNEIDER, R. C., QVIST, J., SNIDER, M. T., CREASY, R. K. & HOCHACHKA, P. W. 1979. Regional blood flow during diving in the conscious Weddell seal. *J. Appl. Physiol.* **47** (5): 968–973.

ZORAB, P. A. 1961. The historical and prehistorical background of ankylosing spondylitis. *Proc. Roy. Soc. Medicine* 415–420.

ØRITSLAND, T. 1964. Klappmysshunnens forplantningsbiologi. *Fisken og Havet* **1**: 1–15.

ØRITSLAND, T. 1970. Sealing and seal research in the south-west Atlantic pack ice Sept–Oct 1964. pp. 367–376 in *Antarctic ecology* **1** Ed. M. W. Holdgate. Academic Press, London.

ØRITSLAND, T. 1973. Walrus in the Svalbard area. pp. 59–68 in *Seals. IUCN Suppl. Paper* No. 39.

ØRITSLAND, T. 1977. Food consumption of seals in the Antarctic pack ice. pp. 749–768. in *Proc. 3rd SCAR Symp. on Ant. Biol. Adaptations within Ant. Ecosystems.* Smithsonian Inst.

ØRITSLAND, T. & BONDØ, G. 1980. Klappmyssunge født pa norskekysten. *Fauna.* **33**: 74–76.

ØYNES, P. 1966. Sel i sør-Norge. *Fiskets Gang* **45**: 834–839.

Further reading

Some of these are popular, some scientific; and some consist of a collection of papers, or chapters by individual authors.

RONALD, K., HANLEY, L. M., HEALY, P. J. & SELLEY, L. J. 1976.
An annotated bibliography on the Pinnipedia. 794pp. DK-2920 Charlottenlund: International Council for the Exploration of the Sea. 'This bibliography is really indispensable for the library of any institution where work is done on pinnipeds, and in view of its price, it also may not lack in the private library of a zoologist working on animals of this suborder. Approximately 9500 references on seals, sea lions and walruses dating from the time of Homer and Aristotle to 1975 are listed and cross-indexed.'

ALLEN, J. A. 1880. *History of North American pinnipeds.* 785pp. US Geol. and Geogr. Survey. Terr. Misc. Publ. Washington.

ANDERSEN, H. J. (ed.) 1969. *The biology of marine mammals.* 511 pp. Academic Press.

ANON. 1975. Biology of the seal. Proceedings of Guelph Symposium 1972. *Rapp. P-v. Réun. Cons. int. Explor. Mer.* **169**:

BACKHOUSE, K. M. 1969. *Seals.* 96pp. Arthur Barker Ltd, London.

BRAZIER HOWELL, A. 1930. *Aquatic mammals.* 350pp. Charles C. Thomas, Baltimore.

BURT, W. H. (ed.) 1971. Antarctic Pinnipedia. *Am. Geophys. Union. Ant. Res. Series* **18**:

FAO FISHERIES SERIES 1979. No. 5. **2**: *Mammals in the seas.*

HARRISON MATTHEWS, L. 1952. *Sea elephant.* 185pp. The Scientific Book Club, London.

HARRISON, R. J. 1972, 1974, 1977. *Functional anatomy of marine mammals.* Vols. 1, 2, 3. Academic Press, London.

HARRISON, R. J., HUBBARD, R. C., PETERSON, R. S., RICE, C. E. & SCHUSTERMAN, R. J. 1968. *The behavior and physiology of pinnipeds.* Appleton-Century-Crofts, New York.

HARRISON, R. J. & KING, J. E. 1980. *Marine mammals.* 2nd ed. 192pp. Hutchinson Univ. Library.

HOLDGATE, M. W. (ed.) 1970. *Antarctic ecology.* Vol. 1. Academic Press, London.

HONACKI, J. H., KINMAN, K. E., & KOEPPL, J. W. (eds.) 1982. *Mammal species of the world. A taxonomic and geographic reference.* 694pp. Allen Press Inc.

HOOKE, N. W. 1964. *The seal summer.* 160pp. Arthur Barker Ltd, London

HURRELL, H. G. 1963. *Atlanta my seal.* 168pp. Wm. Kimber, London.

KING, J. E. 1964. *Seals of the world.* 154pp. British Museum Natural History, London.

LOCKLEY, R. M. 1954. *The seals and the Curragh.* 149pp. J. M. Dent & Sons Ltd, London.

MORGAN, G. 1979. *Flip. The story of a seal.* 80pp. Collins, London.

PEARSON, R. H. 1959. *A seal flies by.* 154pp. Rupert Hart-Davis, London.

PERRY, R. 1967. *The world of the walrus.* 162pp. Taplinger Publ. Co, New York.

RIDGWAY, S. H. (ed.) 1972. *Mammals of the sea. Biology and medicine.* Charles C. Thomas, Springfield, Illinois.

RIDGWAY, S. H. & HARRISON, R. J. (Eds). 1981. *Handbook of marine mammals.* Vol. 1. *The Walrus, sea lions, fur seals and the sea otter.* 235pp. Vol. 2. *Seals.* 359pp. Academic Press, London.

SCHEFFER, V. B. 1958. *Seals, sea lions and walruses. A review of the Pinnipedia.* 189pp. Stanford Univ. Press.

SCHEFFER, V. B. 1970. *The year of the seal.* 205pp. Charles Scribner's Sons, New York.

SCHEFFER, V. B. 1976. *A natural history of marine mammals.* 157pp. Charles Scribner's Sons, New York.

THOMSON, D. 1954. *The people of the sea.* 214pp. Turnstile Press, London.

Index

For each species of seal, as appropriate, the following information is included in the section on the Diversity of Pinnipeds. These headings, when referring to the individual seals, are not listed separately in the index.